WAVES AND IMAGING THROUGH COMPLEX MEDIA

Waves and Imaging through Complex Media

edited by

PATRICK SEBBAH

CNRS,
Laboratoire de Physique de la Matière Condensée,
Université de Nice-Sophia Antipolis, France

KLUWER ACADEMIC PUBLISHERS
DORDRECHT / BOSTON / LONDON

Proceedings of the International Physics School on
Waves and Imaging through Complex Media
Cargese, France
August 30–September 3, 1999

Library of Congress Cataloging-in-Publication Data

International Physics School (1999 : Cargèse, France)
    Waves and imaging through complex media : International Physics School, Cargèse,
France, 30 August-3 September 1999 / edited by Patrick Sebbah.
        p. cm.
    Includes indexes.
    ISBN 0-7923-6814-2 (acid-free paper)
    1. Electromagnetic waves--Scattering--Congresses. 2. Condensed matter--Congresses.
I. Sebbah, Patrick. II. Title.

    QC173.45 .I56 1999
    539.2--dc21
                                                        00-069210

Published by Kluwer Academic Publishers,
P.O. Box 17, 3300 AA Dordrecht, The Netherlands.

Sold and distributed in North, Central and South America
by Kluwer Academic Publishers,
101 Philip Drive, Norwell, MA 02061, U.S.A.

In all other countries, sold and distributed
by Kluwer Academic Publishers,
P.O. Box 322, 3300 AH Dordrecht, The Netherlands.

Drosophila embryo observed by simultaneous optical coherence tomography
and two-photon excited fluorescence microscopy. Both contrast modes are
based on the selection of ballistic photons and provide micron-resolution
imaging in scattering samples. E.Beaurepaire, L.Moreaux & J.Mertz
(INSERM-ESPCI). Embryo provided by E.Farge (CNRS-Curie).

Printed on acid-free paper

Printed in the Netherlands.

Contents

Preface

Recent advances in wave propagation in heterogeneous media are consequences of new approaches to fundamental issues as well as strong interest in potential applications. As an example, the recent introduction of wave chaos and more specifically random matrix theory – an old tool from nuclear physics - to the study of random media has pointed the way to a deeper understanding of wave coherence in complex media. At the same time, efficient new approaches for retrieving information from random media promise to allow wave imaging of small tumors in opaque tissues.

The International Summer School in Physics in Cargèse, France (Aug. 30th – Sept. 3rd 1999), on "Waves and Imaging through Complex Media" was organized by the PRIMA (*PRopagation et Imagerie en Milieu Aléatoire*–CNRS) research group, directed by A. C. Boccara and R. Maynard. This school was intended to bring together students and researchers with either fundamental or applied interests in the multiple scattering of waves. The program was composed of reviews and lectures on basic and advanced theoretical and experimental approaches to multiple scattering, on advances in imaging techniques in random media, and on applications to medical imaging. Aware of the heterogeneity of the audience, the lecturers were concerned with being tutorial and complete at the same time. The quality of the presentations by specialists in diverse domains such as Anderson localization, speckle analysis, semiclassical methods, optical tomography or diffuse wave spectroscopy led to a stimulating interaction between the two communities and to the spontaneous consensus that these lectures be collected in a book. Nonetheless, this book should not be regarded as a collection of isolated research reports. Rather, it is a collective effort to bring to the community a thorough view of this rapidly evolving field.

The book is organized as follows: Fundamental concepts in wave chaos and multiple scattering (O. Legrand) and the state of the art in optical imaging through biological tissues are reviewed (P.M.W. French and C. Boccara) in two introductory chapters. The second section is dedicated to multiple wave scattering with a tutorial presentation of a new approach based on the probability of quantum diffusion (E. Akkermans and G. Montambaux) and a thorough review on the statistical approach to transport in random media and Anderson localization demonstrated in microwave experiments (A.Z. Genack and A.A. Chabanov). The third section is devoted to methods originally developed in the field of quantum chaos, and only recently introduced into the field of multiple scattering, namely the semiclassical approach (D. Delande) and the random matrix theory (J.-L. Pichard). Elastic and sound waves serve as experimental illustrations of these concepts in multiple scattering and wave chaos (R.L. Weaver) or time reversal and enhanced backscattering (M. Fink and J. de Rosny). The next two sections focus on optical imaging and present two categories of techniques recently introduced. (a) Techniques that concentrate on the detection of the minute fraction of light that suffers no scattering are illustrated by recent advances in the domains of fluorescence lifetime imaging and holography (N.P. Barry *et al.*), optical coherence and two-photon fluorescence microscopy (E. Beaurepaire *et al.*), low coherence interferometry (G. Brun *et al.*), laser optical feedback tomography (E. Lacot *et al.*), numerical holography (M. Gross *et al.*) and time-gating methods (E. Lantz *et al.*). (b) Techniques related to the detection of the average radiation density such as inverse problem methods in heterogeneous (S. Arridge) or stratified (J.-M. Tualle *et al.*) media are first introduced by a detailed analysis of the transition from singly scattered to diffuse light (K.K. Bizheva *et al.*) and a comparison between an exact electromagnetic solution and the solution to the radiative transfer equation (J.J. Greffet *et al.*). A first analysis of multi-scale systems is proposed in the context of rough surfaces (C.A. Guérin *et al.*). The two last sections focus on the coherent nature of wave propagation in multiple scattering media. Diffusive wave spectroscopy is reviewed (G. Maret and M. Heckmeier) and applied to dense flowing suspensions (P. Snabre *et al.*). Acoustic waves coupled to light have recently been used to image optically absorbing objects embedded in opaque tissues (S. Lévêque-Fort). From a more fundamental point of view, the role of frequency and spatial correlations is analyzed (R. Pnini) and the role of temporal correlations is demonstrated in original optical experiments (F. Scheffold and G. Maret). Finally, possible applications of these concepts to imaging in turbid media (S.E. Skipetrov) or characterization of non-linear random media (R. Bressoux and R. Maynard) are considered.

We hope that this work will serve as a reference for physicists working in various domains where disorder is an obstacle to information retrieval, as in non-destructive testing, acoustics, lidar, ... On the other hand, the focus on medical imaging, which is a rapidly growing application of this field, will serve as a motivating thrust to the fundamental physics community.

This work has been made possible by financial support from the Centre National de la Recherche Scientifique, the Délégation Générale de l'Armement, the Ministère de l'Education Nationale de la Recherche et de la Technologie and the Ministère des Affaires Etrangères. I would like to thank all the authors for their contributions. Special thanks goes to Jean Michel Tualle for sharing the task of the organizing the school, Olivier Legrand and Fabrice Mortessagne for their valuable assistance in the preparation of the manuscript. We thank E. Dubois-Violette, head of the Institut d'Etudes Scientifiques de Cargèse, C. Ariano, B. Cassegrain, A. Chiche and C. Ubaldi for their valuable help during the preparation of the School in Cargèse, whose charming surroundings favored the genesis of this book.

PATRICK SEBBAH

I

INTRODUCTION

Chapter 1

WAVE CHAOS AND MULTIPLE SCATTERING: A STORY OF COHERENCE

O. Legrand
Laboratoire de Physique de la Matière Condensée
CNRS - UMR 6622
Université de Nice-Sophia Antipolis
Parc Valrose F-06108 Nice Cedex 2 FRANCE
olivier.legrand@unice.fr

Abstract   This paper is an introduction to the theoretical progress of the last two decades concerning wave propagation in complex systems originating in the fields of Quantum Chaos and Multiple Scattering Transport Theory. Common fundamental concepts and mutual interactions of theoretical approaches such as Random Matrix Theory, Diagrammatic and Semiclassical Methods for the study of coherence effects in chaotic or disordered systems are reviewed.

Introduction

The progress in the understanding of wave propagation in complex media has benefitted essentially from two fields originally issued from quantum physics, namely *Quantum Chaos* and the *Mesoscopic Theory of Quantum Transport*. For more than a decade now, both domains of research have been cross-fertilizing each other, yielding a more unified approach of coherent effects in complex wave systems. Gathering contributions by people from Imaging, Quantum Chaos and Multiple Scattering communities, this book tries to fill a gap by showing the numerous connections between these fields and how they share many fundamental and conceptual tools. This introductory chapter will be devoted to shedding some light on these common fundamental concepts, bringing forth historical aspects of their birth and also of their mutual interactions and merging in the fields of Random Matrix Theory, Diagrammatic Perturbation methods for disordered systems, Localization theory, and Semiclassical approaches.

3

P. Sebbah (ed.), Waves and Imaging through Complex Media, 3–14.

About thirty years ago, the primary concern of the field coined Quantum Chaos was to unravel the fundamental grounds of the correspondence principle if any, that is to decipher the subtle encoding of generically chaotic classical motion by quantum wave motion.

The outstanding feature of wave motion in cavities, where ray dynamics is strongly chaotic, is the seemingly contradictory behavior of the spectrum : eigenfrequencies display short and long range order. This is particularly manifested by the tendency neighboring eigenfrequencies have to repel one another. Long range order is related to what is known as spectral rigidity and, again, this property is universally encountered in classically chaotic systems. Historically, the first approach to these universal features of spectral statistics came from nuclear physics with the works by Wigner and the Random Matrix Theory (RMT) applied to Hamiltonian systems due to Dyson and Mehta [1, 2]. Then came the semiclassical approach, concerned with the high-frequency (small wavelength) asymptotics relying on a description of the wave motion based on classical (non-quantal) ray trajectories. Much of the relation between classical chaos and the corresponding signature in quantum or wave systems has been understood through a combined study of the above mentioned approaches. While RMT has essentially been successful in describing the universal properties of spectral fluctuations in Gaussian ensembles with specific fundamental symmetries of the system under study (time reversal (non)invariance, rotational (non)invariance), the semiclassical analysis has allowed not only to relate mean modal densities to phase space terms (the famous Thomas-Fermi or Weyl's law) but, above all, to account for fluctuations in terms of periodic orbits (p.o.'s) through the use of the Gutzwiller and akin trace formulae. Especially, Berry [3] was able to apply the Gutzwiller's approach to recover part of the predictions of RMT (those concerning the spectral rigidity) through the so-called *diagonal approximation* based on a random phase assumption between p.o.'s.

During the same period, parallel to the endeavor of the quantum chaos community, solid state physics was thriving in the exploration of anomalous transport properties of electrons in disordered systems. The phenomenon under study was called *weak localization* and was dealing with the quantum interference of conduction electrons on the defects of a metal. It had first emerged from a more general scaling theory of localization introduced by Abrahams *et al.* [4] for two-dimensional conductors but soon became a field of its own, intensively investigated both theoretically and experimentally (see for instance the review by Bergmann [5] and references therein). After the initial diagrammatic perturbation theory, it produced a series of important works devoted to the elaboration of a semiclassical basis for weak localization [5-7]. Starting from a path integral approach for the quantum electron propagation, the wave amplitude was written in terms of classical paths involved in a Brownian motion. This gave a more intuitive picture of the role of interference effects in the underlying physics than allowed by the diagrammatic perturbation theory. For instance, this interpretation permitted to describe, quantitatively, weak localization as resulting from interference between pairs of time reversed paths. Since these seminal works, other works have appeared, trying to use the physically intuitive trace formula approach in order to understand quantum disordered systems as well [8, 9]. This was essentially motivated by the fact that both classically chaotic systems and disordered systems have the same spectral statistics (given by RMT) in appropriate domains of validity. As opposed to the perturbation diagrammatic methods, the semiclassical methods based on trace formulae may provide a more intuitive insight into phenomena such as localization [9] or universal conductance fluctuations [8]. Trace formulae nevertheless suffer from an organic drawback as asymptotic approaches : the convergence of the sums is not well understood and expansions are still ill-defined. This is obviously the reason why diagrammatic methods are most commonly used to study waves in disordered media as they allow a reliable control of perturbation expansions to compute various averaged quantities.

In electronic mesoscopic disordered systems, the use of RMT was led along essentially two distinct approaches : one by viewing these systems as *scattering systems* implied by the fact that the main experimental information is obtained through transport studies (see the recent excellent review by C. Beenakker [10]), another devoted to their *spectral properties* whose relevance for transport properties was pointed out by Thouless in the 70's [11]. The latter approach was first exploited by Altshuler and Shklovskii [12] to provide a true understanding of universal conductance fluctuations as resulting from quantum coherence effects.

To fully appreciate the novel contributions of this book concerning RMT, diagrammatic methods and semiclassics to study coherence effects in multiple scattering media, the reader is invited on a brief introductory survey of the key concepts and methods that have been devised to describe waves in disordered or chaotic systems for about three decades till now.

1. RANDOM MATRIX THEORY

Wigner-Dyson's RMT (WDRMT) is primarily concerned by ensembles of Hermitian matrices as models of Hamiltonians describing autonomous systems. Applied to wave systems, one must call for the close analogy between the Schrödinger equation for spinless particles and the Helmholtz equation for scalar waves, and therefore simply envisage Hamiltonians as representations of operators. These random matrices are mainly divided among three ensembles, which correspond to three types of repulsion between *levels* (eigenvalues of the matrices). They can be obtained by imposing global symmetries together with a maximal statistical independence of the matrix elements. Among the three

classes of symmetries, two are essentially in order for our purposes, namely
the time-reversal invariance or the absence of it. To these symmetries are respectively associated two types of matrix ensembles : the Gaussian Orthogonal
Ensemble (GOE) consisting of real symmetric matrices (left invariant by the
canonical group of orthogonal transformations), and the Gaussian Unitary Ensemble (GUE) consisting of complex hermitian matrices (left invariant by the
canonical group of unitary transformations).

To compare the statistical spectral properties of random matrices with real
spectra - for instance, resonance frequencies of a bounded medium - it is convenient to evaluate frequency scales in units of average frequency spacing Δ
between neighboring frequencies. In the generic case of a dispersive medium,
this obviously can be done only locally, in a given frequency range. By average, here it is meant averaging over the spectrum or averaging over ensembles
in the case of disordered systems even though the equivalence between both
averages is most often a matter of conjecture. For ensembles of random $N \times N$
matrices, complete distributions have been computed for their eigenvalues or
eigenvectors [1] and an introductory presentation of these calculations is given
in the contribution due to J.-L. Pichard in the present book [13].

One of the main statistical features of the Gaussian ensembles is level repulsion, a *local* tendency neighboring levels have to repel each other. It may
be characterized by the nearest-neighbor spacing distribution, approximately
given by the so-called *Wigner's surmise*

$$P(s)ds = \left(\frac{\pi s}{2}\right) \exp\left(-\frac{\pi s^2}{4}\right) ds \qquad (1)$$

(note that this result is exact for the Gaussian ensemble of real symmetric
2×2 matrices and is numerically close to the result for the GOE of $N \times N$
matrices in the limit $N \to \infty$). To characterize long-range correlations of the
spectra, better suited is a measure of *two-point* correlations, as for instance the
variance $\Sigma^2(L) = \langle (n(L) - L)^2 \rangle$ of the number of levels within a spectral
interval of length $L = \langle n(L) \rangle$ measured in units of mean spacing Δ (see for
instance [2] or [14]). For a GOE spectrum, in the limit of a wide interval,
$\Sigma^2(L) \approx \frac{2}{\pi^2} \ln(L)$, whereas for uncorrelated spectra one has the general result
for the sum of independent random variables $\Sigma^2(L) = L$. This property of
reduced fluctuations of $n(L)$ for GO(U)E spectra with respect to uncorrelated
spectra is called spectral rigidity. The number variance is easily deduced from
the 2-level correlation function defined as

$$R\left(s = |x_1 - x_2|\right) = \langle d(x_1) d(x_2) \rangle - 1 , \qquad (2)$$

where $d(x) = \sum_i \delta(x - x_i)$ is the spectral density, through the double integral [2]

$$\Sigma^2(L) = \iint_{x-L/2}^{x+L/2} R\left(|x_1 - x_2|\right) dx_1 dx_2 . \tag{3}$$

This quantity (or the number variance) will prove to be a key one in what follows since most quantities of interest for the computation of fluctuations of transport properties in chaotic or disordered systems involve quadratic statistical averages.

R. L. Weaver was among the first to clearly demonstrate the relevance of these concepts to acoustics and elastodynamics [15], and more generally to non-quantal waves. Elastic systems are most appropriate, along with microwaves and optics, to explore wave chaos and multiple scattering. Among many interesting features, elastic systems are particularly suitable for studying time-resolved transients, which play such a crucial role in the phenomenon of reverberation. By investigating the elastic response of *ergodic* aluminium blocks, Weaver and co-workers have been able to demonstrate spectral rigidity [16], and the ensuing *quantum echo*, which is observed at times longer than the *break time* (time of the order of the inverse mean frequency spacing) [17]. Mind that this effect somehow goes beyond the standard enhanced backscattering effect associated to weak localization, which is observed at earlier times: it is a coherent effect that reveals the spectral statistics. An interesting consequence of this quantum echo is demonstrated by M. Fink and J. de Rosny to be of practical use for time-reversal of ultrasounds in chaotic billiard-like plates (see [18] and also [19]).

At this stage of our introduction, it is advisable to briefly present the two main historical RMT approaches for scattering off a complex system. These were originally introduced within the framework of nuclear reaction theory to obtain a model for the fluctuations of nuclear cross-section but were adapted, as soon as the 80's, for the theories of quasi one-dimensional (1-D) conductors.

The first approach could be called the *random Hamiltonian approach* following Guhr *et al.* [20] who gave a wonderful comprehensive review of common concepts of RMT in quantum (and also wave) physics. It relies on the hypothesis that the scattering matrix is expressed in terms of the Hamiltonian of the transmitting system. Thereby, an ensemble of scattering matrices is obtained for instance from the GOE, naturally allowing for the frequency dependence of the scattering matrix (or more generally for any parametric dependence). In the beginning of the 80's it became clear that WDRMT could be used to describe spectral fluctuations in disordered metals at low temperature i.e. for the diffusive regime in electronic transport. This was confirmed by Efetov [21] in the so-called zero-dimensional limit of his nonlinear σ model (Efetov's work on *supersymmetry* was greatly motivated by the question of deriving RMT from first principles within this context [22]), and deviations from Wigner-Dyson's statistics have henceforth been seen as manifestations of the spatial extension of the system.

The second approach is generally referred to as the *random S matrix approach* because it is directly concerned with the S matrix elements as stochastic variables and does not require the introduction of an underlying Hamiltonian. A similar approach known under the name of the *random transfer matrix* was also introduced to describe electronic transport through a disordered conducting wire. Transport properties are most easily deduced from the multichannel Landauer formulation for the conductance in terms of transmission amplitudes. In the present book, J.-L. Pichard gives a clear introductory review of the random scattering formalism leading to predictions for the *Universal Conductance Fluctuations* (UCF) which are different in the zero-dimensional case (relevant to ballistic chaotic cavities) and in the quasi 1-D case (relevant to disordered wires) (see references in [13]). In the latter case, a Fokker-Planck equation (the so-called DMPK equation, named after Dorokhov, Mello, Pereira and Kumar) is derived for the distribution function of the transmission eigenvalues, which enables, amongst other things to recover the standard result of UCF (otherwise established through the perturbation diagrammatic method - see below)

$$var(g) = \frac{8}{15\beta} \qquad (\beta = 1 \quad \text{for orthogonal symmetry} \tag{4}$$

$$\beta = 2 \quad \text{for unitary symmetry})$$

where $g \equiv \frac{h}{e^2} G = 2T$ is the dimensionless conductance written in terms of the total transmission T through the Landauer formula.

2.   FROM DIAGRAMMATICS TO A TRAJECTORIES APPROACH : THE WAVE RETURN PROBABILITY

To establish the link between RMT and the diagrammatic approach to transport, one generally relies on the equivalence between the Kubo formula for the conductance at zero temperature and the multichannel Landauer formula. This permits to write the conductivity (its longitudinal part along x), in the k-representation, in terms of the Green functions of the conductor (see for instance [23] and also [24][1])

$$\sigma_{xx} = \frac{2e^2}{h} \frac{\hbar^4}{m^2 L^d} \sum_{\mathbf{k},\mathbf{k}'} k_x k'_x |G^R(\mathbf{k},\mathbf{k}')|^2 \tag{5}$$

---

[1]Note that, in [24], the extra factor of 2 is missing because spinless electrons are considered.

where L is the linear dimension of the sample (d is the dimension of space). The impurity perturbation approach thus deals with products of Green functions and the hard work consists in finding, when averaging over disorder, relevant contributions to increasing orders in the small parameter $(k\ell)^{-1}$ in the so-called *weak disorder* limit (ℓ being the elastic mean free path). In the present book, E. Akkermans and G. Montambaux give a very thorough review of coherent scattering in disordered media and propose a novel insight into the perturbation theory for multiple scattering by relating the dominant diagrams, namely the *diffuson* (also called *particle-hole ladders*) and the *cooperon* (also named *particle-particle ladders*), to the corresponding contributions of the *quantum* (or, more generally, wave) *return probability*. Along this picture, the *diffuson* approximation gives a quantum probability function $P_d(\mathbf{r}_1, \mathbf{r}_2; \omega)$, which obeys a dynamic diffusion equation and has the property to be properly normalized as a true probability should be. At this stage, the diffusion approximation could be assumed to be complete and, indeed, this might account for the observed robustness of the radiative transfer equation, so commonly used for practical purposes to evaluate the averaged transmitted intensity through highly scattering media (see the contribution due to J.-J. Greffet *et al.* in this book). The *cooperon* is however crucial when evaluating the probability to come back at the origin, since its contribution $P_c(\mathbf{r}_1, \mathbf{r}_2; \omega)$ is equal to the diffuson contribution at $\mathbf{r}_1 = \mathbf{r}_2$. This is the now famous *enhanced backscattering* phenomenon. The space integral of the wave return probability, once integrated over time, yields the recurrence time which is shown to be related to the spectral determinant of the problem under study (we will see below that it may also serve to derive the spectral correlations in a closed system). This spectral determinant serves as a generating function to calculate various quantities such as $\langle \delta g \rangle$ or $\langle \delta g^2 \rangle$ and permits to recover the UCF result given in Eq. (4).

This last result is intimately related to infinite-range spatial intensity correlations of speckle in random media as reviewed in this book by R. Pnini who introduces an alternative Langevin approach to evaluate the three dominant contributions (to leading order in $(k\ell)^{-2}$) to the intensity-intensity correlation function, namely the so-called C_1 (short-range correlations), C_2 (long-range correlations), and C_3 (infinite-range correlations. The latter contribution is precisely the one which yields UCF when summing over all (in-and-outgoing) channels (see for instance the review [25] due to Berkovits and Feng). The role of long-range and infinite correlations in the dynamic speckle correlations has been exemplified by F. Sheffold and G. Maret in recent photon correlation spectroscopy experiments where they could show the dynamic equivalent of UCF for the first time in optics [26]. In his contribution to this book, R. Pnini then proceeds to show how spatial correlations may lead to non-Rayleigh intensity distributions, their asymptotic tails depending strongly on the relative position and distance between source and detector [27]. It is worth remarking here that these results obtained from diagrammatics in the $g \gg 1$ limit, are reminiscent of similar results, obtained in the framework of the non-perturbation RMT approach with DMPK equation, by van Langen, Brouwer and Beenakker [28] and also through the supersymmetric σ-model approach (see the recent extensive review by Mirlin [29]). Concerning the related problem of evaluating the distributions of transmission coefficients of a disordered waveguide, one should mention here that they were experimentally studied by Genack's group [30] with findings in good agreement with the above mentioned theoretical predictions [28]. In the present book, A. Z. Genack and A. A. Chabanov take the opportunity to convey an innovative statistical approach of Anderson localization by illustrating how many aspects of localization (transmission, spatial correlation of intensity, various moments of transmission variables, etc.) are specified by the single parameter $\langle g \rangle$, or, alternatively, by the variance of the normalized transmission as absorption comes into play. In the localized regime, which the diagrammatic approach is not supposed to describe, a log-normal distribution for the total transmission is expected [27], [31], which is to be related to the multifractality of localized wave functions [29].

Anderson localization has known an unceasingly growing interest since the beginning of the 70's, when it began to share the same statistical concepts as RMT. Thouless proposed to use scaling ideas to establish a relation between the electrical conductance and the way the energy spectrum is modified when boundary conditions are changed. This led him to express the dimensionless conductance g as a function of the ratio E_c/Δ, where $E_c = \frac{\hbar D}{L^2}$, ($D$ being the diffusion constant and L the typical size of the system) [11] (see also [32]). More specifically, in the diffusive regime where $E_c \gg \Delta$, this relation reads

$$g = \frac{E_c}{\Delta} = \langle N(E_c) \rangle, \tag{6}$$

where $N(E)$ is the number of levels within an energy interval of length E. This relation provides a natural way to evaluate $\langle \delta g^2 \rangle$ in terms of the fluctuations of the spectrum through the variance number Σ^2. Altshuler *et al.* [12] then used Wigner-Dyson statistics (admittedly relevant to the metallic case) to deduce UCF by allowing for a finite width of levels due to the coupling to the exterior of the conductor. From the scaling theory for the change of g with the length L, one deduces that for $d > 2$, there exists a transition from diffusive (or metallic) to localized regimes. The approach to localization can be viewed through its influence on the energy eigenvalue statistics. At the transition, Altshuler *et al.* proposed that Σ^2 should be of the order of $\langle N(E) \rangle$, recalling the expected behavior for uncorrelated spectra. In fact, this vision of the transition metal-insulator was not completely correct, as Aronov, Kravtsov, and Lerner [33] could show by using a semiclassical approach developed by Argaman, Imry, and Smilansky [8]. They considered the *spectral form factor* $K(t)$ defined as the Fourier transform of the two-point correlation $R(s)$ introduced above,

$$K(\tau) = \frac{1}{2\pi} \int R(s) \exp(is\tau)ds, \tag{7}$$

and related it to the classical return probability $P(\tau)$ through the relation [8] :

$$K(\tau) = \frac{2\tau P(\tau)}{4\pi^2 \beta}. \tag{8}$$

In the diffusive regime, $P(\tau) \propto (D\tau)^{-d/2}$. When extending this expression to the critical regime, the scaling theory predicts a τ-dependent diffusion constant yielding $R(s) = 0$ (i.e. an uncorrelated spectrum) if one just allows for the dominant behavior $D_0(\tau) \propto \tau^{-1+2/d}$ deduced from the fact that, at the transition, g reaches a critical value, which is assumed to be independent of L. If the way g approaches its critical value is taken into account, a correction to the behavior of D is found, which recovers a result for the asymptotic (large s) behavior of $R(s)$ previously derived within the diagrammatic approach. This last result was one of the first to confirm the hypothesis of a new universality class of the statistics at the transition. But we will mention more about the ubiquitous character of this new class below. Here, it should be made clear that the true *spectral form factor* is nothing but the space integral of the wave return probability mentioned above [24]. Only, in Eq. (8), the classical (in the geometrical limit) probability is used to approximate the exact form factor, in the same way as Akkermans and Montambaux evaluate the diffuson and cooperon contributions to the wave return probability to recover RMT [24]. This last remark quite naturally leads us to say a few words about the recent findings concerning semiclassical approximations in chaotic Hamiltonian systems and/or in the presence of multiple scattering. This is the subject of a very complete tutorial proposed by D. Delande in the present book.

If a wave system has complicated ray dynamics, the spectral density of states $d(E)$, or, more precisely, the oscillatory part of it, can be expressed, in the high frequency limit, as a sum over the periodic orbits (p.o.'s) of the system, the famous Gutzwiller trace formula :

$$d(E) = \sum_{p.o.'s} A_j \exp[iS_j], \tag{9}$$

where A_p and S_p are the amplitude and phase accumulated along the p.o. at energy E. This formula may be derived from an asymptotic expression for the Green's function of the system (see [14] and references therein). If one substitutes this semiclassical formula into the definition of the 2-level correlation function (2) and takes the Fourier transform, one obtains the semiclassical form factor

$$K_{sc}(\tau) = \sum_{i,j} A_i A_j^* \exp\left[i(S_i - S_j)\right] \delta\left[\tau - \frac{1}{2}(T_i + T_j)\right], \qquad (10)$$

where T_i is the period of the orbit i. One then proceeds, as suggested by Berry [3], by assuming that only $i = j$ terms contribute to the above sum (the so-called *diagonal approximation*). A few more manipulations (see [8]) permit to express the resulting weighted density of p.o.'s in terms of the classical probability of return, leading to the expression in Eq. (8). In the chaotic regime, for times larger than the few shortest p.o.'s, $P(\tau)$ is a constant leading to GOE statistics. If the system is not strictly ballistic as in a billiard with point-like scatterers, diffractive orbits come into play that modify this picture [14] and may lead to a situation where level repulsion is not associated to spectral rigidity, in a way similar to what is observed for certain pseudo-integrable billiards or for the Anderson model at the localization transition as mentioned above. This corresponds to a new general type of spectral statistics recently coined *semi-poisson* by Bogomolny *et al.* [34] because of the linear behavior of the repulsion at small spacings and of its exponential fall-off at large ones.

Conclusion

The present overview wishes to stress the intimate relationship that seems to gather diagrammatics to RMT (including supersymmetry) and semiclassics for a more profound understanding of wave transport in disordered systems. This is especially true when one considers the recent advances of the subtle coherent mechanisms involved at the threshold of localization : spatial and frequency correlations are the key to the characterization of large fluctuations, the latter being probably the main hindrance, but also maybe the best ally if tamed, to extract the information about the heterogeneous medium, encoded in a formidable interplay of interferences.

References

[1] M. L. Mehta, *Random Matrices* (Academic Press, Boston, 1991).

[2] O. Bohigas, in *Chaos and Quantum Physics*, Proceedings of the Les Houches Summer School, Session LII, edited by M. J. Giannoni, A. Voros and J. Zinn-Justin (North-Holland, Amsterdam, 1991).

[3] M. V. Berry, Proc. R. Soc. London A **400**, 229 (1985).

[4] E. Abrahams, P.W. Anderson, D. C. Licciardello, and T. V. Ramakrishnan, Phys. Rev. Lett. **42**, 673 (1979).

[5] G. Bergmann, Phys. Rep. **107**, 1 (1984).

[6] D. E. Khmelnitskii, and A. I. Larkin, Usp. Fiz. Nauk. **136**, 533 (1982) [Sov. Phys. Usp. **25**, 185, (1982)].

[7] S. Chakravarty, and A. Schmid, Phys. Rep. **140**, 193 (1986).

[8] N. Argaman, Y. Imry, and U. Smilansky, Phys. Rev. B **47**, 4440 (1993).

[9] D. Cohen, J. Phys. A: Math. Gen. **31**, 277-287 (1998).

[10] C. W. J. Beenakker, Rev. Mod. Phys. **69**, 731-808 (1997).

[11] D. J. Thouless, Phys. Rep. **13**, 93 (1974); D. J. Thouless, Phys. Rev. Lett. **39**, 1167 (1977).

[12] B. L. Altshuler and B. I. Shklovskii, Zh. Eksp. Teor. Fiz. **91**, 220 (1986) [Sov. Phys. JETP **64**, 127 (1986)].

[13] J.-L. Pichard, contribution in this book.

[14] D. Delande, contribution in this book.

[15] R. L. Weaver, contribution in this book.

[16] R. Weaver, J. Acoust. Soc. Am. **85**, 1005 (1989); D. Delande, D. Sornette, and R. Weaver, J. Acoust. Soc. Am **96**, 1873 (1994).

[17] R. L. Weaver, and O. I. Lobkis, Phys. Rev. Lett. **84**, 4942 (2000).

[18] M. Fink and J. de Rosny, contribution in this book.

[19] J. de Rosny, A. Tourin, and M. Fink, Phys. Rev. Lett. **84**, 1693 (2000).

[20] T. Guhr, A. Müller-Groeling, and H. A. Weidenmüller, Phys. Rep. **299**, 189-425 (1998).

[21] K. B. Efetov, Adv. in Phys. **32**, 53 (1983).

[22] K. B. Efetov, *Supersymmetry in Disorder and Chaos*, Cambridge University Press (1997).

[23] S. Datta, *Electronic Transport in Mesoscopic Systems*, Cambridge University Press (1995).

[24] E. Akkermans and G. Montambaux, contribution in this book.

[25] R. Berkovits, and S. Feng, Phys. Rep. **238**, 135-172 (1994).

[26] F. Sheffold and G. Maret, contribution in this book.

[27] R. Pnini, contribution in this book.

[28] S. A. van Langen, P. W. Brouwer, and C. W. J. Beenakker, Phys. Rev. E **53**, R1344 (1996).

[29] A. D. Mirlin, Phys. Rep. **326**, 259 (2000).

[30] A. Z. Genack and N. Garcia, Europhys. Lett. **21**, 753 (1993); M. Stoychev and A. Z. Genack, Phys. Rev. Lett. **79**, 309 (1997).

[31] A. Z. Genack and A. A. Chabanov, contribution in this book.

[32] E. Akkermans, and G. Montambaux, Phys. Rev. Lett. **68**, 642 (1992).

[33] A. G. Aronov, V. E. Kravtsov, and I. V. Lerner, Phys. Rev. Lett. **74**, 1174 (1995).

[34] E. B. Bogomolny, U. Gerland, and C. Schmit, Phys. Rev. E **59**, R1315 (1999).

Chapter 2

TOWARDS OPTICAL BIOPSY :
A BRIEF INTRODUCTION

A. C. Boccara

Laboratoire d'Optique, CNRS UPR A0005
Ecole Supérieure de Physique et Chimie Industrielle
10 rue Vauquelin, 75005 Paris, France
Tel : +33-1-40 79 46 03
email : boccara@optique.espci.fr

P. M. W. French

Femtosecond Optics Group, Physics Dept.
Imperial College
London SW7 2BZ, U.K.
Tel: +44-171-594 7784, Fax: +44-171-594 7782
email: mj.cole@ic.ac.uk

Abstract   In this introductory chapter, we discuss the problem of optical imaging in strongly scattering biological tissues and review various approaches and techniques recently developed for in depth tissue imaging. The efficiency of high resolution methods is now well established for shallow investigations. However, imaging inhomogeneities deeper in tissues is more difficult. As a result, time or frequency gating methods developed to solve the inverse problem still require a compromise between depth and resolution. New methods based on the coupling of light and ultrasound led to a recent breakthrough in the achievement of good resolution in thick tissues.

Introduction

Medical diagnosis and research require the practitioners to acquire knowledge concerning the state and function of biological tissue that is typically inside the body of the patient. Physicians today must collate the information from a range of different tools and techniques in order to optimise their diagnosis. Of

P. Sebbah (ed.), Waves and Imaging through Complex Media, 15–26.
© *2001 Kluwer Academic Publishers. Printed in the Netherlands.*

these tools, the most established imaging modalities are x-ray and ultrasound imaging, which are invaluable because they present the clinician of an image of the structure of various types of tissue that can be used to establish physical damage and to identify various organs etc by their shape. Unfortunately these modalities typically only detect differences in tissue density and cannot provide *functional* information, other than by monitoring physical movement. The ability to detect physiological changes, e.g. the oxygenation of blood, requires *chemically specific* imaging, as usually does the ability to distinguish between healthy and diseased tissue. Distinguishing and quantifying different chemical species requires some kind of spectroscopic tool. At present the "gold standard" is biopsy followed by histopatholgy, i.e. a sample of tissue is physically removed from a patient, cut into thin sections, stained and then examined under a microscope. It is the staining that provides the spectroscopy in that different colours in the final image contrast different tissues, or states of tissue. This procedure, combining spectroscopy and morphology, is immensely powerful but it is clearly invasive, as well as time-consuming, and there are many perceived benefits in the development of a technique for non-invasive biopsy. A non-invasive diagnostic technique implies using some kind of radiation that penetrates the body and interacts differently with the different chemical species of interest. Optical radiation meets this requirement and "optical biopsy" is a goal for many research groups.

Spectroscopy using optical radiation is a standard tool in laboratory analysis and may indeed be applied to detect and monitor the concentration of analytes such as oxygen by investigating the absorption, fluorescence or Raman spectra. Performing spectroscopy in biological tissue, however, is very different from identifying or quantifying a solution in a cuvette. The strong heterogeneity of tissue means that if one simply irradiates a sample with a light beam and observes the optical "signature", there will typically be many contributions from different types of tissue that have interacted with the beam. It is therefore highly desirable to combine spectroscopy with imaging, so that the spectroscopic signature corresponding to a particular piece of tissue can be identified. Unfortunately tissue heterogeneity also results in strong scattering of optical radiation, making optical imaging in biological tissue extremely difficult. On average a photon typically scatters after propagating only 50 to 100 μm in biological tissue. Thus even simple line-of-sight absorption measurements can represent a significant challenge and all the optical techniques used for imaging through tissues must overcome the scattering problem. These considerations have precluded the widespread application of optics in clinical medicine and currently much work is being done with a different laboratory spectroscopic tool, namely nuclear magnetic resonance, which has been successfully transferred to clinical practice where it is described as Magnetic Resonance Imaging. MRI can be chemically specific and provide high-resolution (functional) images contrasting types of tissue that would have been indistinguishable using x-ray or ultrasound imaging. Unfortunately, in its present manifestations, MRI is relatively expensive, owing to its requirements for high field magnets, and economic considerations prevent it from being deployed everywhere it might be useful. Also, its ability to deliver chemically specific images is limited, as there are important analytes it cannot detect, and there are often trade-offs between chemical specificity and resolution. In particular, imaging at the subcellular level, which is often necessary to diagnose many conditions and diseases, remains a formidable challenge. The same kind of consideration holds for positron imaging even if this functional imaging technique is rapidly growing thanks to the reduction in the price of the positron cameras.

A rapid, safe, non-invasive technique to perform spectroscopic imaging *in vivo* would permit patients to be much more thoroughly screened for disease with reduced cost and discomfort. If one could overcome the problems associated with scattering, optical imaging could meet this requirement and "optical biopsy" would find immediate markets, e.g. screening for cervical and skin cancers. Another area of application for functional imaging is the real-time *in situ* monitoring of tissue during therapy or surgical intervention, in order to aid the clinician in evaluating the status of a therapeutic procedure and minimizing collateral damage. Optical imaging should be able to contribute here. More ambitiously, functional imaging of whole organs, such as the brain, could reveal much about the workings of physiological processes and would be an invaluable diagnostic tool. Thick tissue spectroscopic imaging is an area where functional MRI is currently supreme but into which optical techniques are making inroads, e.g. in mammography and brain imaging.

1. OPTICAL IMAGING THROUGH BIOLOGICAL TISSUE

To realise optical biopsy, one must be able to non-invasively acquire the same information obtained from stained histological sections directly from the patient. In vivo microscopy and functional imaging, particularly via endoscopy, is a rapidly growing field that is promising to produce important new diagnostic tools. One of the main goals is to increase the penetration depth at which these can image below the surface of tissues. To do this one must address the issues of absorption and scattering in biological tissue. Let us underline that the overall damping due to photon absorption is weak and that photons can thus easily propagate through centimeters of tissues if they are chosen in a specific spectral range from red ($\sim$ 650 nm) to near infrared ($\sim$ 1.3 μm). Absorption is indeed rather weak ($\sim$ 0.1 to 1 cm^{-1}) in this range whereas hemoglobin and other chromophores absorb in the visible and water starts to be a strong absorber in the near IR above 1.3 μm. All of us have seen red light emerging through the hand when a flash-light is placed beneath: such an experiment clearly demonstrates that, at these wavelengths, the main effect which prevents from observing in-depth structures is the strong scattering experienced by the light propagating through tissues. In this spectral region of relatively high transmission [0.7–1.3 μm], that is often described as the "therapeutic window" of biological tissue, semiconductor lasers and recently developed solid-state laser technology can now conveniently provide high power tunable optical radiation in the near infra-red (NIR) spectral region. This enabling technology has prompted many researchers to address the challenge of imaging through scattering media such as biological tissue.

2.   ORDERS OF MAGNITUDE

The various scattering parameters which determine the behaviour of the light propagation in scattering media are the following:

- *The scattering mean free path, ℓ:* It is defined as the length between two successive scattering events. In biological tissues, ℓ is of the order of 50 μm. Because the flux ϕ_0 of ballistic photons – the photons who have propagated without loosing their spatial coherence properties – is damped exponentially along the beam direction z:

$$\phi = \phi_0 \exp\left(-z/\ell\right) , \tag{1}$$

  the damping after 1 mm is about a hundred dB.

- *The transport mean free path, ℓ^*:* In fact single scattering events in the tissues are mainly induced by structures whose sizes are larger than the wavelength and the index mismatch with the surrounding medium (typ-ically $\sim 10^{-2}$) is not large enough to induce a complete memory loss of the initial direction after one scattering event. Therefore, each scat-tering event only contributes by a small deviation $\theta \ll \pi/2$ of the light trajectory. For this reason, the transport mean free path $\ell^* = \ell/(1- < \cos\theta >)$, which accounts for the anisotropic factor involved in the scat-tering events, is considered. ℓ^* is of the order of 1 mm in most tissues. This parameter or its inverse ($\mu'_s = 1/\ell^*$) is the pertinent parameter in the light diffusion equation.

The damping of the energy propagation in an absorbing and scattering medium writes

$$\phi = \phi_0 \exp\left[-z\sqrt{3\mu_a\left(\mu'_s + \mu_a\right)}\right] \text{ (1-D model)} . \tag{2}$$

Because in the therapeutical window, $\mu_s \sim 10$ cm^{-1} is much bigger than the absorption damping $\mu_a \sim 0.1$ cm^{-1}, Eq. 2 reduces to

$$\phi \approx \phi_0 \exp\left[-z\sqrt{3\mu_a\mu_s'}\right] . \tag{3}$$

In this case, even a 10 cm thick sample will produce a damping of the photon flux smaller than 100 dB.

3.   SHALLOW TISSUE IMAGING THROUGH SELECTION OF BALLISTIC PHOTONS

When imaging thin ($\lesssim 2$ mm) tissue samples, it is possible to use photons that have not been scattered: these are called the *ballistic* photons and they can be used to form high-resolution images. Their number, however, decreases exponentially with propagation distance (Eq. 1) and this ballistic signal is usually swamped by multiply scattered photons that obscure any image and saturate most detectors as illustrated in Fig. 1. For relatively shallow tissue depths it is

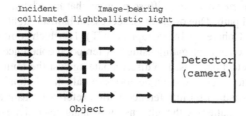

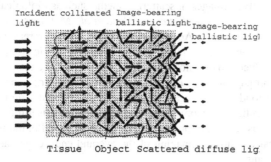

Figure 1.   Imaging through scattering media

possible to use various filtering techniques to block the multiply scattered photons and there is much current research that aims to extend optical imaging and biopsy to depths of a few mm. Confocal microscopy rejects much scattered light through its spatial filtering action and has been shown to form useful images at tissue depths of up to 300-500 μm, both *in vitro* and *in vivo* which corresponds to 40 to 50 dB of damping for the ballistic photons. Two-photon fluorescence confocal microscopy is particularly interesting in this regime since it ensures that the all detected fluorescence photons originate from the desired image plane. This technique also ensures a better penetration (using IR excitation) and a better resolution (non linear response) than single photon confocal microscopes (see the contribution of E. Beaurepaire *et al.* in this book). Scattering does, however, make quantitative imaging very difficult and so wavelength-ratiometric imaging or fluorescence lifetime imaging (FLIM) are likely to be increasingly employed (see the contribution of N. P. Barry *et al.* in this book).

Imaging to much greater tissue depths does not appear to be practical using optical microscopy and spatial filtering alone and more sophisticated techniques must also be employed to discriminate in favour of the ballistic photons. These techniques exploit either the fact that the ballistic signal photons retain spatial *coherence* with the incident light or the fact that the ballistic photons arrive at the detector *earlier* than any diffuse photons that have been scattered back into their original path. One of the most successful ballistic light imaging techniques is Optical Coherence Tomography (OCT) [1], which combines confocal microscopy with coherence gating using low temporal coherence light. Essentially coherent detection is used to discriminate against the multiply scattered photons that scatter back into the trajectories of ballistic photons. The ballistic light signal is measured by interference with a reference beam and detecting the resulting fringe pattern with high sensitivity using heterodyne detection (see Fig. 2. By using low coherence length radiation and matching the time-of-flight of the reference beam signal to that of the desired ballistic light, OCT can acquire depth-resolved images in scattering media such as biological tissue (see E. Beaurepaire *et al.*, E. Lacot *et al.*, G. Brun *et al.*, M. Gross *et al.* and K. K. Bizheva *et al.* in this book). According to Hee *et al.* [2] , for reasonable powers of tissue irradiation in the infrared, the unscattered ballistic component of the light will be reduced to the shot noise detection limit (more than 150 dB attenuation) after propagating through approximately 36 scattering mean free paths ℓ. This corresponds to a 'typical' tissue thickness of about 4 mm (or $\sim$ 2 mm tissue depth in reflection geometry), which is potentially useful for clinical applications such as screening for skin cancer, diagnosing and/or monitoring other dermatological conditions, imaging of the eye and endoscope-based imaging of internal organs.

OCT has proved clinically useful as a means of acquiring depth-resolved images in the eye and is being developed for application in strongly scattering

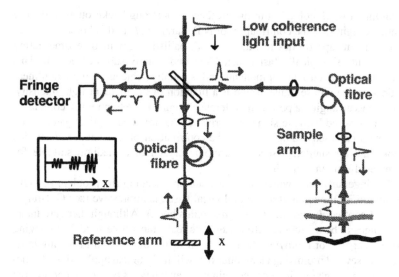

Figure 2.   Optical Coherence tomography

biological tissue [3]. Although the image acquisition time is relatively slow (due to the requirement to scan pixel-by-pixel) the use of new high power superluminescent diodes in the 100 mW range and of ultrafast lasers to provide high average power, low coherence length radiation has resulted in an *in vivo* OCT system providing real-time imaging. This technique readily lends itself to endoscopic application and one exciting development is the use of OCT for intravascular imaging. There are a number of other approaches to coherent imaging including whole-field 2-D acquisition techniques with no requirement to transversely scan pixel by pixel. Some of these techniques exploit heterodyne detection in the time domain, like OCT, and others are based on holography [4], i.e. they use transverse *spatial* interference. Whole-field imaging that is used at ESPCI-Paris using a temporally and spatially incoherence source and a well balanced interferometric microscope intrinsically provides higher image acquisition rates and can take advantage of inexpensive spatially incoherent, broadband sources such as high power LED's [5]. At Imperial College a real-time 3-D imaging system is developed, based on photorefractive holography in Multiple quantum well devices [6], which can potentially acquire depth-resolved images at 1000's of frames per second. The photorefractive effect is sensitive, not to the intensity of incident light but, to the spatial derivative of intensity. This somewhat esoteric physics makes it particularly well suited to recording a weak hologram in the presence of a strong background of diffuse scattered light (see the contribution of N.P. Barry *et al.* in this book).

Incoherent approaches to imaging with ballistic light involve time gating with "shutters" typically faster than 10 ps and usually sub-picosecond. This is too fast for electronic components and is usually achieved using a nonlinear optical device whose transmission is a function of the incident light intensity. It involves arranging a powerful "reference" signal to be temporally coincident with the desired ballistic signal on arrival at the nonlinear optical gate, which may be a parametric amplifier (see the contribution of E. Lantz *et al.* in this book), a Kerr shutter, a second harmonic crystal or a medium suitable for stimulated Raman scattering.

The ingenuity with which physicists are approaching the problem of high-resolution 3-D imaging through biological tissue is impressive (see G. Brun *et al.*, E. Lacot *et al.* and M. Gross *et al.* in this book). Although there are many challenges to be overcome before there is a real-time whole-field 3-D imaging system capable of removing the need for invasive biopsy and histopathology, there are several promising techniques that will find clinical application. Partic-ularly promising areas include screening for several types of cancer, endoscopic inspection of joints and suspect tumours and *in situ* monitoring of tissue during minimally invasive surgery and therapy.

4.   DEEP TISSUE IMAGING THROUGH TIME AND FREQUENCY GATING

At their most successful, these ballistic light techniques can extend the depth of tissue image to $\sim$ 2 mm when imaging in reflection and rather further when imaging in transmission. If it is necessary to image through cm, rather than mm of tissue, however, there will be no detectable ballistic light signal and one must extract useful information from the scattered photons. One approach is to use time gating to select the earliest arriving scattered light, which will provide the most useful image information. For moderate tissue depths one can exploit the fact that biological tissue is highly forward scattering, with most photons being only slightly deviated from their original direction upon each scattering event. After a few cm of tissue, there can be a significant number of photons that have followed a reasonably well-defined *"snake-like"* path about the original direction through the tissue, arriving at the detector after the ballistic light but before the fully *diffuse* photons whose trajectories have become effectively randomised. This early arriving light can be selectively gated using ultrafast lasers and a variety of detectors including streak cameras, time-gated optical image intensifiers and time-gated photon counting systems, which provide picosecond time gates ranging from $\sim$ 1-100 ps. A chapter of this book is dedicated to the investigation of the transition from single scattering to light diffusion (K. K. Bizheva *et al.*).

An alternative cheaper approach to temporal gating is the use of (high) frequency signal discrimination (which is the Fourier Transform of the pulse experiment). The periodic solution of the diffusion equation with a periodic term source gives rise to the so called "photon density waves". Such diffusive waves, like thermal waves, are used in the near field regime (their amplitudes are damped by a factor $\exp(-2\pi)$ after propagating over a single wavelength path) and they can reveal subsurface structures through their refractive and diffusive behaviour [7]. Coherence gating and polarisation gating can be used when the early arriving light still retains some degree of coherence with the incident light. A coherent heterodyne system working in transmission has been demonstrated to tomographically image through biological tissues of several mm thickness, such as human fingers and teeth [8], albeit at lower resolution than is achieved when imaging with ballistic light.

For many important biomedical applications, such as mammography or imaging brain function, it is necessary to penetrate through $\gtrsim 5 - 10$ cm of tissue, after which almost all the detected signal is diffuse. Detecting earlier arriving photons can provide more information but, if the time window is too narrow, the detected signal becomes too weak to use with clinically acceptable data acquisition times. This extremely challenging problem is, however, being addressed with some success using statistical models of photon transport, with different degrees of approximation ranging from full Monte Carlo simulation of photon propagation to the diffusion equation. The approach is to address "*the inverse problem*", i.e. to measure the scattered light signal as comprehensively as is practical and to calculate what distribution of absorption and scattering properties would have produced the measured signal. This provides a means to quantify the optical properties of biological tissue, averaged over a volume corresponding to a particular distribution of photon paths, and to form relatively low-resolution tomographic images (see contributions of S. Arridge and J.-M. Tualle *et al.*).

Multiplying the number of sources and detectors and correlating the various intensity distributions over the whole structure under examination gives, as expected, better precision in localizing the size and position of the local structure to be identified. However, due to the inhomogeneity of biological tissues, these techniques applied in real life situations do not provide yet the level of resolution required to be of practical use in medical imaging. (the typical tumour sizes observed are within the cm-range while the mm-range is desired). One difficulty may come from this very basic approach which relies on the diffusion equation. Indeed, it is not always valid for instance in multi-scale samples such as muscular tissues. Calculations based on anomalous diffusion [9] and fractal geometry (see the contribution of C.A. Guerin *et al.* in this book) may be necessary to handle such complex problems.

One important exception to this lack of success in finding precisely the position, the size and the optical properties of hidden structure is the case of "activation" studies. This corresponds to a "differential" situation in which an unknown structure is locally changing with time. Here the goal is not to reach a full description of the unknown structure but only to characterize the local changes induced by the activation. The main field of application of this approach is certainly the brain activation study. A matrix of light emitters (usually at two wavelengths, e.g. 780 and 840 nm, which discriminate between oxy-hemoglobine and desoxy-hemoglobine) is coupled to a matrix of light detectors. The signal difference between each pair of detectors is electronically balanced before activation takes place. A local activation (e.g. in the brain) induces a supplementary absorption which perturbs the photon distribution and unbalances the detectors. This "triangulation" technique provides, with a cheap and simple setup, rather spectacular results. Sometimes a modulation is added and the phase shift of the photon density wave is enhanced by the differential approach [10]. Recently it has been demonstrated that for weak perturbations (weak variations of the optical parameters), the inverse problem can be rigorously solved [11].

5.   COUPLING OPTICS AND ACOUSTICS TO REVEAL OPTICAL CONTRASTS

The two techniques that will be described now take advantage of a different way of handling the optical information, namely by coupling optics and acoustics. These two approaches are fairly different in their basic principle as well as in the experimental setups although their names could be rather confusing : opto-acoustics (sometimes called photo-acoustics) and acousto-optics.

5.1.   OPTO-(OR PHOTO-)ACOUSTICS

In this technique a pulsed (or modulated) electromagnetic beam irradiates the structure under examination. Because of their tortuous paths, the photons spread in the volume in a typical time of a few nano-seconds. During this time, local absorbing centers absorb the electromagnetic energy, experience a fast thermal expansion and become acoustic sources. Because the speed of sound is much smaller than the speed of light, the optical problem of tomographic reconstruction has been converted into a simpler problem of acoustic reconstruction of source distribution. This approach is rather new compared to the purely optical transillumination method and has led already to a large number of studies in which time gating brings a new piece of information in the accurate determination of the depth of the ultrasonic sources [12].

5.2. ACOUSTO-OPTICS

Here a DC monomode laser is used to illuminate the sample. Due to the good temporal coherence of the source, all the scattered photons are likely to interfere when they emerge from the sample volume. These interferences are seen as a speckle field at the output of the sample [13]. Focusing an ultrasonic beam into the medium will mainly induce a small (smaller than the optical wavelength) displacement of the scatterer and a local variation of the refractive index. We know that for a few cm-thick typical biological sample, the average length of the photons trajectories could reach more than ten times the sample thickness. This means that the probability for a photon to interact with a an ultrasonic beam focused into the sample volume is rather large. This interaction will result in a speckle modulation at the ultrasonic frequency. If the zone irradiated by the ultrasonic field is optically absorbing, less photons will emerge from this zone and will contribute to the modulated signal. A reduced modulation will be observed when scanning the acoustic source across the absorbing center. The modulation can be detected by selecting a single speckle grain and by averaging over a number of random realizations (using for instance latex particles). However, this is not always possible, in particular because the overall geometry is rather stable in a semisolid tissue. A multiple detector and a parallel processing of many speckle grains has been used to strongly improve the signal-to-noise ratio and the acquisition rate (see S. Lévêque-Fort's contribution in this book).

Conclusion and possible trends

The complexity associated to multiple scattering in turbid media such as tissues of the human body, prevents an easy use of light as a routine tool for in depth examination. Nevertheless examination of shallow tissue of retina skin, digestive tract, teeth, ... , start to be used today as complementary diagnostic tools. Optical techniques will obviously progress through to better characterization of optical properties of biological tissues and will benefit from improvement of laser properties and detection techniques as well as new mathematical approaches of the inverse problem.

In the various theoretical and experimental approaches reviewed in this paper, the scattering of light is an obstacle and a limit. Very few methods –like acousto-optics, diffuse wave spectroscopy (see contributions of G Maret and M. Heckmeier, and P. Snabre *et al.*) or speckle correlation analysis (see S. E. Skypetrov's contribution)– take advantage of the short and long range coherence properties of the multiply-scattered light to retrieve "imaging information".

A better understanding of the fundamental physics of wave in random media should stimulate new ideas, new experimental schemes leading to new breakthrough in this delicate problem of using light to image highly scattering media.

References

[1] D. Huang, E. A. Swanson, C. P. Lin, J. S. Schuman, W. G. Stinson, M. R. Hee, T. Flotte, K. Gregory, C. A. Puliafito and J. G. Fujimoto, Science, **254**, 117 (1991).

[2] M. R. Hee, J. A. Izatt, J. M. Jacobson, J. G. Fujimoto and E. Swanson, Opt. Lett., **18**, pp 950-952 (1993).

[3] J. G. Fujimoto, M. E. Brezinski, G. J. Tearney, S. A. Boppart, B. Bouma, M. R. Hee J. F. Southern and E. A. Swanson, Nature Medicine **1**, 970 (1995).

[4] E. Leith, C.Chen, H. Chen, Y. Chen, D.Dilworth, J. Lopez, J. Rudd, P-C. Sun, J.Valdmanis, and G.Vossler: J. Opt.Soc.Am. A, **9**, 1148 (1992).

[5] E. Beaurepaire, A.C. Boccara, M. Lebec, L. Blanchot, and H. Saint-Jalmes: Opt. Lett. **23**, 224 (1998).

[6] R. Jones, S. C. W. Hyde, M. J. Lynn, N. P. Barry, J. C. Dainty, P. M. W. French, K. M. Kwolek, D. D. Nolte and M. R. Melloch, Appl. Phys. Lett. **69**, 1837 (1996).

[7] J. Ripoll, M. Nieto-Vesperinas, and R. Carminati : J. Opt. Soc. Am. A **16**, 1466-1476 (1999).

[8] B. Devaraj, M. Usa, K. P. Chan, T. Akatsuka and H. Inaba, IEEE J STQE-**2**, 1008 (1996).

[9] A. H. Gandjbakhche and G.H. Weiss in *Progress in Optics*, vol. XXXIV, E. Wolf edit. (1995).

[10] B. Chance, E. Anday, S. Nioka, S. Zhou, H. Long, K. Worden, C. Li, T. Turray, Y. Ovetsky, D. Pidikiti, and R. Thomas : Opt. Express, **2** (1998).

[11] V. Ntziachristos, B. Chance and A.G. Yodh, Optics Express, **20**, 230 (1999).

[12] A. A. Oraevsky et al., : Appl. Opt. **36**, 402-415 (1997)

[13] M. Kempe, D. Zaslavsky, and A. Z. Genack, J. Opt. Soc. Am. A **14**, 1151 (1997).

II

MULTIPLE WAVE SCATTERING

Chapter 1

COHERENT MULTIPLE SCATTERING
IN DISORDERED MEDIA

E. Akkermans
Department of Physics, Technion
32000 Haifa, Israel
eric@physics.technion.ac.il

G. Montambaux
Laboratoire de Physique des Solides
Université Paris-Sud
Bât. 510, F-91405 Orsay Cedex, France

Abstract   These notes contain a rapid overview of the methods and results obtained in the field of propagation of waves in disordered media. The case of Schrödinger and Helmholtz equations are considered that describe respectively electrons in metals and scalar electromagnetic waves. The assumptions on the nature of disorder are discussed and perturbation methods in the weak disorder limit are presented. A central quantity, namely the probability of quantum diffusion is defined and calculated in the same limit. It is then shown that several relevant physical quantities are related to the return probability. Examples are provided to substantiate this, which include the average electrical conductivity, its fluctuations, the average albedo and spectral correlations.

Keywords: coherence, multiple scattering, diffuson, cooperon, radiative transfer

Introduction

The study of wave propagation in random media gave rise to a huge amount of work especially during the last twenty years. Today, this field is split in two main subfields. One is concerned with the interplay between coherence and disorder in metallic systems and the second deals with the same problematics but for electromagnetic waves (this includes as well sound or gravity waves).

P. Sebbah (ed.), Waves and Imaging through Complex Media, 29–52.
© 2001 *Kluwer Academic Publishers. Printed in the Netherlands.*

Each one of the subfields has its own specificities and advantages, such as the effect of magnetic fields on transport and thermodynamics, interactions for metals and angular structure (spectroscopy) for waves. Unfortunately, the split between these two subfields has been growing so that their interplay is now certainly too weak in spite of the existence of a number of excellent reviews. These notes represent a very preliminary step towards a unified presentation of this field [1]. We have tried to present a general formalism that can apply to both situations (electrons and waves). It is centered around the existence of a basic quantity *the probability of quantum diffusion* which allows to describe either weak localization effects of the electrical conductivity of metals, spectral quantities of isolated electronic systems or the coherent and incoherent albedo, dynamical effects in multiple scattering of light by suspensions, etc ...

The next three sections contain basic definitions and generalities on the kind of waves we consider and the model of disorder. The section 4 deals with the definition and the calculation of the probability of quantum diffusion in multiple scattering using various approximations. Then, the subsequent sections apply these results to a selection of examples taken either from electronic systems (conductivity and spectral correlations) or multiple scattering of electromagnetic waves (average albedo in optical systems).

1.   MODELS FOR THE DISORDERED POTENTIAL

We shall consider mainly two problems where waves and disorder are involved. The first one corresponds to the study of spinless electrons in disordered metals or semiconductors. For a degenerate gas of free electrons of mass m and charge $-e$, the Schrödinger equation is governed by the Hamiltonian

$$\mathcal{H}\psi(\mathbf{r}) = -\frac{\hbar^2}{2m}(\nabla + \frac{ie}{\hbar}\mathbf{A})^2\psi(\mathbf{r}) + V(\mathbf{r})\psi(\mathbf{r}) , \qquad (1)$$

where $\mathbf{B} = \nabla \times \mathbf{A}$ is the magnetic field. Effects associated to the band structure or to interactions between the electrons in the framework of Fermi liquid theory are accounted for by the replacement of the mass m by an effective mass. The potential $V(\mathbf{r})$ decribes both the scattering by inhomogeneities and the confinement potential.

The second problem we shall consider is the propagation of electromagnetic waves of frequency ω in the *scalar* approximation. The behaviour of the electric field $\psi(\mathbf{r})$ is obtained from the Helmholtz equation

$$-\Delta\psi(\mathbf{r}) - k_0^2\mu(\mathbf{r})\psi(\mathbf{r}) = k_0^2\psi(\mathbf{r}) , \qquad (2)$$

where the function $\mu(\mathbf{r}) = \delta\epsilon/\bar{\epsilon}$ is the relative fluctuation of the dielectric constant, $k_0 = \bar{n}\frac{\omega}{c}$ and $\bar{n}$ is the average optical index. Under this form, the Helmholtz equation has a structure similar to the Schrödinger equation and the waves are scattered by the fluctuations of the dielectric function. It is nevertheless interesting to notice that in the latter case, the strength μ of the potential is multiplied by the frequency ω^2 so that in contrast with electronic systems, a decrease of the frequency ω leads to a weaker effect of the disorder.

To account for the effects of the disorder in either case, we shall consider a random continuous function $V(\mathbf{r})$, of zero spatial average, $\langle V(\mathbf{r}) \rangle = 0$, where $\langle \cdots \rangle$ represents the spatial disorder average. The potential is characterized by its correlations especially by the two-point correlation function

$$\langle V(\mathbf{r})V(\mathbf{r'}) \rangle = B(\mathbf{r} - \mathbf{r'}) \,. \tag{3}$$

For a potential $V(\mathbf{r})$ which is localized enough, there exits a length r_c which describes the fall off of $B(\mathbf{r} - \mathbf{r'})$. In the limit where the wavelength $\lambda \gg r_c$, the scattering events are statistically independent and we can consider the limiting case

$$\langle V(\mathbf{r})V(\mathbf{r'}) \rangle = \gamma \delta(\mathbf{r} - \mathbf{r'}) \,. \tag{4}$$

Such a potential is usually called a white noise. This is the case we shall consider throughout these notes. For the case of the Helmholtz equation, the potential is taken to be $V(\mathbf{r}) = k_0^2 \mu(\mathbf{r})$ so that γ, in that case, has the dimensions of the inverse of a length.

2.   PERTURBATION THEORY FOR MULTIPLE SCATTERING

2.1.   SINGLE SCATTERING - ELASTIC SCATTERING TIME

For a very dilute system, we can first assume that a given incident wave of wavevector $\mathbf{k}$ is scattered only once into a state $\mathbf{k'}$ before leaving the system. The lifetime τ_k of the state $\mathbf{k}$ is then given (using the notations of quantum mechanics) by the Fermi golden rule

$$\frac{1}{\tau_k} = \frac{2\pi}{\hbar} \sum_{\mathbf{k'}} |\langle \mathbf{k}|V|\mathbf{k'} \rangle|^2 \delta(\epsilon_\mathbf{k} - \epsilon_\mathbf{k'}) \,. \tag{5}$$

For the white noise case, we obtain

$$\frac{1}{\tau_e} = \frac{2\pi}{\hbar} \rho_0 \gamma \,, \tag{6}$$

where ρ_0 is the density of states per unit volume. To this time, we can associate a length, the elastic mean free path l_e, defined as $l_e = v\tau_e$ by using the group velocity v. But the Fermi golden rule is valid for short times $t \ll \tau_e$ and in perturbation with the scattering potential V. In order to go beyond these limitations, we need to resort to a more powerful tool namely the formalism of the Green functions.

The Green function associated to a wave equation can be defined as the response to a pulse i.e. to a δ-function perturbation. For the case of the hamiltonian (1), the Green function $G^{R,A}(\mathbf{r}_i, \mathbf{r}, \epsilon)$ is solution of

$$(\epsilon - \mathcal{H} \pm i0)\, G^{R,A}(\mathbf{r}_i, \mathbf{r}, \epsilon) = \delta(\mathbf{r} - \mathbf{r}_i)\,. \tag{7}$$

We can define as well the free Green function $G_0(\mathbf{r}_i, \mathbf{r}, \epsilon)$ in the absence of scattering potential, which is the solution of

$$\left(\epsilon + \frac{\hbar^2}{2m}\Delta_{\mathbf{r}} \pm i0\right) G_0^{R,A}(\mathbf{r}_i, \mathbf{r}, \epsilon) = \delta(\mathbf{r} - \mathbf{r}_i)\,, \tag{8}$$

so that the Green function $G^{R,A}(\mathbf{r}_i, \mathbf{r}, \epsilon)$ can be also expressed as a solution of the integral equation

$$G(\mathbf{r}_i, \mathbf{r}, \epsilon) = G_0(\mathbf{r}_i, \mathbf{r}, \epsilon) + \int G(\mathbf{r}_i, \mathbf{r}', \epsilon) V(\mathbf{r}') G_0(\mathbf{r}', \mathbf{r}, \epsilon) d\mathbf{r}'\,. \tag{9}$$

The solution of (8) is given (for d=3) by

$$G_0^{R,A}(\mathbf{r}_i, \mathbf{r}, \epsilon) = -\frac{m}{2\pi\hbar^2}\frac{e^{\pm ikR}}{R}\,, \tag{10}$$

with $\epsilon = \hbar^2 k^2/2m$.

For the Helmholtz equation (2), we obtain a sequence of similar equations, but attention needs to be paid to the fact that the dispersion of the waves is now linear instead of quadratic for the Schrödinger case. The Green equation is

$$\left(\Delta_{\mathbf{r}} + k_0^2(1 + \mu(\mathbf{r}))\right) G(\mathbf{r}_i, \mathbf{r}, k_0) = \delta(\mathbf{r} - \mathbf{r}')\,, \tag{11}$$

while the solution of the free Green equation obtained for $\mu(\mathbf{r}) = 0$, is

$$G_0^{R,A}(\mathbf{r}_i, \mathbf{r}, k_0) = -\frac{1}{4\pi}\frac{e^{\pm ik_0 R}}{R}\,. \tag{12}$$

2.2.   ELECTROMAGNETIC WAVES

The formalism of Green functions provides an appropriate technical framework for the study of solutions of wave equations for free systems namely without sources. It becomes essential for the study of the propagation of electromagnetic waves from a distribution of sources $j(\mathbf{r})$ of the field. This is the problem of radiative transfer [2]. For a pointlike source, we are back to the previous problem of the Green equation. In the general case, The Helmholtz equation (2) needs to be replaced by

$$\Delta\psi(\mathbf{r}) + k_0^2(1 + \mu(\mathbf{r}))\psi(\mathbf{r}) = j(\mathbf{r})\,, \tag{13}$$

where $\psi(\mathbf{r})$ is indeed a Green function i.e. it depends on the distribution of sources $j(\mathbf{r})$. The equation (13) can also be written in the form of the integral equation

$$\psi(\mathbf{r}) = \int d\mathbf{r}_i j(\mathbf{r}_i) G_0(\mathbf{r}_i, \mathbf{r}, k_0) - k_0^2 \int d\mathbf{r}' \psi(\mathbf{r}') \mu(\mathbf{r}') G_0(\mathbf{r}', \mathbf{r}, k_0) , \quad (14)$$

which allows to consider separately the effects of the source and of the random potential $\mu(\mathbf{r})$.

3.   MULTIPLE SCATTERING EXPANSION

Either the expressions (9) or (14) provide the starting point for a systematic expansion of the Green function in terms of the free Green function. It can be written

$$\begin{aligned} G(\mathbf{r}, \mathbf{r}') &= G_0(\mathbf{r}, \mathbf{r}') + \int d\mathbf{r}_1 G_0(\mathbf{r}, \mathbf{r}_1) V(\mathbf{r}_1) G_0(\mathbf{r}_1, \mathbf{r}') \\ &+ \int d\mathbf{r}_1 d\mathbf{r}_2 G_0(\mathbf{r}, \mathbf{r}_1) V(\mathbf{r}_1) G_0(\mathbf{r}_1, \mathbf{r}_2) V(\mathbf{r}_2) G_0(\mathbf{r}_2, \mathbf{r}') \\ &+ \dots \end{aligned} \quad (15)$$

We can now calculate the average Green function using the white noise potential (4). All the odd terms in the potential V disappear from (15), and it remains

$$\overline{G}(\mathbf{r}, \mathbf{r}') = G_0(\mathbf{r}, \mathbf{r}') + \gamma \int d\mathbf{r}_1 G_0(\mathbf{r}, \mathbf{r}_1) G_0(\mathbf{r}_1, \mathbf{r}_1) G_0(\mathbf{r}_1, \mathbf{r}') + \dots \quad (16)$$

where we denote from now on the disorder average by $\overline{\cdots}$. By averaging over the disorder, the medium becomes again translational invariant and the Green function $\overline{G}(\mathbf{r}, \mathbf{r}') = \overline{G}(\mathbf{r} - \mathbf{r}')$. The average over the disorder generates all possible diagrams. Among them, there is a subclass called irreducible diagrams that cannot be split into two already existing diagrams without cutting an impurity line. It is possible to rewrite the average Green function or its Fourier transform in terms of the contribution of these diagrams only. We then obtain the so called Dyson equation

$$\overline{G}(\mathbf{k}) = G_0(\mathbf{k}) + G_0(\mathbf{k}) \Sigma(\mathbf{k}, \epsilon) \overline{G}(\mathbf{k}) , \quad (17)$$

where the function $\Sigma(\mathbf{k}, \epsilon)$ is called the self-energy. It should be emphasized that, although the self-energy contains only the irreducible diagrams, there is an infinity of them. Thus, the calculation of Σ is a difficult problem. For the white noise potential (4), Σ can be expanded in powers of γ. To first order, for the Schrödinger equation, we obtain

$$\Sigma_1^{R,A}(\mathbf{k}, \epsilon) = \frac{\gamma}{\Omega} \sum_\mathbf{q} G_0^{R,A}(\mathbf{q}) , \quad (18)$$

where Ω is the volume of the system. The real part of Σ corresponds to an irrelevant shift of the origin of the energies that we shall ignore. The imaginary part is

$$\mathrm{Im}\Sigma_1^R(\mathbf{k}, \epsilon) = -\pi\rho_0(\epsilon)\gamma \,, \tag{19}$$

while for the Helmholtz equation, it is

$$\mathrm{Im}\Sigma_1^R(\mathbf{k}, \epsilon) = -\frac{\gamma k_0}{4\pi} \,. \tag{20}$$

We emphasize again that the difference between these two expressions results from the two distinct dispersions of respectively the Schrödinger and Helmholtz equations. Higher orders terms in the expansion of Σ are proportional to $\mathrm{Im}\Sigma_1^R$ times some power of the dimensionless parameter $\frac{1}{kl_e}$. The contribution Σ_1 describes the multiple scattering of the wave as a series of independent effective collisions. The higher corrections include interference effects between those successive scattering events. The weak disorder limit $kl_e \gg 1$ amounts to neglecting these interferences. We deal then with the so called self-consistent Born approximation. We shall, from now on, consider only this limit.

It is then a straightforward calculation to get an expression for the average Green function at this approximation:

$$\overline{G}^{R,A}(\mathbf{r}_i, \mathbf{r}, k_0) = G_0^{R,A}(\mathbf{r}_i, \mathbf{r}, k_0) \ e^{-|\mathbf{r}-\mathbf{r}_i|/2l_e} \,. \tag{21}$$

To conclude this section, we would like to notice that although this expression has been obtained for the case of an infinite system, this restrictive assumption can be relieved and we need to consider, for this relation to be valid, only systems of sizes $L \gg l_e$.

4.   PROBABILITY OF QUANTUM DIFFUSION

The quantities of physical interest are usually not related to the average Green function but instead to the so called *probability of quantum diffusion* which describes the probability for a quantum particle (or a wave) to go from the point $\mathbf{r}$ to the point $\mathbf{r}'$ in a time t. Once we average over the disorder, we shall see that this probability $P(\mathbf{r}, \mathbf{r}', t)$ contains mainly three contributions:

- i. The probability to go from $\mathbf{r}$ to $\mathbf{r}'$ without scattering.

- ii. The probability to go from $\mathbf{r}$ to $\mathbf{r}'$ by an incoherent sequence of multiple scattering, which is called the *diffuson*.

- iii. The probability to go from $\mathbf{r}$ to $\mathbf{r}'$ by a coherent multiple scattering sequence. We shall calculate one such coherent process, called the *cooperon*.

We shall first define the probability for the Schrödinger case. Please notice that in the rest of the paper, we shall take $\hbar = 1$. To that purpose, we consider a gaussian wavepacket of energy ϵ_0. We shall also assume that around ϵ_0 the density of states is constant. Then, we can write for the Fourier transform $P(\mathbf{r}, \mathbf{r}', \omega)$ of the probability $P(\mathbf{r}, \mathbf{r}', t)$ the expression:

$$P(\mathbf{r}, \mathbf{r}', \omega) = \frac{1}{2\pi\rho_0} \overline{G^R(\mathbf{r}, \mathbf{r}', \epsilon_0) G^A(\mathbf{r}', \mathbf{r}, \epsilon_0 - \omega)} \, . \tag{22}$$

This probability is normalized to unity which means that either

$$\int P(\mathbf{r}, \mathbf{r}', t) d\mathbf{r}' = 1 \tag{23}$$

or

$$\int P(\mathbf{r}, \mathbf{r}', \omega) d\mathbf{r}' = \frac{i}{\omega} \, . \tag{24}$$

4.1. FREE PROPAGATION

In the absence of disorder, the Green functions in (22) take their free expression (10) and it is straightforward to obtain for the three dimensional case

$$P(\mathbf{r}, \mathbf{r}', t) = \frac{\delta(R - vt)}{4\pi R^2} \, , \tag{25}$$

where $R = |\mathbf{r}' - \mathbf{r}|$ and v being the group velocity. This probability is indeed normalized.

4.2. DRUDE-BOLTZMANN APPROXIMATION

In the presence of disorder, we need, in order to calculate the probability, to evaluate the average of the product of the two Green functions that appear in (22). The simplest approximation is to replace the average by the product of the two averaged Green functions. Here again, since we have calculated in the weak disorder limit (21) the expression of $\overline{G}$, we obtain

$$P_0(\mathbf{r}, \mathbf{r}', \omega) = \frac{e^{i\omega R/v - R/l_e}}{4\pi R^2 v} \tag{26}$$

so that

$$\int P_0(\mathbf{r}, \mathbf{r}', \omega) d\mathbf{r}' = \frac{\tau_e}{1 - i\omega\tau_e} \, . \tag{27}$$

At this approximation, the probability is not normalized, but instead

$$\int P_0(\mathbf{r}, \mathbf{r}', t) d\mathbf{r}' = e^{-t/\tau_e} \, . \tag{28}$$

It is then clear that some part is missing in the probability. The Drude-Boltzmann approximation overlooks a large part of the probability; since after a time t, it predicts that the wavepacket disappears.

4.3.   THE DIFFUSON

There is another contribution associated to the multiple scattering which can be calculated in the weak disorder limit $kl_e \gg 1$ in a semiclassical way. Using the description we obtained previously for the calculation of the average Green function, we can associate [3] to each possible sequence $\mathcal{C}$, of independent effective collisions a complex amplitude $A(\mathbf{r}, \mathbf{r}', \mathcal{C})$. Then, using a generalization of the Feynman path integral description, we can in principle write the Green function as a sum of such complex amplitudes.

Then, in order to evaluate the product of two Green functions, we notice the following two points.

- i. Due to the short range of the scattering potential, the set of scatterers entering in the sequences for both G^R and G^A must be identical.

- ii. For the effective collisions, the mean distance between them is set by the elastic mean free path $l_e \gg \lambda$. Therefore, if any two scattering sequences differ by even one collision event, the phase difference between the two complex amplitudes, which measures the difference of path lengths in units of λ will be very large and then the corresponding probability will vanish on average.

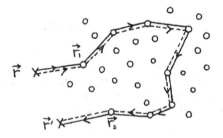

Figure 1.

We shall therefore retain only contributions of the type represented on Fig. 1 for which the corresponding probability $P_d(\mathbf{r}, \mathbf{r}', \omega)$ is

$$P_d(\mathbf{r}, \mathbf{r}', \omega) = \frac{1}{2\pi\rho_0} \int \overline{G}_\epsilon^R(\mathbf{r}, \mathbf{r}_1) \overline{G}_{\epsilon-\omega}^A(\mathbf{r}_1, \mathbf{r}) \overline{G}_\epsilon^R(\mathbf{r}_2, \mathbf{r}') \overline{G}_{\epsilon-\omega}^A(\mathbf{r}', \mathbf{r}_2) \times$$
$$\times \; \Gamma_\omega(\mathbf{r}_1, \mathbf{r}_2) d\mathbf{r}_1 d\mathbf{r}_2 \tag{29}$$

It is made of two multiplicative contributions. The first one is

$$\overline{G}_\epsilon^R(\mathbf{r}, \mathbf{r}_1)\overline{G}_\epsilon^R(\mathbf{r}_2, \mathbf{r}')\overline{G}_{\epsilon-\omega}^A(\mathbf{r}_1, \mathbf{r})\overline{G}_{\epsilon-\omega}^A(\mathbf{r}', \mathbf{r}_2) \ .$$

It describes the mean propagation between whatever two points $\mathbf{r}$ and $\mathbf{r}'$ in the medium and the first ($\mathbf{r}_1$) (respectively the last ($\mathbf{r}_2$)) collision event of the sequence of scattering events. The second contribution defines the quantity $\Gamma_\omega(\mathbf{r}_1, \mathbf{r}_2)$ which we shall call the *structure factor* of the scattering medium. In a sense, it generalizes to the multiple scattering situation the usual two-point correlation function in the single scattering case. We now use once again the assumption of independent collisions in order to write for $\Gamma_\omega(\mathbf{r}_1, \mathbf{r}_2)$ the integral equation

$$\Gamma_\omega(\mathbf{r}_1, \mathbf{r}_2) = \gamma\delta(\mathbf{r}_1 - \mathbf{r}_2) + \gamma\int \overline{G}_\epsilon^R(\mathbf{r}_1, \mathbf{r})\overline{G}_{\epsilon-\omega}^A(\mathbf{r}, \mathbf{r}_1)\Gamma_\omega(\mathbf{r}, \mathbf{r}_2)d\mathbf{r} \ . \quad (30)$$

This equation can be solved exactly in some geometries. For the infinite three dimensional space, we can make use of the translational invariance and get for the structure factor the expression

$$\Gamma_\omega(\mathbf{q}) = \frac{\gamma}{1 - P_0(\mathbf{q}, \omega)/\tau_e} \ . \quad (31)$$

where $P_0(\mathbf{q}, \omega)$ is the Fourier transform of (26) and is given by $\frac{1}{qv}\arctan\frac{ql_e}{1-i\omega\tau_e}$ with $q = |\mathbf{q}|$. Then, the probability rewrites

$$P_d(\mathbf{q}, \omega) = P_0(\mathbf{q}, \omega)\frac{P_0(\mathbf{q}, \omega)/\tau_e}{1 - P_0(\mathbf{q}, \omega)/\tau_e} \ . \quad (32)$$

Using this expression of P_d, the normalization of the total probability $P = P_0 + P_d$ can be readily checked namely $P(\mathbf{q} = 0, \omega) = \frac{i}{\omega}$. For the semi-infinite space with a point source, it is also possible to obtain a closed analytical expression for the probability P_d using the Wiener-Hopf method. But beyond these two cases, for simple finite geometries, it is not possible to obtain the solution of (29) without resorting to numerical calculations. We are then led to look for some approximate solutions. An excellent one is the diffusion approximation obtained for large times $t \gg \tau_e$ and large spatial variations $r \gg l_e$. It is obtained by expanding the structure factor under the form

$$\Gamma_\omega(\mathbf{r}, \mathbf{r}_2) = \Gamma_\omega(\mathbf{r}_1, \mathbf{r}_2) + (\mathbf{r} - \mathbf{r}_1).\nabla_{\mathbf{r}_1}\Gamma_\omega + \frac{1}{2}[(\mathbf{r} - \mathbf{r}_1).\nabla_{\mathbf{r}_1}]^2\Gamma_\omega \ , \quad (33)$$

which together with the integral equation (30) gives

$$[-i\omega - D\Delta_{\mathbf{r}_1}]\Gamma_\omega(\mathbf{r}_1, \mathbf{r}_2) = \frac{1}{2\pi\rho_0\tau_e^2}\delta(\mathbf{r}_1 - \mathbf{r}_2) \ , \quad (34)$$

where the diffusion coefficient is $D = \frac{1}{d}\frac{l_e^2}{\tau_e} = \frac{1}{d}v^2\tau_e$. At this approximation, we have between P_d and Γ_ω the following relation

$$P_d(\mathbf{r}, \mathbf{r}', \omega) \simeq 2\pi\rho_0\tau_e^2\Gamma_\omega(\mathbf{r}, \mathbf{r}') \tag{35}$$

so that P_d, as well, obeys a diffusion equation. It is interesting to check the validity of the diffusion approximation. For an infinite system, and for $\frac{r}{l_e} = 1$ the relative correction between the exact solution and diffusion approximation is 0.085 while for $\frac{r}{l_e} = 2.5$, it is less than 5.10^{-3}.

4.4.   THE COOPERON

With the normalized expression of the probability we have just obtained, it seems that we fulfilled the demand of evaluating all the relevant processes that contribute to the probability. But it could be, and it is certainly the case, that there are many other contributions that sum up to zero.

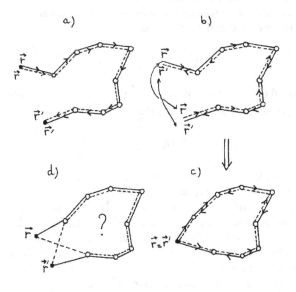

Figure 2.

For instance, we may consider the possibility represented on Fig. 2. It corresponds to the product of Green functions such as we considered before, but where now the two identical trajectories are time reversed one from the other. It is clear that if these trajectories are closed on themselves, there is no phase difference left between them. This requires that the system has time-reversal invariance namely that $G^{R,A}(\mathbf{r}, \mathbf{r}', t) = G^{R,A}(\mathbf{r}', \mathbf{r}, t)$. This relation does not hold anymore in the presence of a magnetic field for electronic systems.

The contribution, we shall call P_c, of this process to the total probability can be evaluated as we did before for the diffuson. Thus we have instead of (29)

$$
P_c(\mathbf{r}, \mathbf{r}', \omega) = \frac{1}{2\pi\rho_0} \int \overline{G}_\epsilon^R(\mathbf{r}, \mathbf{r}_1) \overline{G}_\epsilon^R(\mathbf{r}_2, \mathbf{r}') \overline{G}_{\epsilon-\omega}^A(\mathbf{r}', \mathbf{r}_1) \overline{G}_{\epsilon-\omega}^A(\mathbf{r}_2, \mathbf{r}) \times
$$
$$
\times \ \Gamma_\omega'(\mathbf{r}_1, \mathbf{r}_2) d\mathbf{r}_1 d\mathbf{r}_2 \, , \tag{36}
$$

where the new structure factor Γ_ω' is solution of the integral equation

$$
\Gamma_\omega'(\mathbf{r}_1, \mathbf{r}_2) = \gamma\delta(\mathbf{r}_1 - \mathbf{r}_2) + \gamma \int \overline{G}_\epsilon^R(\mathbf{r}_1, \mathbf{r}'') \overline{G}_{\epsilon-\omega}^A(\mathbf{r}_1, \mathbf{r}'') \Gamma_\omega'(\mathbf{r}'', \mathbf{r}_2) d\mathbf{r}'' \, .
$$
$$
\tag{37}
$$

Notice that unlike the structure factor associated to the diffuson, the new combination $\overline{G}_\epsilon^R(\mathbf{r}_1, \mathbf{r}'') \overline{G}_{\epsilon-\omega}^A(\mathbf{r}_1, \mathbf{r}'')$ that appears here cannot be simply written in terms of the probability P_0. But as before, we can evaluate $P_c(\mathbf{r}, \mathbf{r}', \omega)$ in the diffusion approximation (i.e. for slow variations) and we obtain

$$
P_c(\mathbf{r}, \mathbf{r}', \omega) \simeq \frac{\Gamma_\omega(\mathbf{r}, \mathbf{r})}{2\pi\rho_0} [\int \overline{G}_\epsilon^R(\mathbf{r}, \mathbf{r}_1) \overline{G}_\epsilon^A(\mathbf{r}', \mathbf{r}_1) d\mathbf{r}_1]^2 \, . \tag{38}
$$

Since in the presence of time reversal invariance, we have

$$
\overline{G}_\epsilon^R(\mathbf{r}_1, \mathbf{r}) \overline{G}_\epsilon^A(\mathbf{r}_1, \mathbf{r}) = \overline{G}_\epsilon^R(\mathbf{r}_1, \mathbf{r}) \overline{G}_\epsilon^A(\mathbf{r}, \mathbf{r}_1) \, , \tag{39}
$$

then, $\Gamma_\omega'(\mathbf{r}_1, \mathbf{r}_2) = \Gamma_\omega(\mathbf{r}_1, \mathbf{r}_2)$ and finally,

$$
P_c(\mathbf{r}, \mathbf{r}', \omega) = P_d(\mathbf{r}, \mathbf{r}, \omega) g(\mathbf{r} - \mathbf{r}')^2 \, , \tag{40}
$$

where in $3d$ we have the relation

$$
g(R) = \frac{\sin kR}{kR} e^{-R/2l_e} \tag{41}
$$

and $R = |\mathbf{r} - \mathbf{r}'|$. For $R = 0$, i.e. for $\mathbf{r} = \mathbf{r}'$, we have

$$
P_c(\mathbf{r}, \mathbf{r}, \omega) = P_d(\mathbf{r}, \mathbf{r}, \omega) \, , \tag{42}
$$

namely, the probability to come back to the initial point is *twice* the value given by the diffuson. The contribution of the cooperon P_c to the total probability is given by

$$
\int P_c(\mathbf{r}, \mathbf{r}', \omega) d\mathbf{r}' = P_d(\mathbf{r}, \mathbf{r}, \omega) \frac{\tau_e}{\pi\rho_0} \tag{43}
$$

for any space dimensionality. How does this contribution compare with the diffuson contribution ? We have found that $P_d(q = 0, t) \simeq 1$ while $P_c(q = 0, t) \simeq \frac{\tau_e}{\pi \rho_0} \frac{1}{(Dt)^{3/2}}$ for small enough times. Then, $P_c(q = 0, t)$ is maximum for $t \simeq \tau_e$ and given by $P_c(q = 0, \tau_e) = \frac{1}{(kl_e)^{d-1}}$. Thus, the contribution of P_c to the total probability is vanishingly small for $kl_e \gg 1$ and for a space dimensionality $d \geq 2$. But although it is very small, P_c must be compensated by another contribution in order to restore the normalization of the probability. The additional contributions result from other irreducible diagrams. But the subsequent terms in this series are not known.

We would like to conclude this section on the cooperon by emphasizing that although Γ'_ω obeys a diffusion equation, it would be meaningless and incorrect to state that $P_c(\mathbf{r}, \mathbf{r}', \omega)$ obeys it as well. The exact statement is that for $\mathbf{r} = \mathbf{r}'$, P_c and P_d are proportional and that P_d obeys a diffusion equation.

5.   RADIATIVE TRANSFER - LOCAL INTENSITY AND CORRELATION FUNCTION

In the previous sections, we have defined the quantum probability for electronic systems. The probability P is directly related to quantities that are physically measurable like the electrical conductivity or the magnetic response e.g. the magnetization [4, 5]. For the study of the propagation of electromagnetic waves in disordered media, the quantity which is usually measured is the local intensity of the field or its correlation function [6]. As we discussed previously, and by definition of the Green function, the radiative solution $\psi_\epsilon(\mathbf{r})$ of the Helmholtz equation (13) with a localized source at point $\mathbf{R} = \mathbf{0}$ is $\psi_\epsilon(\mathbf{r}) = G_\epsilon(\mathbf{0}, \mathbf{r})$. The correlation function of the field is then

$$\overline{\psi_\epsilon(\mathbf{r})\psi^*_{\epsilon-\omega}(\mathbf{r}')} = \overline{G^R_\epsilon(\mathbf{0}, \mathbf{r})G^A_{\epsilon-\omega}(\mathbf{r}', \mathbf{0})} . \qquad (44)$$

It is not directly related to the probability of quantum diffusion $P(\mathbf{r}, \mathbf{r}', \omega)$. But the radiated intensity $I(\mathbf{r})$ defined by

$$\begin{aligned} I(\mathbf{r}) &= \frac{4\pi}{c}|\psi_\epsilon(\mathbf{r})|^2 \\ &= \frac{4\pi}{c}G^R_\epsilon(\mathbf{0}, \mathbf{r})G^A_\epsilon(\mathbf{r}, \mathbf{0}) \end{aligned} \qquad (45)$$

is indeed related on average to the probability P and

$$\bar{I}(\mathbf{r}) = \frac{4\pi}{c}\overline{G^R_\epsilon(\mathbf{0}, \mathbf{r})G^A_\epsilon(\mathbf{r}, \mathbf{0})} . \qquad (46)$$

Using for the probability the following relation which is the counterpart, for the Helmholtz equation, of the relation (22)

$$P_d(\mathbf{r}, \mathbf{r}') = \frac{4\pi}{c}\overline{G^R_\epsilon(\mathbf{r}, \mathbf{r}')G^A_\epsilon(\mathbf{r}', \mathbf{r})} , \qquad (47)$$

we obtain $\overline{I}(\mathbf{r}) = P_d(\mathbf{0}, \mathbf{r})$. From now on, we shall denote $I(\mathbf{r})$ the average intensity.

We can rephrase what we did before in order to calculate the various contributions to the intensity that come respectively from the Drude-Boltzman, the diffuson and the cooperon approximations. The first contribution is given by

$$I_0(R) = \frac{1}{4\pi R^2 c} e^{-R/l_e} . \tag{48}$$

It corresponds to the contribution to the radiative intensity of waves that did not experience any collision on a distance R from the source.

The diffuson contribution is given by

$$I_d(\mathbf{r}) = \frac{4\pi}{c} \int d\mathbf{r}_1 d\mathbf{r}_2 |\overline{\psi}_\epsilon(\mathbf{r}_1)|^2 \Gamma_{\omega=0}(\mathbf{r}_1, \mathbf{r}_2) |\overline{G}_\epsilon^R(\mathbf{r}_2, \mathbf{r})|^2 \tag{49}$$

and finally, the contribution of the cooperon to the intensity is

$$I_c(\mathbf{r}) = \frac{4\pi}{c} \int d\mathbf{r}_1 d\mathbf{r}_2 \overline{\psi}_\epsilon(\mathbf{r}_1) \overline{\psi}_\epsilon^*(\mathbf{r}_2) \Gamma(\mathbf{r}_1, \mathbf{r}_2) \overline{G}_\epsilon^R(\mathbf{r}_2, \mathbf{r}) \overline{G}_\epsilon^A(\mathbf{r}, \mathbf{r}_1) , \tag{50}$$

where we used the notation $\Gamma_{\omega=0} = \Gamma$. In the diffuson approximation, the intensity I_d rewrites

$$I_d(\mathbf{r}) = P_d(\mathbf{0}, \mathbf{r}) = \frac{l_e^2}{4\pi c} \Gamma(\mathbf{0}, \mathbf{r}) \tag{51}$$

and like P_d, it obeys the diffuson equation

$$-D\Delta I_d(\mathbf{r}) = \delta(\mathbf{r}) , \tag{52}$$

whose solution in the $3d$ free space is

$$I_d(R) = \frac{1}{4\pi D R} . \tag{53}$$

We shall now apply all the considerations developed in this section to the calculations of physical quantities in some specific situations both for metallic systems and for the propagation of electromagnetic waves in suspensions.

6.   EXAMPLE 1. THE ELECTRICAL CONDUCTIVITY OF A WEAKLY DISORDERED METAL

We previously defined a weakly disordered metal as a non interacting and degenerate (spinless) electron gas (at $T = 0$), moving in the field of defects and impurities described by the white noise potential (4).

The average electrical conductivity $\sigma(\omega)$ calculated in the framework of the linear response theory [7] is given by the Kubo formula:

$$\sigma(\omega) = \frac{e^2 \hbar^3}{2\pi m^2 \Omega} \sum_{\mathbf{k},\mathbf{k}'} k_x k'_x \ \overline{G_\epsilon^R(\mathbf{k},\mathbf{k}')G_{\epsilon-\omega}^A(\mathbf{k}',\mathbf{k})} \ . \tag{54}$$

Then, we see from this definition, that the structure of the conductivity is up to the product $k_x k'_x$ very similar to those of the quantum probability P. Therefore, and just as we did before, the very definition of $\sigma(\omega)$ leads us to the following set of approximations.

6.1.   THE DRUDE-BOLTZMAN APPROXIMATION

It is obtained by replacing the average of the product of two Green functions by the product of the averages, namely

$$\overline{G_\epsilon^R(\mathbf{k},\mathbf{k}')G_{\epsilon-\omega}^A(\mathbf{k}',\mathbf{k})} \simeq \overline{G}_\epsilon^R(\mathbf{k},\mathbf{k}')\overline{G}_{\epsilon-\omega}^A(\mathbf{k}',\mathbf{k}) \ , \tag{55}$$

where the Fourier transform of the averaged Green functions (21) is

$$\overline{G}_\epsilon^{R,A}(\mathbf{k},\mathbf{k}') = \overline{G}_\epsilon^{R,A}(\mathbf{k})\delta_{\mathbf{k},\mathbf{k}'} = \frac{\delta_{\mathbf{k},\mathbf{k}'}}{\epsilon - \epsilon(\mathbf{k}) \pm i\frac{\hbar}{2\tau_e}} \ . \tag{56}$$

Then, we obtain for the conductivity $\sigma_0(\omega)$ at this approximation the following expression

$$\sigma_0(\omega) = \frac{ne^2}{m} \int d\mathbf{r}' P_0(\mathbf{r},\mathbf{r}',\omega) \ , \tag{57}$$

where P_0 is the quantum probability calculated at the same approximation (26). Using the expression (27), we obtain

$$\sigma_0(\omega) = \frac{ne^2}{m} \frac{\tau_e}{1 - i\omega\tau_e} \ , \tag{58}$$

which is the well-known Drude expression. It must be noticed that because of the Kronecker delta function that appears in the average Green functions, the scalar product $k_x k'_x$ reduces simply to k_x^2 and eventually, after averaging, to $\frac{k_F^2}{d}$ where k_F is the Fermi wavevector.

6.2.   THE CONTRIBUTIONS OF THE DIFFUSON AND THE COOPERON

Here again, we approximate the average product in the relation (54), using the same scheme we used for the diffuson. The contribution $\sigma_d(\omega)$ of the diffuson to the conductivity is thus

$$\sigma_d(\omega) = \frac{e^2\hbar^3}{2\pi m^2 \Omega^2} \frac{1}{2\pi \rho_0 \tau_e^2} P_d(0,\omega) \sum_{\mathbf{k},\mathbf{k'}} k_x k'_x \tilde{P}_0(\mathbf{k},0,\omega) \tilde{P}_0(\mathbf{k'},0,\omega) , \quad (59)$$

where the function $\tilde{P}_0(\mathbf{k},\mathbf{q},\omega) = \frac{1}{2\pi\rho_0}\overline{G}_\epsilon^R(\mathbf{k}+\frac{\mathbf{q}}{2})\overline{G}_{\epsilon-\omega}^A(\mathbf{k}-\frac{\mathbf{q}}{2})$. Since the function $\tilde{P}_0(\mathbf{k},\mathbf{q}=0,\omega)$ depends only on the modulus of the wavevector $\mathbf{k}$ and not on his direction, the angular integral in the previous expression gives a vanishing contribution namely $\sigma_d(\omega) = 0$. Then, it is interesting to notice that although the diffuson gives the main contribution to the quantum probability, its contribution to the conductivity which indeed measures such a probability, vanishes identically.

We evaluate, the contribution $\sigma_c(\omega)$ of the cooperon using the relation (36) so that

$$\sigma_c(\omega) = -\frac{e^2\hbar^3}{2\pi m^2}\frac{k_F^2}{d}[\frac{1}{\Omega}\sum_{\mathbf{k}}\tilde{P}_0^2(k,\mathbf{q}=0,\omega)]\frac{1}{\Omega}\sum_{\mathbf{Q}}P_d(\mathbf{Q},\omega) . \quad (60)$$

The sum in the brackets is straightforward so that in the diffusion approximation, we obtain

$$\sigma_c(\omega) = -\frac{e^2 D}{\pi\hbar}P_d(\mathbf{r},\mathbf{r},\omega) . \quad (61)$$

Using now the relation (43) between the cooperon and the diffuson, $\sigma_c(\omega)$ rewrites

$$\sigma_c(\omega) = -\frac{ne^2}{m}\int d\mathbf{r'} P_c(\mathbf{r},\mathbf{r'},\omega) \quad (62)$$

and the total conductivity at this order is now given by

$$\sigma(\omega) = \frac{ne^2}{m}\int d\mathbf{r'} \left(P_0(\mathbf{r},\mathbf{r'},\omega) - P_c(\mathbf{r},\mathbf{r'},\omega)\right) . \quad (63)$$

The conductivity $\sigma(\omega)$ is reduced by the coherent (cooperon) contribution. This correction is called the *weak localization correction* to the conductivity [8, 9, 10]. Relatively, this contribution of P_c is much larger than the normalization correction to the total probability. This is because P_c is now compared to P_0 and not to P_d which represents the main contribution.

6.3.   THE RECURRENCE TIME

It is of some interest to study the dc conductivity $\sigma(\omega = 0)$ using a slightly different point of view [1]. From the relation (63), and using the dc expression

$\sigma_0 = \frac{ne^2\tau_e}{m}$, we obtain for the relative correction to the conductivity,

$$\frac{\delta\sigma}{\sigma_0} = -\frac{1}{\pi\hbar\rho_0} \int_0^\infty dt P_c(\mathbf{r}, \mathbf{r}, t) , \qquad (64)$$

where ρ_0 is the density of states per unit volume.

Consider now the quantity $Z(t)$ defined by

$$Z(t) = \int_\Omega P_d(\mathbf{r}, \mathbf{r}, t) d\mathbf{r} , \qquad (65)$$

where the integral is over the volume Ω of the system. It represents the return probability to a point $\mathbf{r}$ averaged over all those points. This quantity which characterizes the solutions of the diffusion equation is sometimes called *the heat kernel* in the literature. The time integral of $Z(t)$ defines the characteristic time T_R

$$T_R = \int_\Omega d\mathbf{r} \int_0^\infty P_d(\mathbf{r}, \mathbf{r}, t) dt = \int_0^\infty Z(t) \, dt \qquad (66)$$

called the *recurrence time*. It measures the space average of the time spent by a diffusive particle within each infinitesimal volume. T_R diverges, as stated by the Polya theorem [11], for a random walk in the free space of dimensionality $d \leq 2$. Then, we define a regularized expression for T_R given by the Laplace transform

$$T_R(s) = \int_0^\infty Z(t)e^{-st} \, dt . \qquad (67)$$

This expression of the recurrence time may be interpreted by saying that $T_R(s)$ selects the contribution of all the diffusive trajectories of lengths smaller than $L_s = \sqrt{D/s}$.

Then, the expression (64) of the relative contribution of the cooperon to the conductivity rewrites

$$\frac{\delta\sigma}{\sigma_0} = -\frac{\Delta}{\pi\hbar} \int_0^\infty dt Z(t)e^{-st} = -\frac{\Delta}{\pi\hbar} T_R(s) , \qquad (68)$$

where the energy defined by $\Delta = \frac{1}{\Omega\rho_0}$ is the mean level spacing measured at the Fermi level between the energy levels of an electron gas confined in a box of volume Ω. The weak localization correction to the dc conductivity can be essentially expressed in terms of a purely classical quantity, namely the recurrence time T_R. When it diverges, we need to use its regularized version (67), where now the length L_s can be given a physical meaning. The cooperon correction results from the absence of any relative phase between two time reversed multiple scattering trajectories. Any interaction of an electron with an external perturbation may destroy this phase coherence and then the contribution of the cooperon. For a large class of such perturbations which includes inelastic collisions at finite temperature, Coulomb interactions or coupling to other excitations, we can define a phenomenological length L_ϕ called the phase coherence length and a phase coherence time τ_ϕ such that $L_\phi^2 = D\tau_\phi$. Thus, L_ϕ is the length L_s used to regularize the recurrence time. We shall see later other examples of physical quantities that can be expressed using either the heat kernel or the recurrence time.

6.4.   CONDUCTANCE FLUCTUATIONS- SPECTRAL DETERMINANT

We could as well calculate the so-called conductance fluctuations using the properties of the heat kernel. The conductance G is related to the conductivity σ by Ohm's law $G = \sigma L^{d-2}$. We define the dimensionless conductance $g = \frac{h}{e^2}G$. Its fluctuations, defined by $\langle \delta g^2 \rangle = \langle g^2 \rangle - \langle g \rangle^2$, can be expressed in terms of $Z(t)$ through

$$\langle \delta g^2 \rangle = \frac{12}{\tau_D^2} \int_0^\infty dt\, t\, Z(t)e^{-st} , \tag{69}$$

where $\tau_D = L^2/D$ is the diffusion time. For a one-dimensional system (in the sense of the diffusion equation), we have

$$\langle \delta g^2 \rangle = \frac{2}{15} . \tag{70}$$

It is interesting to recover the last two results on the average conductivity (or conductance) and its fluctuation using a systematic expansion of the recurrence time. To that purpose, we define the spectral determinant $S(s)$ by $S(s) = det(-\Delta + s)$ where Δ is the Laplacian operator. Then, by definition of $Z(t)$, we have the relation

$$T_R(s) = \int_0^\infty Z(t)e^{-st}\, dt = \frac{\partial}{\partial s}\ln S(s) , \tag{71}$$

up to a regularization independent of the Laplace variable s. This relation is valid for all space dimensionality. Consider now as a working example [5] the case of a $1d$ diffusive wire of length L. In order to describe the case of a perfect coupling of the wire to the reservoir, we demand Dirichlet boundary conditions for the diffusion equation. This is in contrast to the Schrödinger equation for which this choice corresponds to Neumann boundary conditions. Then, the spectral determinant $S(s)$ of the diffusion equation can be readily calculated and it is given by

$$S(x) = \frac{\sqrt{x}}{\sinh \sqrt{x}} , \tag{72}$$

with $x = s\tau_D$. The average conductance δg deduced from (68) and the fluctuation $\langle \delta g^2 \rangle$ can be written generally as

$$\delta g = 2\frac{\partial}{\partial x}\ln S(x) = -\frac{1}{3} \tag{73}$$

$$\langle \delta g^2 \rangle = 12\frac{\partial^2}{\partial x^2}\ln S(x) = \frac{2}{15} . \tag{74}$$

In the limit $x \to 0$, we can expand the spectral determinant namely $\ln S(x) \simeq_{x\to 0}$ $-\frac{x}{6} + \frac{x^2}{180}$ and recover the previous result.

7.   EXAMPLE 2. MULTIPLE SCATTERING OF LIGHT: THE ALBEDO

We shall now give an example of the use of the quantum probability taken from the multiple scattering of electromagnetic waves in disordered suspensions. To that purpose, consider the scattering medium as being the half-space $z \geq 0$. The other half-space is a free medium which contains both the sources of the waves and the detectors. We also assume that both incident and emergent waves are plane waves with respective wavevectors $\mathbf{k_i} = k\hat{\mathbf{s}}_i$ and $\mathbf{k_e} = k\hat{\mathbf{s}}_e$, where $\hat{\mathbf{s}}_i$ and $\hat{\mathbf{s}}_e$ are unit vectors. The waves experience only elastic scattering in the medium so that after a collision, only the direction $\hat{\mathbf{s}}$ changes while the amplitude $k = \frac{\omega_0}{c}$ remains constant. The reflection coefficient $\alpha(\hat{\mathbf{s}}_i, \hat{\mathbf{s}}_e)$ for this geometry is called the *albedo* and is proportional to the intensity $I(\mathbf{R}, \hat{\mathbf{s}}_i, \hat{\mathbf{s}}_e)$ emerging from the medium per unit surface and per unit solid angle measured at a point $\mathbf{R}$ at infinity. If F_{inc} defines the flux of the incident beam related to the incident intensity I_{inc} by $F_{inc} = cSI_{inc}$ where S is the illuminated surface in the plane $z = 0$, then the albedo is given by

$$\alpha(\hat{\mathbf{s}}_i, \hat{\mathbf{s}}_e) = \frac{R^2 c}{F_{inc}}I(\mathbf{R}, \hat{\mathbf{s}}_i, \hat{\mathbf{s}}_e) . \tag{75}$$

In order to calculate the intensity $I(\mathbf{R}, \hat{\mathbf{s}}_i, \hat{\mathbf{s}}_e)$, we need to calculate first the radiative solutions of the Helmholtz equation at the Fraunhoffer approximation. It is given by

$$\psi_{\omega_0}(\hat{\mathbf{s}}_i, \hat{\mathbf{s}}_e) = \int d\mathbf{r}d\mathbf{r}' e^{ik(\hat{\mathbf{s}}_i.\mathbf{r} - \hat{\mathbf{s}}_e.\mathbf{r}')} G(\mathbf{r}, \mathbf{r}', \omega_0) \tag{76}$$

and the intensity is then $I(\mathbf{R}, \hat{\mathbf{s}}_i, \hat{\mathbf{s}}_e) = |\psi_{\omega_0}(\hat{\mathbf{s}}_i, \hat{\mathbf{s}}_e)|^2$. Before averaging over the disorder, the albedo looks like a random pattern of bright and dark spots. This is a speckle pattern [12, 13] whose long range correlations are a specific feature of multiple scattering. By averaging over the disorder (either liquid suspension [14-16] or rotating solid sample [17]), the speckle is washed out and the two surviving contributions are respectively associated to the diffuson and the cooperon.

7.1.   THE INCOHERENT ALBEDO: THE DIFFUSON

We have calculated previously the various contributions to the intensity. The incoherent part is given by the diffuson contribution, namely the relation (49), where we take as the source term the incident plane wave

$$\overline{\psi}_i(\mathbf{r}_1) = \sqrt{\frac{cI_{inc}}{4\pi}} e^{-|\mathbf{r}_1-\mathbf{r}|/2l_e} e^{-ik\hat{\mathbf{s}}_1 \cdot \mathbf{r}_1} , \tag{77}$$

where $\mathbf{r}$ is the impact point on the surface $z = 0$ and $\mathbf{r}_1$ is the position of the first collision. The albedo rewrites

$$\alpha_d = \frac{R^2}{S} \int d\mathbf{r}_1 d\mathbf{r}_2 |\overline{G}^R(\mathbf{r}_2, \mathbf{R})|^2 \Gamma(\mathbf{r}_1, \mathbf{r}_2) e^{-|\mathbf{r}_1-\mathbf{r}|/l_e} . \tag{78}$$

The Fraunhoffer approximation consists in taking the limit $|\mathbf{R} - \mathbf{r}_2| \to \infty$, so that the Green function $\overline{G}^R(\mathbf{r}_2, \mathbf{R})$ can be expanded as

$$\begin{aligned}
\overline{G}^R(\mathbf{r}_2, \mathbf{R}) &= e^{-|\mathbf{r}'-\mathbf{r}_2|/2l_e} \frac{e^{ik|\mathbf{R}-\mathbf{r}_2|}}{4\pi|\mathbf{R} - \mathbf{r}_2|} \\
&\simeq e^{-|\mathbf{r}'-\mathbf{r}_2|/2l_e} e^{-ik\hat{\mathbf{s}}_e \cdot \mathbf{r}_2} \frac{e^{-ikR}}{4\pi R} .
\end{aligned} \tag{79}$$

By defining the respective projections μ and μ_0 of the vectors $\hat{\mathbf{s}}_i$ and $\hat{\mathbf{s}}_e$ on the z axis, we obtain finally

$$\alpha_d = \frac{1}{(4\pi)^2 S} \int d\mathbf{r}_1 d\mathbf{r}_2 e^{-\frac{z_1}{\mu_0 l_e}} e^{-\frac{z_2}{\mu l_e}} \Gamma(\mathbf{r}_1, \mathbf{r}_2) . \tag{80}$$

This expression calls for a number of remarks. We first notice that it does not contain any dependence on the direction namely on the incident and emerging vectors $\hat{\mathbf{s}}_i$ and $\hat{\mathbf{s}}_e$. Therefore, at this approximation, the albedo is flat i.e. does not have any angular structure. Second, if we compare this expression to its counterpart for the conductivity, we see that although both of them describe transport, the conductivity vanishes for the diffuson while it does not for the albedo.

The structure factor Γ is related to the diffusion probability through the relation

$$P_d(\mathbf{r}_1, \mathbf{r}_2) = \frac{l_e^2}{4\pi c} \Gamma(\mathbf{r}_1, \mathbf{r}_2) , \tag{81}$$

so that we obtain

$$\alpha_d = \frac{c}{4\pi l_e^2} \int_0^\infty dz_1 dz_2 e^{-\frac{z_1}{\mu_0 l_e}} e^{-\frac{z_2}{\mu l_e}} \int_S d^2\rho P_d(z_1, z_2, \rho) , \tag{82}$$

where now due to the geometry of the medium, the function P_d depends on z_1, z_2 and the projection ρ of the vector $\mathbf{r}_1 - \mathbf{r}_2$ onto the plane $z = 0$. Within the diffusion approximation, P_d can be calculated for this geometry using the image method [18, 19]. This gives

$$P_d(\mathbf{r}, \mathbf{r}') = \frac{1}{4\pi D}\left[\frac{1}{|\mathbf{r} - \mathbf{r}'|} - \frac{1}{|\mathbf{r} - \mathbf{r}'^*|}\right], \tag{83}$$

where $\mathbf{r}'^* = (r'^*_{\perp}, -z' + z_0)$ and $z_0 = \frac{2}{3}l_e$. Then, it appears that P_d does not vanish on the plane $z = 0$ but instead for $z = -z_0$. It is possible for that geometry to compare the validity of the diffusion approximation with the known exact solution for the semi-infinite problem without sources (the so called Milne problem) that can be solved using the Wiener-Hopf method [20]. There, we obtain instead that P_d vanishes on the plane $z = -0.7104l_e$. This justifies the use of the diffusion approximation. But, it is important at this point to make two important remarks. First, it would be a mistake to believe that on the basis of the two examples, infinite and semi-infinite spaces, for which the diffusion approximation works well that it is indeed always the case. The generalization and the validity of the diffusion equation in a restricted geometry is a difficult problem. Second, all the expressions we have obtained so far are valid for isotropic scatterers for which there is no difference between the elastic and the transport mean free paths. When relaxing this approximation, namely dealing with anisotropic scattering, we have to include both of them. Although it appears clearly from simple physical considerations that in the expression (82), we must remove the elastic mean free path l_e in the exponential factors that describe the first and last collisions, and replace it by the transport mean free path in the expression of P_d, this point needs further investigation using for instance transport theory [6, 19].

Finally, by replacing (83) into (82), we obtain

$$\alpha_d = \frac{3}{4\pi}\mu\mu_0\left(\frac{z_0}{l_e} + \frac{\mu\mu_0}{\mu + \mu_0}\right). \tag{84}$$

7.2.   THE COHERENT ALBEDO: THE COOPERON

Along the same lines, we can now evaluate the contribution of the cooperon to the albedo. Using the expression (50) for the intensity, we obtain

$$\alpha_c(\hat{\mathbf{s}}_i, \hat{\mathbf{s}}_e) = \frac{1}{(4\pi)^2 S}\int d\mathbf{r}_1 d\mathbf{r}_2 e^{-\frac{1}{2}(\frac{1}{\mu} + \frac{1}{\mu_0})\frac{z_1 + z_2}{l_e}}\Gamma(\mathbf{r}_1, \mathbf{r}_2)e^{ik(\hat{\mathbf{s}}_i + \hat{\mathbf{s}}_e)\cdot(\mathbf{r}_2 - \mathbf{r}_1)}. \tag{85}$$

Unlike the previous case, there is now a phase term present in this expression which is at the origin of a new angular structure of the albedo. It is straightforward to check that, assuming $\hat{\mathbf{s}}_i + \hat{\mathbf{s}}_e = \mathbf{0}$, i.e. measuring the albedo right in the backscattering direction, we have

$$\alpha_c(\theta = 0) = \alpha_d , \tag{86}$$

where the angle θ is between the directions $\hat{s}_i$ and $\hat{s}_e$. Then, the total albedo $\alpha(\theta) = \alpha_d + \alpha_c(\theta)$ is such that

$$\alpha(\theta = 0) = 2\alpha_d . \tag{87}$$

Using the expression (83) for the probability and integrating, we obtain

$$\alpha_c(\theta) = \frac{3}{8\pi} \frac{\mu\mu_0}{(1 + \mu k_\perp l_e)(1 + \mu_0 k_\perp l_e)} \left(\frac{1 - e^{-2k_\perp z_0}}{k_\perp l_e} + 2 \frac{\mu\mu_0}{\mu + \mu_0} \right) . \tag{88}$$

In this expression, $k_\perp = (k_i + k_e)_\perp = k(\hat{s}_i + \hat{s}_e)_\perp$ is the projection onto the plane xOy of the vector $k_i + k_e$. At small angles, we have $k_\perp \simeq \frac{2\pi}{\lambda}\theta$, while at large angles, $\alpha_c(\theta \to \infty) = 0$. Then, the coherent contribution to the albedo is finite only in a cone of angular aperture $\frac{\lambda}{2\pi l_e}$ around the backscattering direction. By expanding near $\theta = 0$ we obtain the expression

$$\alpha_c(\theta) \simeq \alpha_d - \frac{3}{4\pi} \frac{(l_e + z_0)^2}{l_e} k_\perp + O(k_\perp)^2 . \tag{89}$$

Thus, the albedo shows a cusp near $\theta = 0$ namely, its derivative is discontinuous at this point.

8.   EXAMPLE 3: SPECTRAL CORRELATIONS

In the previous two examples, we discussed transport coefficients. They give a description of the system when it is connected to another "reference" medium. There is another characterization independent of the coupling to another system, which focuses on *spectral properties* i.e. properties of the energy spectrum of the solutions of the wave equations. For electronic systems, they are related to the equilibrium thermodynamic properties like the magnetization of a metallic sample. For disordered systems, it also raises an important issue: these systems are part of the larger class of *complex systems*, usually non integrable. The complete set of correlation functions of the energy levels provides a complete description of the thermodynamic properties. For instance, it is possible just by inspection of the energy spectrum to determine whether or not an electronic system is a good metal or an Anderson insulator.

Let us start by defining the set of eigenenergies ϵ_α of the Schrödinger (or Helmholtz) equation in a confined geometry (e.g. a box). The density of states per unit volume is given by

$$\rho(\epsilon) = \frac{1}{\Omega} \sum_\alpha \delta(\epsilon - \epsilon_\alpha) . \tag{90}$$

It can be written as well in terms of the Green function for the Schrödinger equation, using the equality

$$\rho(\epsilon) = -\frac{1}{\pi\Omega}\text{Im} \int d\mathbf{r} G^R(\mathbf{r}, \mathbf{r}, \epsilon) \,. \tag{91}$$

By averaging over the disorder, we define the various correlation functions. For instance, the two-point correlation function is

$$\overline{\rho(\epsilon_1)\rho(\epsilon_2)} - \overline{\rho}(\epsilon_1)\,\overline{\rho}(\epsilon_2) \,, \tag{92}$$

where the average density of states coincides with its free value ρ_0 and therefore is related to the mean level spacing Δ through $\overline{\rho} = \frac{1}{\Delta\Omega}$. It is important at this point to make the following remark. The average we consider here is over the random potential (4). We could have taken as well the average over different parts of the spectrum as is done for instance in the quantum description of classically chaotic billiards where there is no random potential to average over. In the diffusion approximation, it has been shown numerically that these two ways to average are equivalent [4, 21] but this not need to be the general case. We then define the dimensionless correlation function

$$K(\omega) = \frac{\overline{\rho(\epsilon)\rho(\epsilon - \omega)}}{\overline{\rho}(\epsilon)^2} - 1 \,. \tag{93}$$

It can be expressed in terms of the Green function under the form

$$K(\omega) = \Delta^2 \int d\mathbf{r}d\mathbf{r}' K(\mathbf{r}, \mathbf{r}', \omega) \,, \tag{94}$$

with

$$K(\mathbf{r}, \mathbf{r}', \omega) = \frac{1}{2\pi^2}\text{Re}\,\overline{G^R(\mathbf{r}, \mathbf{r}, \epsilon)G^A(\mathbf{r}', \mathbf{r}', \epsilon - \omega)}^c \,, \tag{95}$$

where $\overline{\cdots}^c$ represents the cumulant average. The diffuson and cooperon contributions to the local function $K(\mathbf{r}, \mathbf{r}', \omega)$ can be expressed in terms of P_d and the structure factor as [22]

$$K_d(\mathbf{r}, \mathbf{r}', \omega) = \frac{1}{2\pi^2}\text{Re}\,\left[P_d(\mathbf{r}, \mathbf{r}'\omega)P_d(\mathbf{r}', \mathbf{r}, \omega)\right] \,, \tag{96}$$

while

$$K_c(\mathbf{r}, \mathbf{r}', \omega) = 2\rho_0^2\tau_e^4\text{Re}\,\left[\Gamma_\omega'(\mathbf{r}, \mathbf{r}')\Gamma_\omega'(\mathbf{r}', \mathbf{r})\right] \,, \tag{97}$$

when the system has time-reversal invariance. Finally, we define the Fourier transform $\tilde{K}(t)$ which is often called the *form factor*. By collecting the two previous contributions, it can be written

$$\tilde{K}(t) = \frac{\Delta^2}{4\pi^2}|t|Z(|t|) = \frac{\Delta^2}{4\pi^2}|t| \int_\Omega P_d(\mathbf{r}, \mathbf{r}, |t|)d\mathbf{r} \,, \tag{98}$$

which is precisely the form that is obtained assuming the Random Matrix Theory [4, 21] provided we consider the regime for which $\hbar/\Delta \gg t \gg \tau_D$ where $\tau_D = L^2/D$. The first inequality which involves the Heisenberg time $\hbar/\Delta$ enforces the condition of a continuous spectrum while the second one indicates that we must be in the ergodic limit where the diffusing particle explored the whole system.

Conclusion

We have presented in this short review a very partial selection of highlights in the field of multiple scattering of waves in disordered media. Our aim was more to give a feeling of the profound unity of the physical phenomena and therefore of the methods to handle them rather than discussing an extensive list of effects. Among them, we should mention the study of fluctuations of the intensity for waves both in the weak disorder regime and near the Anderson transition that will be covered in details by the reviews of R. Pnini, A. Z. Genack and A. A. Chabanov, and F. Scheffold and G. Maret. This is certainly a problem for which the recent developments have been very spectacular. We have studied the coherent albedo in the weak disorder case. It should be extended to the strong disorder limit. Finally, our last example on spectral correlations should be considered as another way to obtain the semiclassical results reviewed in the contribution of D. Delande.

References

[1] These notes are a partial and preliminary account of the monography *Cohérence et diffusion dans les milieux désordonnés* by E. Akkermans and G. Montambaux, to be published by CNRS Intereditions for the french version.

[2] S. Chandrasekhar, *Radiative transfer* (Dover, N.Y. 1960).

[3] S. Chakraverty and A. Schmid, Phys. Rep. **140**,193 (1986)

[4] G. Montambaux, in *Quantum fluctuations*, proceedings of the Les Houches Summer School, Session LXIII, ed. by S. Reynaud et al. (North Holland, Amsterdam, 1996), p.387.

[5] M. Pascaud and G. Montambaux, Phys. Rev. Lett. **82** (1999), 4512 and Phys. Uspekhi **41**,182 (1998).

[6] A. Ishimaru, *Wave propagation and scattering in random media*, Vol.I,II (Academic, N.Y. 1978).

[7] S. Doniach et E.H. Sondheimer, *Green's Functions for Solid State Physicists*, Frontiers in Physics, W.A. Benjamin, (1974)

[8] G. Bergmann, Phys. Rep. **107**, 1 (1984)

[9] J. Rammer, *Quantum transport theory*, Frontiers in Physics, Perseus books (1998)

[10] *Mesoscopic quantum physics*, proceedings of the Les Houches Summer School, Session LXI, ed. by E.Akkermans, G. Montambaux, J.L. Pichard and J. Zinn-Justin (North Holland, Amsterdam, 1995).

[11] C. Itzykson and J.M. Drouffe, *Statistical field theory*, Vol. 1 (Cambridge, 1989)

[12] C. Dainty ed., *Laser speckle and related phenomena*, Topics in a applied physics, Vol.9 (Springer, Berlin, 1984).

[13] G. Maret, in *Mesoscopic quantum physics*, proceedings of the Les Houches Summer School, Session LXI, ed. by E.Akkermans, G. Montambaux, J.L. Pichard and J. Zinn-Justin (North Holland, Amsterdam, 1995).

[14] Y. Kuga and A. Ishimaru, J.Opt.Soc. Am., **A8**, 831 (1984)

[15] M.P. van Albada et A. Lagendijk, Phys. Rev. Lett. **55** (1985), 2692.

[16] P.E. Wolf et G. Maret, Phys. Rev. Lett. **55** (1985), 2696.

[17] M. Kaveh, M. Rosenbluh, I. Edrei et I. Freund, Phys. Rev. Lett. **57** (1986), 2049

[18] E. Akkermans, P.E. Wolf, and R. Maynard, Phys. Rev. Lett. **56** (1986), 1471-1474

[19] E. Akkermans, P.E. Wolf, R. Maynard and G. Maret, J. de Physique (France), **49** (1988), 77-98

[20] H. C. van de Hulst, *Multiple light scattering* (Dover, N.Y. 1981).

[21] O. Bohigas, in *Chaos and Quantum Physics, Proceedings of the Les Houches Summer School, Session LII*, ed. by M.J. Giannoni, A. Voros and J. Zinn-Justin (North Holland, Amsterdam, 1991), p.91

[22] B.L. Al'tshuler and B. Shklovskiĭ, Sov. Phys. JETP **64**, 127 (1986).

Chapter 2

STATISTICAL APPROACH
TO PHOTON LOCALIZATION

A. Z. Genack and A. A. Chabanov
Department of Physics, Queens College
City University of New York
Flushing, New York 11367, USA
genack@qc.edu

Abstract   A statistical description of wave propagation in random media is necessary to characterize large fluctuations found in these samples. The nature of fluctuations is determined by the closeness to the localization transition. In the absence of inelastic processes, this can be specified, in many circumstances, by a single parameter - the ensemble average of the dimensionless conductance. As a result, the extent of localization can be determined by any of a wide variety of related statistical measurements. Among the quantities that most directly reflect key aspects of localization are the following: (i) the ensemble average of the dimensionless conductance, (ii) the degree of spatial correlation of intensity, (iii) the variances of the probability distribution of transmission quantities, such as the intensity, the total transmission, and the dimensionless conductance, and (iv) the ratio of the width to the spacing of modes of an open sample. We will emphasize the relationships between key statistical aspects of propagation in quasi-one-dimensional samples and the different impact of absorption upon these.

   We find that even in the presence of absorption, the extent of localization can be characterized by a single parameter - the variance of the total transmission normalized by its ensemble average. Measurements of fluctuations in intensity and total transmission of microwave radiation allow us to study photon localization in collections of dielectric spheres and in periodic metallic wire meshes containing metallic scatterers. We find in low-density collections of alumina spheres contained in a copper tube, at frequencies near the first Mie resonance, that the variance of normalized total transmission scales exponentially once it becomes greater than unity. When this parameter is large, transmission spectra are observed to be a series of narrow lines with widths that are smaller than the separation between peaks. These spectra have an extraordinarily wide intensity distribution and correspondingly large variance. These results demonstrate that

P. Sebbah (ed.), Waves and Imaging through Complex Media, 53–84.
© 2001 *Kluwer Academic Publishers. Printed in the Netherlands.*

the variance of normalized transmission serves as a powerful guide in the search
for and characterization of photon localization.

Keywords: statistics, photon localization, microwave

Introduction

Waves are the means by which we probe our environment and communicate
with one another. Their study has expanded from primordial fascination with
everyday observations of ripples in a pond and the ocean's waves to the systematic study of sound, light and the full gamut of electromagnetic radiation,
and quantum mechanical waves. The field of electromagnetic propagation has
grown with the development of the laser to encompass the exploration of nonlinear and quantum optical phenomena in an expanding array of new materials
including optical fibers and photonic band gap structures. The joint application of optics and electronics in communications has made photonics a rapidly
expanding aspect of modern life. The study of transport in random media has
grown apace. It has been spurred in recent years by advances in imaging, by the
interchange with electronic mesoscopic physics, and by the expanding range of
statistical aspects of propagation that can be measured and computed.

Wave transport in random media is essentially a statistical problem [1-5].
Since the precise structure of a random sample is not known, the fine-grained
variation in intensity of reflected and transmitted light cannot be predicted. An
example of a random intensity pattern produced in reflection of a helium-neon
laser beam from a sheet of paper is shown in Fig. 1. Similar speckle patterns

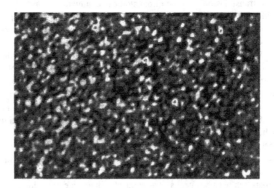

Figure 1.   The far-field speckle pattern of reflected light is seen in this CCD image of a
helium-neon laser beam scattered from a sheet of white paper. Lighter regions have higher
intensity.

are produced in transmission. Despite the apparently haphazard character of scattered light, the nature of transport in a random medium can be obtained from a statistical characterization of such complex random speckle patterns for all incident frequencies and incident spatial modes for an ensemble of statistically equivalent random realizations. An example of a spectral "speckle" pattern of microwave intensity at a point on the output surface of a random sample is shown in Fig. 2. The sample is composed of polystyrene spheres randomly

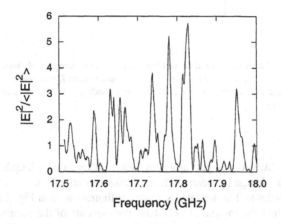

Figure 2. Normalized spectrum of the intensity of a single polarization component of the microwave field transmitted through a random collection of 1.27-cm-diam. polystyrene spheres contained in a 100-cm-long waveguide. Only a single polarization component is detected using a wire antenna positioned at the sample output surface. Its amplitude is squared and divided by the ensemble average of this quantity.

positioned in a long copper tube. An even more detailed "fingerprint" of the sample is contained in the spatial and spectral variation of the field itself.

The field is the sum of randomly phased partial waves associated with all possible paths from the source to the point of detection. The field fluctuates with position because of the changing set of partial waves that contribute to the field at different points. The linear superposition of such partial waves corresponds to the sum of phasors shown schematically in Fig. 3. Each partial wave may be represented by an arrow or phasor with length proportional to the field amplitude and with phase equal to that of the partial wave. The component of the phasor along the x-axis (y-axis) gives the component of the field, which is in-phase (out-of-phase) with the incident wave. The fluctuations in the field with frequency shift at a given point are primarily due to changes in the phase difference,

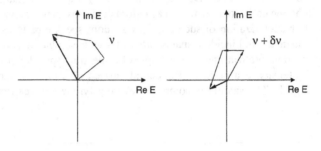

Figure 3.   Partial waves associated with different trajectories are shown as phasors and summed to give the total field. Each phasor is a complex-valued contribution E with real part corresponding to the in-phase component and imaginary part corresponding to the out-of-phase component of the partial wave.

$\Delta\varphi = 2\pi\Delta s/\lambda$, between partial waves, which differ in path length by Δs, as a result of the change in wavelength λ. Measurements of the microwave field at the same point and in the same sample configuration as in Fig. 2 are shown in Fig. 4a. Both the in- and out-of-phase components of the transmitted field are measured with use of a Hewlett-Packard network analyzer. The analyzer provides the ratio of the in- and out-of-phase components of the transmitted wave to a reference field, which is the field at the input plane of the sample. The intensity in Fig. 2 is calculated as the sum of the squares of the in- and out-of-phase components. The magnitude of the field $|E|$ and the phase modulus 2π, ϕ, are computed from these components and the resultant spectra are displayed in Fig. 4b.

In order to describe the nature of transmission in a random medium, the full probability distributions of random variables should be given rather than their ensemble averages. These distributions reflect correlation within the medium, whose dependence upon position and frequency give key static and dynamic aspects of average transmission. Let us consider, for example, correlation functions of the field. The field correlation function with displacement on the sample surface for an ensemble of samples is the Fourier transform of the variation of the ensemble average of the far-field intensity with angle [6]. Similarly, the field correlation function with frequency shift is the Fourier transform of the time-of-flight distribution of radiation reaching a point [7, 8]. Thus considerable insight is gained by dealing with field correlation functions, which involve both the amplitude and phase of the wave.

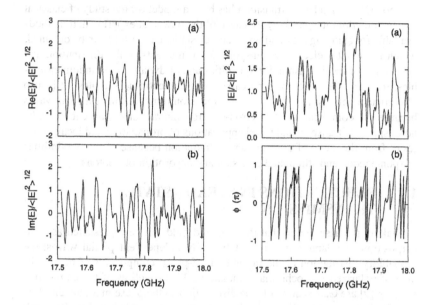

Figure 4a. The real and imaginary parts of the microwave field at the same point and in the same sample configuration as in Fig. 2.

Figure 4b. The magnitude $|E|$ normalized to its ensemble average value and the phase, modulus 2π, ϕ, of the microwave field calculated using the data in Fig. 4a.

It is often convenient, however, to consider these variables separately [9]. Key aspects of the statistics of steady-state transmission with monochromatic sources can be obtained from measurements of distribution [1, 2, 10, 11, 13-16, 21-23, 26, 28, 29, 32-39, 41-45] and correlation [15, 17-20, 22-25, 27, 30, 31, 42, 43, 44, 45] of the local intensity, $|E|^2$, or of the spatially averaged flux. At the same time, basic aspects of the statistics of dynamics are revealed by measuring the probability distribution of the frequency derivative of the cumulative phase φ [47-51, 9, 52-54]. The cumulative phase is obtained by incrementing ϕ of Fig. 4b by 2π each time the phase passes through the upper bound of ϕ of π rad. The derivative $d\varphi/d\omega$ equals the average photon transit time of a narrow band pulse from the source to the detector. Large fluctuations in steady-state and dynamic transmission quantities are a distinguishing feature of transport in random media. Here we will focus on the statistics of static transmission quantities.

The study of statistics provides a platform for the characterization of the localization transition. Though initially proposed in the electronic context

[55, 56, 57], the Anderson transition has been a model for the study of classical waves in random media [3, 5]. Because of the powerful experimental methods available for studying classical waves, a rich statistical portrait can be obtained, which illuminates electronic transport. Measurements of the distribution and correlation of local transmission quantities and of dynamics in random ensembles, for example, have not been carried out in electronic samples. It is also of particular interest to investigate the Anderson transition for classical waves because the complication of electron-electron interactions is absent. Finally, new aspects of transport and novel applications of propagation and localization result from the study of electromagnetic radiation because of the possibilities of spontaneous and stimulated emission and absorption of photons.

1.   FLUCTUATIONS IN STEADY-STATE TRANSMISSION

Making the assumption that the scattered field can be represented as a superposition of a large number of statistically independent partial waves, the probability distribution of the in- and out-of-phase components of the amplitude of polarized monochromatic radiation [37] as well as of the cumulative phase φ [9] are Gaussian, while the distribution of the phase modulus 2π, ϕ, is flat [2]. The corresponding intensity distribution was calculated by Rayleigh [10] and bears his name. For a given polarization component of the field, the probability distribution of the associated intensity in the absence of correlation is a negative exponential. When the intensity is normalized to its ensemble average value for an ensemble of equivalent samples, its distribution is a universal function independent of the physical dimensions or scattering strength of the sample, $P(s_{ab} = T_{ab}/\langle T_{ab}\rangle) = \exp(-s_{ab})$. Here T_{ab} is the transmission coefficient from an incident mode a into an outgoing mode b, and $\langle...\rangle$ represents the average of the quantity between the brackets over an ensemble of statistically equivalent samples. Depending upon the experiment under consideration, these modes may correspond to different transverse momentum states of the far field, or to different modes of a microwave cavity, or to the fields within different coherence areas on the sample surface. A schematic representation of key transmission quantities is presented in Fig. 5. For this negative exponential distribution for a single polarization component of the radiation, the variance of the intensity is equal to its average value, $var(s_{ab}) = 1$. Since the exponential distribution obtained in the absence of intensity correlation is universal, its measurement provides no information regarding wave propagation beyond the confirmation of the reasonableness of the model, which may apply to a lesser or greater extent in the diffusive limit. However, the average transmission, $\langle T_{ab}\rangle$, which is the first moment of the transmission distribution, depends upon the absorption coefficient and scattering strength of the medium,

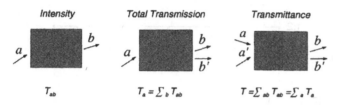

Figure 5.   Transmission coefficients in random media in order of increasing spatial averaging. The incident and outgoing modes a and b may be in any complete representation of the field.

and both the transport mean free path ℓ and the diffusive absorption length L_a can be determined from the scaling of $\langle T_{ab} \rangle$. In particular, when absorption is absent, $\langle T_{ab} \rangle \sim \ell/L$ [1, 7].

As a result of spatial correlation, however, waves are not statistically independent and deviations from the Rayleigh distribution are observed. The intensity distribution then provides essential information regarding the nature of transport and intensity correlation within a medium. The cumulant correlation function of normalized intensity is $C = \langle \delta s_{ab} \delta s_{ab'} \rangle$, where $\delta s_{ab} = s_{ab} - 1$ is the fluctuation from the ensemble average value of s_{ab}. Short-range intensity correlation is associated with field correlation and reflects the inability of a wave to change on a scale much shorter than the wavelength [15]. This short-range component of the normalized intensity cumulant correlation function C is denoted C_1 and is obtained by factorizing the field [15, 17, 42]. Its dependence upon displacement ΔR is given by $|\langle E(R)E(R + \Delta R) \rangle|^2 / \langle |E(R)|^2 \rangle \langle |E(R + \Delta R)|^2 \rangle$ and can be computed directly from measurements of the field [46]. It is shown in Fig. 6. This term is immediately evident in the speckle pattern of laser radiation scattered from an object, such as that shown in Fig. 1. For an illuminated area A and wavelength λ, the number of transverse modes for a vector wave is equal to the number of speckle spots, $N = 2\pi A/\lambda^2$. Since the number of partial waves associated with distinct paths within the medium, which contribute to the field at a point, greatly exceeds N, these waves must be correlated and departures from Rayleigh statistics can be expected. In addition to short-range correlation, scattering in the medium and subsequent diffusion induce intensity correlation across the entire sample. The difference $C - C_1$ gives longer-range contributions to C. Aside from an oscillatory contribution of the same form as seen in Fig. 6, $C - C_1$ has a long-range component with displacement. In quasi-one-dimensional samples, with length L much greater than the transverse dimensions of the sample, the wave is confined by reflecting walls so that the modes are completely mixed. In this case, the long-range contribution to C is

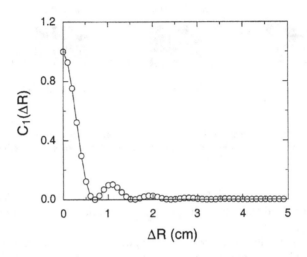

Figure 6.   The square of the field correlation function with displacement. This is the field factorization contribution C_1 to the normalized cumulant correlation functions with displacement of the intensity, C, for the sample described in Fig. 2 and in the text.

constant between points in the transmitted wave [17, 20, 22, 24, 27]. The degree of long-range correlation may be represented by $\langle \delta s_{ab} \delta s_{ab'} \rangle$, where b and b' are points separated by several intervening speckle spots. The main part of the oscillatory component arises from the correlation in the field. The constant background contribution to C is dominated by a term, denoted by C_2, which is calculated by including a single crossing of fields in the medium, represented by a Hikami box in a perturbation expansion of the Green function [17, 20]. In the absence of absorption or gain, the degree of long-range correlation is proportional to $L/N\ell$ [17, 20, 22, 24, 27].

The impact of spatial correlation upon distributions of transmission quantities is perhaps most transparent for the distribution of the total transmission for a given incident mode a, normalized by its ensemble average value, $P(s_a = T_a/\langle T_a \rangle)$, where $T_a = \Sigma_b T_{ab}$ is the sum of transmission coefficients over all output modes, as shown schematically in Fig. 5. If the intensity at different points were uncorrelated, transmission would be the sum of statistically independent fluxes at an unbounded number of points. The total transmission distribution would then be a delta function. However, as a result of short-range correlation in the intensity, there can be no more than N coherence areas in the transmission speckle pattern. Thus, if there were no correlation in the flux in distinct coherence areas, we would expect that the variance of the normalized total transmission resulting from the sum of N contributions, each of order $1/N$, would be $N(1/N)^2 = 1/N$. Instead, for diffusing waves in the absence of absorption, the variance is larger by a factor of L/ℓ, giving $var(s_a) \sim L/N\ell$, which equals the degree of long-range correlation. Thus enhanced fluctuations in transmission are the result of extended spatial correlation and the distribution of the normalized total transmission reflects the nature of scattering in the medium.

For quasi-one-dimensional samples, Kogan and Kaveh [34] found that the distribution $P(s_{ab})$ can be expressed in terms of the distribution $P(s_a)$,

$$P(s_{ab}) = \int_0^\infty \frac{ds_a}{s_a} P(s_a) \exp(-s_{ab}/s_a). \tag{1}$$

This relationship was found using random matrix theory. It has been confirmed in microwave measurements [41] in random polystyrene samples both with and without the influence of absorption, for diffusive waves and at the threshold of localization, as will be shown below. It is seen from Eq. (1), that negative exponential intensity statistics would only be obtained if the distribution of the total transmission were a delta function. But, since spatial correlation is always present to some degree, deviations from Rayleigh statistics are present as well. The intensity distribution displays the extent of this correlation and hence displays the nature of transport within the sample. Random matrix theory gives the relation between the moments of two distributions, [34]

$$\langle s_{ab}^n \rangle = n! \langle s_a^n \rangle. \tag{2}$$

Again, measurements establish that this relation holds independent of the degree of spatial correlation or of absorption of the wave [44].

The existence of intensity correlation at separations well beyond a field coherence length was first recognized in the analysis of fluctuations observed in the electrical conductance of micron-sized samples at low temperatures [12-14]. Such samples are intermediate in size between the atomic and macroscopic scales and so are termed mesoscopic. The electronic wave function is coherent on a time scale longer than the diffusion time within mesoscopic samples. Hence the field inside the sample is temporally coherent with the incident field. As a result of extended spatial correlation of the current transmission coefficients between any pairs of incident and outgoing modes, the size of conductance fluctuations in mesoscopic conducting samples is enhanced by a factor of $(L/\ell)^2$ and is independent of the scattering strength and sample size [13, 14, 40]. Expressing the conductance as $G = (e^2/h)g$, where g is the dimensionless conductance, the variance of g for mesoscopic samples is 2/15. The Landauer relation [60] gives g as the sum of transmission coefficients connecting all incident and outgoing modes, $g \equiv T = \Sigma_{ab}T_{ab}$, as illustrated in

Fig. 5. In the following, we will also use the notation g to denote its ensemble average.

This incoherent sum is appropriate for physical measurements of conductance, in which temporal averages are taken over times long compared to the phase coherence time of incident current modes. Expressing the dimensionless conductance g, or equivalently the transmittance T, in terms of the total transmission, $T = \Sigma_a T_a$, one obtains $g = \langle T \rangle = N\ell/L$, for diffusive waves in the absence of absorption. Thus, $var(s_a) = \langle \delta s_{ab} \delta s_{ab'} \rangle \sim 1/g$, suggesting a relationship between fluctuations of total transmission, spatial correlation of intensity, and average transport. Universal conductance fluctuations for diffusive waves corresponds to $var(g) \simeq 2/15$ or, equivalently, $var(s = T/\langle T \rangle) \simeq 2/15g^2$. This enhancement arises from the equal extent of correlation of transmission coefficients T_{ab} for all input and output modes associated with a term in perturbation theory which includes two Hikami boxes. The contribution to C of this term is proportional to $1/g^2$ and is referred to as C_3 or infinite-range correlation [20, 42]. Unlike the short-range C_1 term whose spatial variation is multiplicative in displacement of the source and detector, and the C_2 term which is additive in this regard, the C_3 term is independent of displacement of either the source or detector. The cumulant intensity correlation function is given by $C = C_1 + C_2 + C_3$. Higher order terms involving odd numbers of Hikami boxes have the spatial dependence of C_2, whereas terms involving even numbers of Hikami boxes have no spatial variation.

2.   SIGNATURES OF LOCALIZATION

A sharp divide exists in the nature of wave propagation [61]. On one side, average transport is not appreciably influenced by wave interference. The wave moves a distance, $\ell \gg \lambda$, before being randomized in direction and, on scales much larger than ℓ, the intensity or charge density follows a diffusion equation. This leads to a transmission coefficient, $T_a \sim \ell/L$, on length scales shorter than the diffusive absorption length, $L_a = \sqrt{D\tau_a}$ [7], where $D = \frac{1}{3}v_E\ell$ is the diffusion coefficient, v_E is the energy transport velocity [62, 63], and τ_a is the absorption time. For $L > L_a$, the wave is attenuated exponentially. The scale dependence of transmission through a wedge-shaped sample of random titania particles dispersed in polystyrene is shown in Fig. 7 [7]. The curve gives the fit of the envelope of the data to diffusion theory. The correlation function of intensity with frequency shift measured in the far field is found to coincide with the square of the Fourier transform of the measured time-of-flight distribution [8]. Hence it corresponds to the C_1 contribution to C. For diffusive waves, the degree of intensity or current correlation is small, and so $var(s_{ab}) \sim 1$, and $var(s_a) \ll 1$.

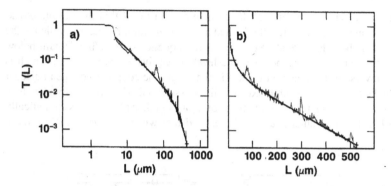

Figure 7.   (a) Log-log and (b) semilog plots of normalized optical transmission through a wedge of rutile titania powder in a polystyrene matrix. The plots show the inverse and exponential attenuation of transmission with L for L smaller than and greater than L_a, respectively.

On the other side of the divide, the constructive interference of waves returning to a point leads to a suppression of transport. The enhancement of intensity in the backscattered direction in a cone of angular width $1/k\ell$, where k is the wave number, may be viewed as a precursor to localization [64-67]. The coherent backscattered cone is the Fourier transform of the point spread function on the input surface. Thus it broadens as the wave is more strongly confined. Such backscattering evidently becomes large when $k\ell \sim 1$. The wave then cannot be envisioned as propagating with well defined wavelength between scattering centers, and the particle diffusion picture breaks down [68].

Localization by disorder was first discussed in the electronic context. Anderson showed that electrons on an atomic lattice with disorder in the site energy become localized when this disorder exceeds the off-diagonal coupling between neighboring sites [55]. The wave functions then fall exponentially in space, and transport is absent in unbounded samples. In bounded samples, however, electrons may flow through the sample as a result of exponential coupling to the boundary. The Ioffe-Regel criterion [68] for the breakdown of diffusive, particle-like propagation, $k\ell \sim 1$, suggests the wave origin of the localization transition. This was further elaborated by John *et al.* [58, 59], who argued that classical waves as well as quantum mechanical waves could be localized.

Edwards and Thouless [56] argued that electrons would be localized when electron states in a block of material could not be shifted into resonance with a neighboring block as a result of interactions at the sample boundary. One measure of the coupling at the boundary is the level width induced by the coupling of the wave to the boundary, δE. Equivalently, the level width may be expressed in frequency units as $\delta\nu = \delta E/h$. The level width $\delta\nu$ may be

identified with the inverse of the particle dwell time within the sample block.
It may also be quite naturally defined as the correlation frequency, which is the
width of the field correlation function with frequency shift. When $\delta\nu$ falls below
the frequency spacing between the levels in equivalent blocks, $\Delta\nu$, which is
the inverse of the density of states in the block, the coupling between adjacent
blocks of material is inhibited. The lines associated with consecutive levels in
two blocks of material that may be brought together are shown schematically
in Fig. 8. Abrahams *et al.* [57] reasoned that when the dimensionless ratio,

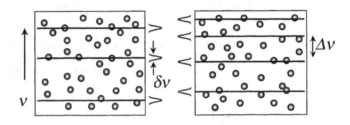

Figure 8.   Schematic representation of the (mis)matching of modes in two adjacent blocks of
a random medium. Here the level width $\delta\nu$ is smaller than the typical level spacing $\Delta\nu$.

$\delta = \delta\nu/\Delta\nu$, is below unity, this quantity will fall exponentially as the sample
is scaled up in size [57]. Using the Einstein relation, $\sigma = e^2 D(dn/dE)$, where
σ is the conductivity, D is the diffusion coefficient, and dn/dE is the density
of states per unit volume, it can be shown that $g = \sigma A/L = \delta$ [57]. Since
the scaling of $g = \delta$ depends upon nothing but the degree of overlap of levels
in adjacent blocks of material, which is given by δ, it was argued that g is a
universal scaling parameter whose scaling depends only upon the value of g
[57]. For localized waves, average transmission quantities scale exponentially.
Because fluctuations in conductance may be significant, especially for localized
waves, scaling relations cannot refer to the value of g for some particular sample
realization. Rather they must relate to the ensemble average value of g.

In contrast to the exponential scaling of g for localized waves, the conductance scales as A/L in conducting samples. Both g and δ increase with this
scaling factor, as the sample size increases proportionately in three dimensions.
Thus g increases as all dimensions are increased for conducting samples and
falls for localized samples. The threshold for localization in three dimensions
occurs at the fixed point for the scaling of conductance, at which point the level
width remains equal the level spacing, $\delta = g = 1$, as the sample size increases.
The corresponding transmission for a cube at the localization threshold scales
as $T_a = g/N \sim 1/(kL)^2 \sim (\ell/L)^2$ [57].

From the foregoing, it is clear that the nature of transport in electronic samples is reflected in the scaling of conductance. The scaling of conductance, at the localization threshold and for localized waves, requires that the wave function be coherent throughout the sample. However, the A/L scaling of conductance holds in conducting samples, even in the presence of inelastic scattering. The situation is less clear cut, however, for classical waves because the number of particles associated with the field is not conserved [59, 69,70-72, 76]. The particle number may be reduced by absorption or augmented by gain. We have seen that in the presence of absorption, for example, transmission falls exponentially for $L > L_a$. Thus exponential decay of transmission is not an unambiguous sign of localization.

An example of transmission of a classical wave in a strongly scattering sample is shown in Fig. 9 [73, 74]. The figure gives the scale dependence of microwave transmission in mixtures of aluminum and Teflon spheres with diameters of 0.47 cm at 19 GHz. At an aluminum sphere volume filling fractions of $f = 0.20$, transmission decays inversely with length, from $L = 1$ cm to $L = 7$ cm, indicating diffusive transport. At $f = 0.30$, transmission falls inversely with L^2, suggesting that the localization threshold is reached. An accurate absolute determination of transmission was not made, but g was of the order of unity in this sample for $L < 7$ cm. At a higher concentration of $f = 0.35$, transmission falls exponentially suggesting the wave is localized. But the possibility that absorption influences transport could not be ruled out.

In more recent experiments, Wiersma *et al.* [75] found a transition from $1/L$ to $1/L^2$ to exponential in the thickness dependence of infrared transmission in a wedge of GaAs particles, as the size of particles was changed. The wavelength was in the band gap of GaAs, so that low absorption was expected. A broadening of the coherent backscattering peak indicated a small value for the transport mean free path. However, even small absorption rates can dramatically influence the point spread function on the input face, and the possibility that these measurements were influenced by absorption has been raised [76]. One interesting feature of the measurements is that, in a sample where $1/L^2$ scaling of transmission was observed, the value of g for a cubic region of the sample with sides equal to the sample thickness L is significantly larger than unity. This is seen by multiplying the number of modes in an area L^2 by the measured transmission coefficient.

Exponential decay in optical transmission in a sample with low absorption has also been observed recently by Vlasov *et al.* [77] in a nearly periodic opal structure. Again, the value of g was much larger than unity for cubic regions of the sample for small values of L. The exponential fall-off of transmission in this case might be related to the evanescent nature of the wave in the nearly periodic structure.

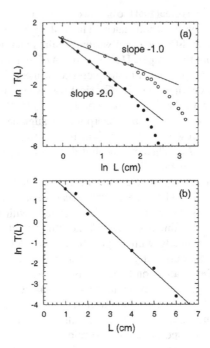

Figure 9.   Scale dependence of relative transmission for three volume fractions f of aluminum in mixture with Teflon spheres: (a) in samples with $f = 0.20$ (circles), transmission falls inversely with L, whereas it falls inversely with L^2 in samples with $f = 0.30$ (filled circles); (b) in samples with $f = 0.35$, transmission falls exponentially. The variation of the scaling of transmission is the same as that expected for a transition from diffusive to critical to localized waves in the absence of absorption.

  Recently van Tiggelen *et al.* [78] have shown that transport, which is renormalized by constructive interference of waves returning to a given point, may be described by an effective position-dependent diffusion coefficient, $D(\mathbf{r})$. In contrast to the conventional description of scaling [57] that treats the scale dependence of transport phenomenologically, via a scale dependent diffusion coefficient, $D(L)$, this approach explicitly treats the variation of the affect of interference upon position. Thus, because the probability of return of a wave to a point near the surface is smaller than for a point in the middle of the sample, $D(\mathbf{r})$ increases from the boundary to the middle of the sample. This approach preserves the scaling of transmission for large L. It may, however, lead to enhanced transmission, particularly in the case of samples with internal reflection.

In this case, photons that are scattered back to the front surface are reinjected into the sample and transmission is thereby enhanced.

Because of the existence of a variety of explanations for the observations of exponential decay in transmission of electromagnetic radiation, the scaling of the average transmission taken by itself may not be an unambiguous indication of localization. Below we will consider other measures of localization and, in particular, statistical measures of fluctuations in transmission. We should note, however, that indications of localization from measurements of average transport and of fluctuations in transport are not necessarily equivalent, even in the absence of absorption. Until recently, it was believed that the value of the conductance, or equivalently the Lyapunov exponent, $\gamma = -\langle \ln(g)/L \rangle$, can give the magnitude of relative fluctuations in conductance [79, 80]. In single-parameter-scaling theory, the magnitude of fluctuations is given in terms of average conductance via the relation, $var(\ln(g)/L) = -2\gamma/L$. But this relation can be violated, as it is found in one-dimensional random systems when the typical distance between localization centers is smaller than the localization lengths of these states, allowing their wave functions to overlap [81, 82].

We have seen that an important aspect of transport, which is not present for electrons, is absorption of the wave. The role of absorption is quite different from dephasing of electrons, since absorption does not lead to a reduction of interference of backscattered waves but simply reduces the amplitude of these waves. Weaver [69] showed in simulations of acoustic waves on a two-dimensional lattice that a localized acoustic excitation remains localized in space when dissipation is introduced. The impact upon localization of absorption is not accurately represented by its affect on the conductance. Absorption suppresses transmission and hence transmittance or conductance, at the same time it weakens the affect of localization. But a reduction of conductance is ordinarily seen as a sign of the strengthening of localization. We therefore seek an indicator of localization, which accurately gives the proximity to the localization threshold in samples both with and without absorption. Similarly, the presence of gain may change the relation between average transport and fluctuations, and an appropriate localization parameter is required in this case.

We also note that, in the presence of absorption, g and δ are affected differently. Whereas g falls, δ increases since the lines broaden due to absorption. Thus these parameters do not reflect the state of transport in the same way. Though we expect that an additional parameter is added when absorption or gain are present, we still seek a parameter that indicates the extent of localization affects.

3.   STATISTICS OF DIFFUSIVE WAVES

The first measurements of the distribution of total transmission were carried out by De Boer *et al.* [32] in optical studies in slabs of titania particles. Samples with $g > 10^3$ were studied and the distribution was found to be a Gaussian to within 1%. A measure of the deviation of the distribution from a Gaussian is the value of the third cumulant $\langle s_a^3 \rangle_c$, which gives the skewness of the distribution and vanishes for a Gaussian distribution. For the samples studied, $\langle s_a^3 \rangle_c$ was of order 10^{-6}.

Greater deviations were observed by Stoytchev and Genack [38] in measurements of the probability distribution of total transmission of microwave radiation in quasi-one-dimensional random waveguides. The samples consisted of 1.27-cm-diam. polystyrene spheres randomly positioned in a copper tube at a volume filling fraction $f = 0.55$. Spectra of total transmission over the frequency range 16.8-17.8 GHz were taken at tube diameters of 7.5 and 5.0 cm and at various sample lengths using an integrating sphere. The distributions of normalized total transmission, $P(s_a)$, for three ensembles of samples with different external dimensions are shown in Fig. 10. In the absence of absorption, the dimensionless conductance, $g = N\ell/L$, would be approximately 15.0, 9.0, and 2.25 for samples a, b, and c, respectively. The distributions are markedly non-Gaussian, and the deviations from Gaussian become more pronounced as either the sample length increases or the tube diameter decreases. A value of $\langle s_a^3 \rangle_c$ as large as 0.112 ± 0.003 was found for sample c. Deviations from Gaussian in the tail of the distributions can be seen in the semi-logarithmic plot in Fig. 10b. For large s_a, the distribution has an exponential tail.

Since the distributions of intensity and of total transmission are affected by absorption, $P(s_a)$ cannot be simply related to g. But, it was found in these strongly absorbing samples that the distribution $P(s_a)$ can be well expressed in terms of a single parameter, $var(s_a)$, or equivalently $g' = 2/3var(s_a)$. In the absence of absorption in the limit $g \gg 1$, the full transmission distribution is given by

$$P(s_a) = \int_{-i\infty}^{i\infty} \frac{dx}{2\pi i} \exp\left(xs_a - \Phi(x)\right), \tag{3}$$

where

$$\Phi(x) = g\ln^2(\sqrt{1 + x/g} + \sqrt{x/g}) \tag{4}$$

is the generating function [33, 34]. It follows from Eqs. (3) and (4), that $var(s_a) = 2/3g$, and hence $P(s_a)$ can be expressed in terms of $var(s_a)$. Plots of $P(s_a)$ of (3), obtained after substituting measured values of $var(s_a)$ into (4), are shown as the solid lines in Fig. 10b. They are in excellent agreement with the measured transmission distributions. This suggests that $var(s_a)$ is the

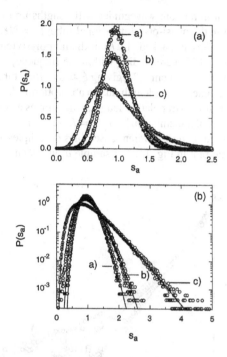

Figure 10.   Linear (a) and semi-logarithmic (b) plot of the distribution function of the normalized transmission $P(s_a)$ for three samples with dimensions: a), $d = 7.5$ cm, $L = 66.7$ cm; b), $d = 5.0$ cm, $L = 50$ cm; c), $d = 5$ cm, $L = 200$ cm. Solid lines in (b) represent theoretical results obtained from Eq. (3), with measured values of g' substituted for g in (4).

essential parameter describing transmission fluctuations in random media, even in the presence of absorption. Since the distribution $P(s_{ab})$ of the normalized intensity is given by the transform (1), $var(s_a)$ is characteristic for statistics of transmitted intensity as well.

The assumptions underlying Rayleigh statistics for transmitted intensity in non-absorbing samples can be expressed by the condition, $g \gg 1$. Deviations in the tail of the distribution $P(s_{ab})$ were observed in microwave experiments in samples with $g \approx 10$ [23, 29] and related to the degree of spatial intensity correlation. Corrections to the Rayleigh distribution were calculated by Nieuwenhuizen and van Rossum [33] and by Kogan and Kaveh [34]. They found that the distribution has a stretched exponential tail, $P(s_{ab}) \sim \exp(-2\sqrt{g s_{ab}})$, for $s_{ab} \gg g = \xi/L$. Similar behavior, but with g' substituted for g, was found in strongly absorbing samples. Measurements of the intensity distribution of microwave radiation in random waveguides with lengths up to L equal to the localization length in the quasi-one-dimensional tube, $\xi = N\ell$, were reported in [41]. Samples of loosely packed, 1.27-cm-diam. polystyrene spheres with a filling fraction $f = 0.52$ were studied within the frequency range 16.8-17.8 GHz. In these samples, $\ell \approx 5$ cm [83], giving $\xi \approx 5$ m for a tube diameter of 5 cm. The exponential attenuation length due to absorption is $L_a = 0.34 \pm 0.02$ m [38], and the diffusion extrapolation length, which gives an effective sample length for the statistics of transmission [84], $\tilde{L} = L + 2z_b$, is $z_b \approx 6$ cm [83]. The distributions measured in three different ensembles of samples are presented in the semi-logarithmic plot in Fig. 11. For sample a, in which $L/\xi = 1/15$, the

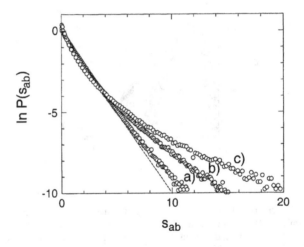

Figure 11.   Normalized intensity distribution $P(s_{ab})$ for three samples of dimensions: a), $d = 7.5$ cm, $L = 66.7$ cm; b), $d = 5.0$ cm, $L = 200$ cm; c), $d = 5.0$ cm, $L = 520$ cm. The Rayleigh distribution is shown as the short-dashed line.

distribution is close to a negative exponential. But, as L/ξ increases, deviations from Rayleigh statistics increase. To study the tail of the normalized intensity distribution $P(s_{ab})$ for $s_{ab} \gg \xi/L$, the measured distributions were fit to the theoretical expression, using g' as a free parameter. The result of the fit for sample b is presented as a dotted line in Fig. 12. For samples, for which $P(s_a)$ was also measured, the values of the fitting parameter g' were found to be within 10% of the values of $g' = 2/3var(s_a)$. The intensity distribution was also compared to the transform (1) of the measured transmission distribution. The transform for sample b is shown as a smooth solid line in Fig. 12, which essentially overlaps with the measured distribution. These measurements, therefore, confirm that $P(s_{ab})$ has a stretched exponential tail to the power of 1/2 and

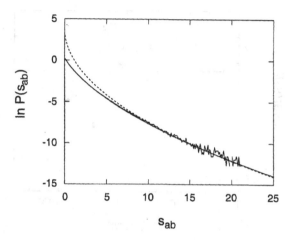

Figure 12.   Comparison of experimental and theoretical results obtained for sample b. The smooth solid curve represents the transform (1) of the measured total transmission distribution b) of Fig. 10. The short-dashed line gives the fit to the tail with the theoretical result, $P(s_{ab}) \sim \exp(-2\sqrt{g's_{ab}})$. The value of the fitting parameter, $g' = 3.20$, is close to the value, $g' = 3.06$, obtained from the total transmission measurement for the same sample.

is given by the transform (1) of $P(s_a)$. Since $var(s_a) = 2/3g$ for $L \ll \xi$, L_a, and because the localization threshold occurs at $g = 1$ in the absence of absorption, we make the conjecture that localization is achieved when $g' = 1$, or equivalently when $var(s_a) = 2/3$, whether absorption is present or not.

In strongly absorbing quasi-one-dimensional dielectric samples, $var(s_a)$ was found to increase sublinearly with length for diffusive waves [38]. This raised the possibility that the value of $var(s_a)$ might saturate with length and that absorption might introduce a cutoff length for the growth of relative fluctuations. Here we show that, though the presence of absorption leads to a decrease in $var(s_a)$, this appropriately reflects a lessening of localization effects. The threshold for localization occurs at $g' = 1$, and for smaller values, g' falls exponentially with length.

We now consider the scaling of $var(s_a)$ in the polystyrene samples and its connection to localization. The role of absorption is investigated by comparing measurements to an analysis of the data that statistically eliminates the influence of absorption. For $L \ll \xi$, L_a, diffusion theory gives $var(s_a) = 2\tilde{L}/3\xi$ [33, 34]. This result is shown as the horizontal short-dashed line in Fig. 13. As $g \to 1$ and localization threshold is approached, the scaling theory of localization suggests that g falls more rapidly [57], and hence $var(s_a)$ should increase

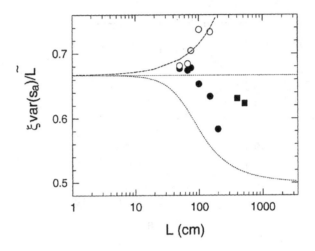

Figure 13.   Influence of absorption and localization, separately and together, on $var(s_a)$ in random polystyrene samples. A semi-logarithmic plot of $\xi var(s_a)/\tilde{L}$ is presented to illustrate the measured scaling of $var(s_a)$ over a large range of L, as well as various theoretical predictions. The upper and lower short-dashed lines represent the two limits of diffusion theory: $L \ll \xi$, L_a and $L_a \ll L \ll \xi$, respectively. The filled circles are obtained from measurements of total transmission, while the filled squares are obtained from measurements of intensity. The circles are the results of an analysis that eliminates the affect of absorption, as explained in the text. The upper, long-dashed curve is a fit of these results to an expression incorporating the first-order localization correction to diffusion theory.

superlinearly with sample length. Measurements of fluctuations in spectra of total transmission in ensembles of polystyrene samples give the results shown as the filled circles in Fig. 13. These results indicate that $var(s_a)$ increases sublinearly with length up to $L = 2$ m, which was the largest length at which accurate measurements of the total transmission could be made.

To extend these studies of statistics in random waveguides to samples of greater lengths, we use Eq. (2). This allows us to relate the variance of the normalized total transmission to the variance of the normalized intensity, which is more readily measured in microwave experiments,

$$2var(s_a) = var(s_{ab}) - 1. \tag{5}$$

Transmitted field spectra are measured in an ensemble of 2,000 polystyrene samples with use of a Hewlett-Packard 8772C network analyzer. The calculated intensity spectra yield $var(s_{ab})$ which gives the corresponding values of $var(s_a)$ using Eq. (5). Values of $var(s_a)$ obtained in this way for $L \leq 2$ m agree within 3% with those shown as the filled circles in Fig. 13. The results for $L > 2$ m are shown in the figure as the filled squares. They indicate a more rapid, superlinear increase in $var(s_a)$ relative to the data for $L \leq 2$ m.

In these measurements, the affect of developing localization and absorption are intertwined. In order to obtain the values of $var(s_a)$ that would be measured in the absence of absorption, we use a procedure which is illustrated in Fig. 14. The measured field spectra are multiplied by the Fourier transform of a Gaussian pulse in time to give the transmission spectra for this pulse. The spectrum is Fourier transformed to give the temporal response to a short incident Gaussian pulse. To compensate for losses due to absorption, the time dependent field is multiplied by $\exp(t/2\tau_a)$, where t is the time delay from the incident pulse and $1/\tau_a$ is the absorption rate determined from measurements of the field correlation function with frequency shift [23]. Since the intensity is the square of the field, the decay rate of the field is one half that of the intensity. The intensity of the modified field is shown as the short-dashed line in Fig. 14. This modified field is then transformed back to the frequency domain. Intensity spectra and the distribution and variance of intensity are then computed. The intensity distributions are in excellent agreement with calculations for diffusive waves [33, 34], which are described in terms of a single parameter g. The values of $var(s_a)$ found in this way are shown as the circles in Fig. 13. A fit of the leading order localization correction [42], $var(s_a) = 2\tilde{L}/3\xi + 4\tilde{L}^2/15\xi^2$, to the data gives the upper long-dashed curve in Fig. 13 with $\xi = 5.51 \pm 0.18$ m and $z_b = 5.25 \pm 0.31$ cm. The results are consistent with independent determination of these parameters [83]. The difference between the circles and the filled circles represents the amount by which $var(s_a)$ is reduced by absorption, and hence represents the extent to which absorption suppresses localization.

For diffusing waves, $var(s_a)$ was predicted to fall from $2\tilde{L}/3\xi$ for $L \ll L_a$ to $\tilde{L}/2\xi$ for $L \gg L_a$ [30, 39], following the lower short-dashed curve in Fig. 13. Notwithstanding the initial drop of $var(s_a)$ from $2\tilde{L}/3\xi$, our measurements rise above this curve as a result of enhanced intensity correlation, as $L \to \xi$. At $L = 5.2$ m, $var(s_a) = 0.6$, which is close to the critical value 2/3.

4.   STATISTICS OF LOCALIZED WAVES

To study the statistics of transmission quantities for localized waves, we examine fluctuations of intensity in strongly scattering quasi-one-dimensional samples of alumina (Al_2O_3) spheres (see Fig. 15). In order to have control over the alumina concentration, the alumina spheres (diameter $d_A = 0.95$ cm, dielectric constant $\epsilon_A = 9.86$) are embedded in Styrofoam spheres of $\epsilon_S = 1.05$. The scattering efficiency of the alumina spheres can be inferred from measurements of microwave transmission in optically thin, low concentration ($f < 0.001$) alumina samples. Extinction of the microwave radiation measured in such samples over the frequency range 11.8-19.0 GHz is shown in Fig. 16.

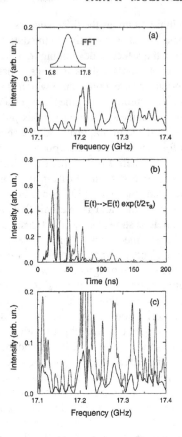

Figure 14. Statistical cancellation of absorption. (a) Intensity spectrum in a sample from the same ensemble as in Fig. 2. The field spectrum, from which the intensity spectrum is obtained, is multiplied by the spectrum of the pulse shown in the inset of (a) and Fourier transformed to the time domain. The result of the transform is squared to give the intensity pulse shown in (b) as the solid curve. The influence of absorption is removed in a statistical sense by multiplying field of the pulse by a factor $\exp(t/2\tau_a)$. The intensity of the modified pulse is shown as the short-dashed line in (b). The modified field is then Fourier transformed to the frequency domain. The field spectrum is then squared to give the transmitted intensity spectrum shown in (c) as the short-dashed curve and compared to the original spectrum. Intensity statistics computed using the transformed spectra are the same as predicted for a medium with the same real part of the dielectric function but without absorption.

We also used Mie theory to obtain the scattering efficiency Q_{sc} for the alumina sphere over the frequency range 7-19 GHz. In calculations, we assumed the alumina sphere to be lossless, and the numerical value of $\epsilon_A = 9.86$ was

Figure 15.   Alumina samples are composed of $\frac{3}{8}$-in-diam. alumina (Al_2O_3) spheres embedded in Styrofoam spheres. Styrofoam is nearly transparent for microwave and serves to control over alumina concentration. In the picture, spheres with diameter of $\frac{3}{4}$ in. are shown. An inch ruler is also shown.

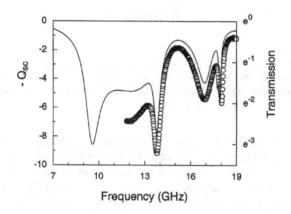

Figure 16.   Measured transmission of the microwave radiation in an optically thin, low concentration ($f < 0.001$) alumina sample is shown over the frequency range 11.8-19.0 GHz. Plotted on semi-logarithmic scale, extinction of microwave radiation indicates frequency dependence of scattering efficiency of the alumina sphere. The scattering efficiency Q_{sc} for the alumina sphere was obtained from Mie theory, and the negative of Q_{sc} is shown as the solid curve over the frequency range 7-19 GHz.

adjusted until the positions of the Mie resonances were located under the dips in the experimental data in Fig. 16.

We measure microwave field transmitted through an ensemble of alumina scatterers randomly positioned in a 7.3-cm-diam. copper tube at an alumina

volume fraction $f = 0.068$. Large values of $var(s_a)$ indicate that the wave is
localized near the first Mie resonance of the alumina sphere. A typical spectrum of s_{ab} obtained in the localization region is shown in Fig. 17. The sharp

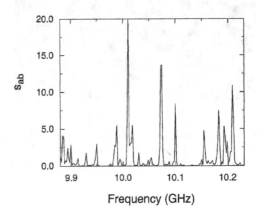

Figure 17.   A typical spectrum of the normalized intensity s_{ab} near the first Mie resonance of
the alumina spheres in a 80-cm-long alumina sample. The sharp and narrow line spectra and giant
fluctuations shown have been predicted for localized waves and are unlike the corresponding
spectrum in diffusive sample shown in Fig. 2.

peaks in s_{ab} appear to be related to resonant transmission through localized photonic states in the sample. We expect that when the frequency of the incident
wave is tuned to resonance with localized states in the medium, transmission
is appreciable. Between the peaks, transmission is exponentially small and the
incident wave is almost completely reflected from the medium. The distribution function $P(s_{ab})$ calculated for an ensemble of 5,000 samples is shown in
Fig. 18 and compared to the Rayleigh distribution. The measured distribution
is remarkably broad with $var(s_{ab}) = 23.5$, and fluctuations greater than 300
times the average value are observed. The scaling of $var(s_a)$ determined using
Eq. (5) at a number of frequencies near the first resonance is shown in Fig. 19.
We find that $var(s_a)$ increases exponentially once it becomes of order unity, as
expected for a localization parameter.

The availability of a measurable localization parameter makes it possible to
determine the existence and the extent of localization in a variety of samples.
This is illustrated in measurements of localization in periodic metallic wire
meshes containing metallic scatterers. John [86] has proposed that photon
localization could be achieved by introducing disorder in a periodic structure

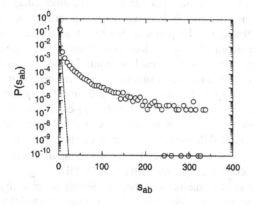

Figure 18.   Semi-logarithmic plot of the distribution $P(s_{ab})$ for the alumina sample. The distribution is obtained in an ensemble of 5,000 spectra within the frequency range 9.88-10.24 GHz, in which statistical parameters do not change substantially. The circle on the horizontal axis represents bins in which there was no measured intensity value. The broken line represents the Rayleigh distribution.

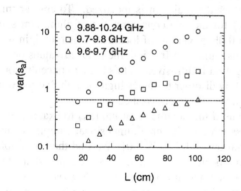

Figure 19.   Scaling of $var(s_a)$ in alumina samples. The values of $var(s_a)$ averaged over the indicated frequency intervals are obtained using Eq. (5). Above a value of order unity, $var(s_a)$ increases exponentially. In the interval 9.88-10.24 GHz, $var(s_a) \sim \exp(L/L_{exp})$, with $L_{exp} \approx 42$ cm.

possessing a photonic band gap. These band gaps exist in the electromagnetic spectrum of a variety of periodic structures in analogy with the electronic band gaps in crystals [87]. In the photonic band gap, electromagnetic waves are evanescent. When disorder is introduced in such structures, localized states are created in the gap. The periodic structure that we examine is a simple cubic lattice made up of copper wires, with a lattice constant of 1 cm. The lattice has eight unit cells along each side and is enclosed in a section of a square waveguide. Measurements of microwave transmission in an empty wire mesh sample show a low-frequency gap with a cut-off frequency of 9.33 GHz (Fig. 20a) [88]. The network is filled with 0.47-cm-diam. Teflon spheres in order to float various scatterers within the structure. The mean free path of randomly positioned Teflon spheres greatly exceeds the length of the structure, so that their only influence is to reduce the wavelength of radiation within the structure. The cut-off frequency shifts to 7.58 GHz (Fig. 20b). This is the same fractional shift as the ratio of the wavelength in air to that in a random sample of Teflon spheres, as determined by a measurement of the coherent field transmitted through the sample. Thus the ratio of the cut-off wavelength to the length of the unit cell is a constant in the structure. When aluminum spheres with the same diameter replace some of the Teflon spheres, transmission peaks appear in the gap just below the band edge. As the scatterer density is increased, the gap fills in. Spectra of the average transmission measured in ensembles of 200 sample configurations at two aluminum sphere volume fractions, $f = 0.05$ and $f = 0.10$, are shown in Fig. 20c. But such measurements leave open the question of whether the radiation is localized. To answer this question, we compute $var(s_a)$ for the two concentrations of aluminum spheres shown in Fig. 21. At $f = 0.05$, a window of localization is found, in which $var(s_a) \geq 2/3$. At twice this aluminum fraction, the reduced values of $var(s_a)$ indicate that wave propagation is diffusive. At this higher concentration, the density of states introduced by disorder becomes high enough that the wave is no longer localized. The gap is effectively washed out.

We have also used measurements of $var(s_a)$ to examine the claim that localization can be achieved in three-dimensional samples of metal spheres at various concentrations [73, 74, 89, 90]. We find that in samples of 0.47-cm-diam. aluminum spheres of length $L = 8.2$ cm and diameter $d = 7.5$ cm, with various volume fraction from 0.1 to 0.475, $var(s_a)$ never rises above the localization threshold of 2/3. A maximum value of 0.29 is reached at $f = 0.45$. Thus we conclude that *three-dimensional* localization is not achieved in these aluminum samples.

Summary

In conclusion, we have demonstrated that the variance of the normalized transmission is a robust localization parameter. It serves as a powerful tool in the search for and characterization of photon localization, even for absorb-

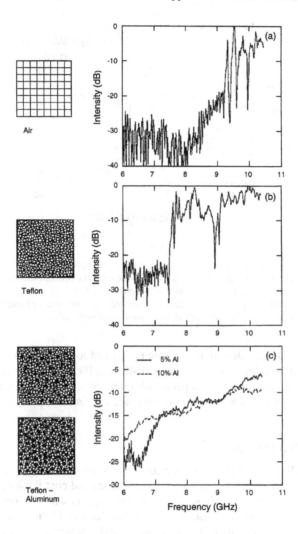

Figure 20. Schematic diagram of the wire mesh lattice filled with different media and associated transmission spectra. The structure has a lattice constant of 1 cm and is 8 unit cells on each side. In (a), the structure is in air. In (b), it is filled with Teflon spheres. In (c), the structure is filled with Teflon-aluminum mixtures at filling fractions of aluminum spheres of $f = 0.05$ and $f = 0.10$. The average transmission spectra obtained from 200 configurations for each value of f are shown.

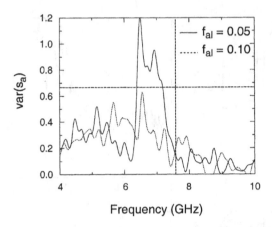

Figure 21. $Var(s_a)$ vs. frequency in a metallic wire mesh containing aluminum scatterers. The broken vertical line indicates the position of the band edge in a periodic structure filled only with Teflon spheres. At an aluminum sphere volume fraction $f = 0.05$, $var(s_a)$ is markedly higher near the edge, rising above the localization threshold of 2/3 shown as the broken horizontal line. At $f = 0.10$, $var(s_a)$ is reduced and wave propagation is diffusive.

ing samples. It has allowed us to locate regimes of localization in quasi-one-dimensional dielectric samples and in periodic metallic wire lattices containing metallic scatterers, and to rule it out in collections of aluminum spheres. These results show that the statistics of transmission are essential aspects of localization.

Acknowledgments

We would like to thank Patrick Sebbah for suggesting the topic and title of this presentation and for stimulating discussions and contributions. We are grateful to Piet Brouwer, Eugene Kogan, Victor Kopp, Edward Kuhner, Alexander Lisyansky, Zdzislaw Ozimkowski, Mark van Rossum, Marin Stoytchev, and Bart van Tiggelen for stimulating discussions and technical assistance without which this work would not have been possible. This research was supported by the National Science Foundation under grants DMR-9973959 and INT-9512975 and by a PSC-CUNY award.

References

[1] A. Ishimaru, *Wave Propagation and Scattering in Random Media* (Academic Press, New York, 1978).

[2] J. W. Goodman, *Statistical Optics* (Wiley, New York, 1985).

[3] *Scattering and Localization of Classical Waves in Random Media*, edited by P. Sheng (World Scientific Press, Singapore, 1990).

[4] *Mesoscopic Phenomena in Solids*, edited by B. L. Altshuler, P. A. Lee, and R. A. Webb (Elsevier, Amsterdam, 1991).

[5] *Diffuse Waves in Complex Media*, edited by J.-P. Fouque (Kluwer Academic Publishers, Dordrecht, 1999).

[6] I. Freund and D. Elyahu, Phys. Rev. A **45**, 6133 (1992).

[7] A. Z. Genack, Phys. Rev. Lett. **58**, 2043 (1987).

[8] A. Z. Genack and J. M. Drake, Europhys. Lett. **11**, 331 (1990).

[9] P. Sebbah, O. Legrand, B. A. van Tiggelen, and A. Z. Genack, Phys. Rev. E **56**, 3619 (1997).

[10] J. W. Strutt (Lord Rayleigh), Proc. Lond. Math. Soc. **3**, 267 (1871).

[11] J. C. Dainty, *Laser Speckle and Related Phenomena* (Springer-Verlag, Berlin, 1975).

[12] R. A. Webb, S. Washburn, C. P. Umbach, and R. B. Laibowitz, Phys. Rev. Lett. **54**, 2696 (1985).

[13] B. L. Altshuler and D. E. Khmelnitskii, JETP Lett. **42**, 359 (1985).

[14] P. A. Lee and A. D. Stone, Phys. Rev. Lett. **55**, 1622 (1985).

[15] B. Shapiro, Phys. Rev. Lett. **57**, 2168 (1986).

[16] B. Shapiro, Phil. Mag. B **56**, 1031 (1987).

[17] M. J. Stephen and G. Cwilich, Phys. Rev. Lett. **59**, 285 (1987).

[18] G. Maret and P. E. Wolf, Z. Phys. B **65**, 409 (1987).

[19] P. A. Mello, E. Akkermans, and B. Shapiro, Phys. Rev. Lett. **61**, 459 (1988).

[20] S. Feng, C. Kane, P. A. Lee, and A. D. Stone, Phys. Rev. Lett. **61**, 834 (1988).

[21] I. Edrei, M. Kaveh, and B. Shapiro, Phys. Rev. Lett. **62**, 2120 (1989).

[22] R. Pnini and B. Shapiro, Phys. Rev. B **39**, 6986 (1989).

[23] N. Garcia and A. Z. Genack, Phys. Rev. Lett. **63**, 1678 (1989).

[24] A. Z. Genack, N. Garcia, and W. Polkosnik, Phys. Rev. Lett. **65**, 2129 (1990).

[25] M. P. van Albada, J. F. de Boer, and A. Lagendijk, Phys. Rev. Lett. **64**, 2787 (1990).

[26] N. Shnerb and M. Kaveh, Phys. Rev. B **43**, 1279 (1991).

[27] E. Kogan and M. Kaveh, Phys. Rev. B **45**, 1049 (1992).

[28] E. Kogan, M. Kaveh, R. Baumgartner, and R. Berkovits, Phys. Rev. B **48**, 9404 (1993).

[29] A. Z. Genack and N. Garcia, Europhys. Lett. **21**, 753 (1993).

[30] N. Garcia, A. Z. Genack, R. Pnini, and B. Shapiro, Phys. Lett. A **176**, 458 (1993).

[31] R. Berkovits and S. Feng, Phys. Rep. **238**, 135 (1994).

[32] J. F. de Boer, M. C. W. van Rossum, M. P. van Albada, Th. M. Nieuwenhuizen, and A. Lagendijk, Phys. Rev. Lett. **73**, 2567 (1994).

[33] M. C. W. van Rossum and Th. M. Nieuwenhuizen, Phys. Rev. Lett. **74**, 2674 (1994).

[34] E. Kogan and M. Kaveh, Phys. Rev. B **52**, R3813 (1995).

[35] V. Fal'ko and K. Efetov, Phys. Rev. B **52**, 17413 (1995).

[36] S. A. van Langen, P. W. Brouwer, and C. W. J. Beenakker, Phys. Rev. E **53**, 1344 (1996).

[37] A. A. Chabanov and A. Z. Genack, Phys. Rev. E **56** R1338 (1997).

[38] M. Stoytchev and A. Z. Genack, Phys. Rev. Lett. **79** 309 (1997).

[39] P. W. Brouwer, Phys. Rev. B, **57**, 10526 (1998).

[40] F. Scheffold and G. Maret, Phys. Rev. Lett. **81**, 5800 (1998).

[41] M. Stoytchev and A. Z. Genack, Opt. Lett. **24**, 262 (1999).

[42] M. C. W. van Rossum and Th. M. Nieuwenhuizen, Rev. Mod. Phys. **71**, 313 (1999).

[43] M. Stoytchev, Ph. D. Thesis (City University of New York, 1998).

[44] A. A. Chabanov, M. Stoytchev, and A. Z. Genack, Nature **404**, 850 (April, 2000).

[45] R. Pnini in this book.

[46] P. Sebbah, R. Pnini, and A. Z. Genack, Phys. Rev. E (2000).

[47] E. Wigner, Phys. Rev. **98**, 145 (1955).

[48] F. T. Smith, Phys. Rev. **118**, 349 (1960); **119**, 2098 (1960).

[49] M. Buttiker, H. Thomas, and A. Pretre, Phys. Lett. A **180**, 364 (1993).

[50] J. G. Muga and D. M. Wardlaw, Phys. Rev. E **51**, 5377 (1995).

[51] P. W. Brouwer, K. M. Frahm, and C. W. J. Beenakker, Phys. Rev. Lett. **78**, 4737 (1997).

[52] P. Sebbah, O. Legrand, and A. Z. Genack, Phys. Rev. E **59**, 2406 (1999).

[53] A. Z. Genack, P. Sebbah, M. Stoytchev, and B. A. van Tiggelen, Phys. Rev. Lett. **82**, 715 (1999).

[54] B. A. van Tiggelen, P. Sebbah, M. Stoytchev, and A. Z. Genack, Phys. Rev. E **59**, 7166 (1999).

[55] P. W. Anderson, Phys. Rev. **109**, 1492 (1958).

[56] J. T. Edwards and D. J. Thouless, J. Phys. **C5**, 807 (1972).

[57] E. Abrahams, P. W. Anderson, D. C. Licciardello, and T. V. Ramakrishnan, Phys. Rev. Lett. **42**, 673 (1979).

[58] S. John, H. Sompolinsky, and M. J. Stephen, Phys. Rev. B **27**, 5592 (1983).

[59] S. John, Phys. Rev. Lett. **53**, 2169 (1984).

[60] R. Landauer, Phil. Mag. **21**, 863 (1970).

[61] B. A. van Tiggelen, in *Diffuse Waves in Complex Media*, edited by J.-P. Fouque (Kluwer Academic Publishers, Dordrecht, 1999).

[62] M. P. van Albada, B. A. van Tiggelen, A. Lagendijk, and A. Tip, Phys. Rev. Lett. **66**, 3132 (1991).

[63] A. Lagendijk and B. van Tiggelen, Phys. Rep. **270**, 143 (1996).

[64] Y. Kuga and A. Ishimaru, J. Opt. Soc. Am. **A1**, 831 (1984).

[65] M. van Albada and A. Lagendijk, Phys. Rev. Lett. **55**, 2692 (1985).

[66] P. F. Wolf and G. Maret, Phys. Rev. Lett. **55**, 2696 (1985).

[67] E. Akkermans, P. E. Wolf, and R. Maynard, Phys. Rev. Lett. **56**, 1471 (1986).

[68] A. F. Ioffe and A. R. Regel, Prog. Semicond. **4**, 237 (1960).

[69] R. L. Weaver, Phys. Rev. B **47**, 1077 (1993).

[70] V. Freilikher, M. Pustilnik, and I. Yurkevich, Phys. Rev. Lett. **73**, 810 (1994).

[71] M. Yosefin, Europhys. Lett. **25**, 675 (1994).

[72] J. C. J. Paasschens, T. Sh. Mizirpashaev, and C. W. J. Beenakker, Phys. Rev. B **54**, 11887 (1996).

[73] N. Garcia and A. Z. Genack, Phys. Rev. Lett. **66**, 1850 (1991).

[74] A. Z. Genack and N. Garcia, Phys. Rev. Lett. **66**, 2064 (1991).

[75] D. S. Wiersma, P. Bartolini, A. Lagendijk, and R. Righini, Nature, **390**, 671 (1997); D. S. Wiersma, J. G. Rivas, P. Bartolini, A. Lagendijk, and R. Righini, R. Nature **398**, 207 (1999).

[76] F. Scheffold, R. Lenke, R. Tweer, and G. Maret, Nature **398**, 206 (1999).

[77] Ya. A. Vlasov, M. A. Kaliteevski, and V. V. Nikolaev, Phys. Rev. B **60**, 1555 (1999).

[78] B. A. van Tiggelen, A. Lagendijk, and D. S. Wiersma, Phys. Rev. Lett. **84**, 4333 (2000).

[79] P. W. Anderson, D. J. Thouless, E. Abrahams, and D. S. Fisher, Phys. Rev. B **22**, 3519 (1980).

[80] P. A. Mello, J. Math. Phys. (N. Y.) **27**, 2876 (1986).

[81] L. I. Deych, D. Zaslavsky, and A. A. Lisyansky, Phys. Rev. Lett. **81**, 5390 (1998).

[82] L. I. Deych, A. A. Lisyansky, and B. L. Altshuler, Phys. Rev. Lett. **84**, 2678 (2000).

[83] N. Garcia, A. Z. Genack, and A. A. Lisyansky, Phys. Rev. B **46**, 14475 (1992).

[84] A. Lagendijk, R. Vreeker, and P. de Vries, Phys. Lett. A **136**, 81 (1989).

[85] A. A. Chabanov and A. Z. Genack, submitted (2000).

[86] S. John, Phys. Rev. Lett. **58**, 2486 (1987).

[87] J. D. Joannopolos, R. D. Meade, and J. N. Winn, *Photonic Crystals* (Princeton University Press, Princeton, 1995).

[88] M. Stoytchev and A. Z. Genack, Phys. Rev. B **55**, R8617 (1997).

[89] K. Arya, Z. B. Su, and J. L. Birman, Phys. Rev. Lett. **57**, 2725 (1986).

[90] C. A. Condat and T. R. Kirkpatrick, Phys. Rev. Lett. **58**, 226 (1987).

III

WAVE CHAOS AND MULTIPLE SCATTERING

Chapter 1

THE SEMICLASSICAL APPROACH
FOR CHAOTIC SYSTEMS

D. Delande

Laboratoire Kastler-Brossel
Tour 12, Etage 1, Université Pierre et Marie Curie
4, place Jussieu, F-75252 Paris Cedex 05, France.
delande@spectro.jussieu.fr

Abstract   Chaotic systems have a very complex classical dynamics. In these lectures, I discuss their quantum properties and show how a semiclassical approach is possible. I also show how the presence of single or multiple scattering by a small object may induce a chaotic-like behaviour and how to calculate the quantum properties using a semiclassical approach.

Keywords: quantum chaos, semiclassical approximation, trace formula

1.   WHAT IS QUANTUM CHAOS?

1.1.   CLASSICAL CHAOS

Chaos is usually defined for classical systems, i.e. systems whose dynamics can be described by deterministic equations of evolution in some phase space. In the specific case of a time-independent Hamiltonian system for a single particle, the phase space coordinates are the position $\mathbf{q}$ and momentum $\mathbf{p}$, and the equations of motion can be expressed using the Hamilton function $H(\mathbf{q}, \mathbf{p})$ as [1]:

$$\frac{dq_i}{dt} = \frac{\partial H}{\partial p_i} \tag{1}$$

$$\frac{dp_i}{dt} = -\frac{\partial H}{\partial q_i}, \quad \text{for } 1 \leq i \leq d, \tag{2}$$

87

P. Sebbah (ed.), Waves and Imaging through Complex Media, 87–124.
© 2001 *Kluwer Academic Publishers. Printed in the Netherlands.*

where d is the number of degrees of freedom. Basically, classical chaos is exponential sensitivity on initial conditions: two neighbouring trajectories diverge exponentially with time, i.e. the distance between the two trajectories generically increases as $\exp(\lambda t)$ where λ is called the Lyapounov exponent of the system [2]. Sensitivity on initial conditions is responsible for the decrease of correlations over long times, loss of memory of the initial conditions and ultimately for deterministic unpredictability of the long time behaviour of the system. Most often, when the system is sensitive on initial conditions, it is also mixing and ergodic [2], i.e. a typical trajectory uniformly fills up the entire phase space at long time.

For low-dimensional systems, the dynamics is often a mixed regular-chaotic one, depending on the initial conditions; also, when a parameter is changed in the Hamilton function, the transition from regularity to chaos is usually smooth with intermediate mixed dynamics. Such mixed systems are rather complicated and not too well understood and we will here restrict to the simple situation where the motion is almost fully chaotic.

1.2.   QUANTUM DYNAMICS

In quantum mechanics, there is neither any phase space, nor anything looking like a trajectory. Hence, the notion of classical chaos cannot be simply extended to quantum physics. Quantum mechanics uses completely different notions, like the state vector $|\psi\rangle$ belonging to some Hilbert space, which describes all the physical properties of the system. Its evolution is given by the Schrödinger equation:

$$i\hbar \, \frac{d|\psi(t)\rangle}{dt} = H(t) \, |\psi(t)\rangle, \qquad (3)$$

where $\hbar$ is the Planck's constant and $H(t)$ the Hamiltonian.

$|\psi\rangle$ is not directly observable in quantum mechanics. In general – according to the standard interpretation of quantum mechanics – the result of a measure is some diagonal element of an Hermitian operator, something like $\langle\psi|O|\psi\rangle$. The physical processes involved in an experimental measurement are quite subtle, difficult and interesting, but beyond the subject of these lectures. It is also the subject of a vast literature [3]. I will not consider this problem and restrict to a purely Hamiltonian evolution.

The time-evolution operator $U(t', t)$ is the linear operator mapping the state $|\psi(t)\rangle$ onto the state $|\psi(t')\rangle$. It obeys the following equation (which is equivalent to Schrödinger equation):

$$i\hbar \, \frac{\partial U(t', t)}{\partial t'} = H(t') \, U(t', t). \qquad (4)$$

Because H is an Hermitian operator, $U(t', t)$ is a linear unitary operator. An immediate consequence is that the overlap between two states is preserved during the time evolution. Indeed, one has:

$$\langle \psi_1(t') | \psi_2(t') \rangle = \langle \psi_1(t) | U^\dagger(t', t) U(t', t) | \psi_2(t) \rangle = \langle \psi_1(t) | \psi_2(t) \rangle, \quad (5)$$

which implies that two "neighbouring" states will remain neighbors forever. Because of linearity and unitarity, quantum mechanics cannot display any sensitivity on initial conditions, hence cannot be chaotic in the ordinary sense!

However, the previous statement must be considered with care. Indeed, classical mechanics can also be seen as a linear theory if one considers the evolution of a classical phase space density $\rho(\mathbf{q}, \mathbf{p}, t)$ given by the Liouville equation [1]:

$$\frac{\partial \rho}{\partial t} = \sum_i \left(\frac{\partial \rho}{\partial p_i} \frac{\partial H}{\partial q_i} - \frac{\partial \rho}{\partial q_i} \frac{\partial H}{\partial p_i} \right). \quad (6)$$

The fact that we obtain both in classical and quantum mechanics a linear equation of evolution in some space just implies that the above argument on linearity in quantum mechanics is irrelevant. Discussions on subjects like "Is there any quantum chaos?" are in my opinion completely uninteresting because they focus on the formal aspects of the mathematical apparatus used.

We will here define quantum chaos as the study of quantum systems whose classical dynamics is chaotic. The questions we would like to answer are thus:

- What are the appropriate observables to detect the regular or chaotic *classical* behavior of the system?

- More precisely, how the chaotic or regular behaviour expresses in the energy levels and eigenstates of the quantum system?

- What kind of semiclassical approximations can be used?

These are the questions discussed in these lectures. I will only present selected topics, forgetting lots of interesting questions and relevant references.

These questions of course go towards an intrinsic definition of quantum chaos not referring to the classical dynamics [4]. Thus, the problem of quantum chaos is essentially related to the correspondence between classical and quantum dynamics, the subject of *semiclassical* physics.

More generally, classical chaos is an example of *complexity*. The dynamics of complex systems is usually so complicated that a fully detailed analysis is of no relevance for a global understanding. For example, for classically chaotic systems, we are not interested in the knowledge of each individual trajectory, but rather in the global, i.e. statistical, properties of the motion like the Lyapounov exponents. The same is true for the quantum properties. A simple example is coherent back-scattering, which typically exists both in disordered complex quantum system and chaotic quantum systems, and is robust versus any change in the detail of the dynamics.

In chaotic systems, the phase structure – although rather complicated – is partly well known which makes it possible to derive and use some sophisticated semiclassical approaches, as will be shown in section 1.3. It is reasonable to think that similar treatments are possible for complex systems. I show in section 4.7 how this leads to some interesting results in the case of a single object which scatters again and again a trapped quantum wave.

1.3.   SEMICLASSICAL DYNAMICS

The whole idea of a semiclassical analysis is to obtain approximate solutions of the quantum equation of motion (the Schrödinger equation) using only classical ingredients (trajectories...) and the Planck's constant $\hbar$. For a macroscopic object, our common knowledge is that an approximate semiclassical solution should be very accurate. Technically, this is true because $\hbar$ is much smaller than any classical quantity of interest (such as the classical action of the particle). One often refers to the "correspondence principle" as an explanation. However, this is a *very vague* concept which is usually not clearly stated. Part of these lectures are devoted to a serious scientific discussion of this issue, using the modern knowledge on classical chaos.

In order to make the connection between classical and quantum quantities, it is useful to define the Wigner representation defined as [5]:

$$W(\mathbf{q},\mathbf{p}) = \frac{1}{(2\pi\hbar)^N} \int \psi\left(\mathbf{q}-\frac{\mathbf{x}}{2}\right) \psi^*\left(\mathbf{q}+\frac{\mathbf{x}}{2}\right) \exp\left(i\frac{\mathbf{p}.\mathbf{x}}{\hbar}\right) d\mathbf{x}. \quad (7)$$

This is a real phase space density probability, or rather quasi-probability because it can be either positive or negative. Its evolution equation is simple to compute [5]:

$$\frac{\partial W}{\partial t} = -\frac{2}{\hbar}H(\mathbf{q},\mathbf{p})\sin\left(\frac{\hbar\Lambda}{2}\right)W(\mathbf{q},\mathbf{p}), \quad (8)$$

with:

$$\Lambda = \sum_i \frac{\overleftarrow{\partial}}{\partial p_i}\frac{\overrightarrow{\partial}}{\partial q_i} - \frac{\overleftarrow{\partial}}{\partial q_i}\frac{\overrightarrow{\partial}}{\partial p_i}, \quad (9)$$

where the left (resp. right) arrow indicates action on the quantity on the left (resp. right) side.

An explicit power expansion of the sine function is possible. This is in fact a power expansion in $\hbar$, hence well suited for a semiclassical approximation. At lowest (zeroth) order, one finds *exactly* the classical Liouville equation, thus establishing a link between the quantum and classical dynamics. At next order in $\hbar$ (actually $\hbar^2$), one finds terms involving third partial derivatives of the Hamiltonian. For harmonic systems, these terms vanish, proving that the classical and quantum phase space dynamics completely coincide.

For non-harmonic systems, the corrective terms produce an additional spreading of initially localized wavepackets. For chaotic systems, the classical solutions of the Liouville equations tend to stretch and fold along (exponentially) unstable directions and – because of conservation of volume in phase space – to shrink along (exponentially) attractive directions. This rapidly creates "whorls" and "tendrils" in the classical phase space density, which in turn implies more and more rapid spatial changes of the density. Thus, as time goes on, one expects some higher order partial derivatives to grow exponentially. Although the corresponding terms in the quantum equation of evolution are multiplied by $\hbar^2$, they will unavoidably grow and overcome the classical Liouville term. Hence, after some "break time", the detailed quantum evolution will differ from the classical one. The estimation of this break time is a difficult question, and different answers are possible, depending on which aspect of the dynamics is under study (local, global...). I will not discuss this important point here, see [3, 4].

Of course, for smaller $\hbar$, the higher order terms are smaller and it requires a longer time for them to perturb the dynamics. Hence, the break time has to tend to infinity in the semiclassical limit $\hbar \to 0$. For a fixed time interval, one can always find a sufficiently small $\hbar$ such that the quantum and classical dynamics are almost identical. In other words, over a finite time range, the quantum dynamics tends to the classical one as $\hbar \to 0$. However, this limit is not *uniform*. For fixed $\hbar$, there is always a finite time after which the quantum dynamics differs from the classical one. In other words, the two limits $t \to \infty$ and $\hbar \to 0$ do not commute. Taking first $\hbar \to 0$, then $t \to \infty$ is studying the long time classical dynamics, i.e. classical chaos. The other limit $t \to \infty$, then $\hbar \to 0$, is what we are interested in, namely quantum (and semiclassical) chaos.

1.4.   PHYSICAL SITUATIONS OF INTEREST

Simple equations of motion may produce a chaotic behaviour. A rather non-intuitive result is that chaos may take place in low dimensional systems. On the other hand, classical chaos can only exist in systems where different degrees of freedom are *strongly* coupled [2]. This implies that a small perturbation added to a regular system cannot make it chaotic.

The simplest possible chaotic systems are thus time-independent 2-dimensional systems. It is also simpler to consider bound systems with a discrete energy spectrum. Various model systems have been studied, among which billiards are the simplest ones. A billiard is a compact area in the plane containing a point particle bouncing elastically on the walls. Depending on the shape of the boundary, the motion may be regular or chaotic. From the quantum point of view, one has to find the eigenstates of the Laplace operator whose wavefunction vanishes on the boundary [6].

In the context of (multiple) scattering, open (i.e. not bound) systems are of primary importance. Two viewpoints are possible: in the scattering point of view, one puts the emphasis on the coupling of the system with the external world and uses a S-matrix which describes how an incoming wave is scattered by the complex medium. In the "internal" point of view, the emphasis is put on the inner properties of the quantum system: there are no longer bound states, but rather resonances with complex energies, the imaginary part being minus half the width (inverse of the lifetime) of the resonance. Of course, the two points of view are strictly equivalent: the resonances, which are complex poles of the Green's function, are also poles of the S-matrix. If the internal medium is sufficiently complex, the classical dynamics of a particle bouncing off the various scatterers may be chaotic. In this sense, there is a strong and deep connection between multiple scattering and quantum chaos. A key issue is localization: in an open system, the classical dynamics is not bound and often shows some similarity with a diffusive motion (although it is perfectly deterministic) with the average excursion increasing as the square root of time. Quantum interferences between various paths – which can be seen as scattering paths in the multiple scattering language or as classical trajectories in the quantum chaos language – may lead to partial (weak) or complete (strong) localization. Whether such a phenomenon can be understood on a semiclassical basis is not clear, but certainly an interesting issue [7].

If we now turn to "experimental" systems, it is obvious that quantum effects are likely to be noticeable only for microscopic systems. The dynamics of nucleons in an atomic nucleus might be chaotic – at least at sufficiently large energy – and the experimental results on highly excited states played a major role in the early development of quantum chaos [6]. The drawback is the existence of complex collective effects and the fact that the interaction is not perfectly well known.

Atoms are among the best available prototypes for studying quantum chaos. Compared with other microscopic complex systems (nuclei, atomic clusters, mesoscopic devices...), they have the great advantage that all the basic components are well understood : these are essentially point particles (electrons and nucleus) interacting through a Coulomb static field, and interacting with the external world through electromagnetic forces. Hence, it is possible to write down an explicit expression of the Hamiltonian. Another crucial advantage of atomic systems is that they can be studied theoretically and *experimentally*. The word "experiment" must here be understood as traditional laboratory experiments, but also as "numerical experiments". Indeed, currently available computers make it possible to numerically compute properties of complex systems described by simple Hamiltonians. During the last fifteen years, the constant interaction between the experimental results and the numerical simulations led to major advances in the field of quantum chaos.

Depending on the energy scale involved, different parts of the atomic dynamics are relevant. At "large" energy – of the order of 1eV – it is the internal dynamics of the atomic electrons (their motion around the nucleus) which may be chaotic. The best example is the hydrogen atom exposed to a strong external uniform magnetic field. Because of the azimuthal symmetry around the magnetic field axis, the dynamics of the electron is effectively two-dimensional, which is the simplest possible situation for observing a chaotic behaviour [2]. This makes this system an almost ideal prototype for the study of quantum chaos [8]. As a crude criterion, chaos is most developed when the Lorentz and Coulomb forces acting on the electron have the same order of magnitude. This can be realized in a laboratory experiment with Rydberg states $n \simeq 40 - 150$ [9, 10, 11]. Another advantage of this system is that the classical dynamics has a scaling property, which makes it possible to effectively study the semiclassical limit $\hbar \to 0$. This possibility has revealed extremely important for understanding the classical-quantal correspondence [10]. From the theoretical point of view, the use of group theory allows extremely efficient numerical experiments [12, 8, 13], making the computation of very accurately highly excited energy levels and wavefunctions possible. The calculated quantities are found in exact agreement with the (less accurate) experimental measurement (see [9])!

At a much lower energy scale – 1 μeV – it is the external dynamics of the center of mass of the atom (considered as a single particle) which may display a chaotic behavior under the influence of an external electromagnetic field [14]. The latter case has been made possible because of the impressive recent improvements on the control of ultra-cold atomic gases using quasi-resonant laser beams [15].

In molecules, the dynamics of the electrons may also be chaotic. In some cases, the motion of the nuclei in the effective potential created by the electrons (which follow the nuclei adiabatically) is chaotic. Some interesting results on the NO_2 molecules have been obtained [16].

At the microscopic level, the dynamics of electrons in a solid state sample may present a chaotic dynamics in, for example, suitable combinations of external fields. This has lead to dramatic results showing very clearly the relevance of periodic orbits for understanding the quantum chaotic dynamics [17].

Another possibility exists for experimental study of quantum chaos. One can consider wave equations describing some other physical phenomena, which have a structure very similar to the Schrödinger equation. As what we are interested in is in fact "wave chaos" (properties of eigenmodes for example) whatever the waves themselves are, this opens a wide variety of possible exper-

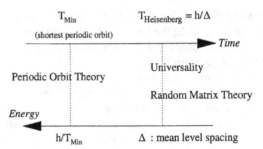

Figure 1.   The important time and energy scales for a chaotic quantum system. The shortest relevant time scale is T_{Min}, the period of the shortest periodic orbit. The most important quantum time scale is T_{H}, associated the mean energy level spacing. In the semiclassical limit, T_{H} is much larger than T_{Min}. One expects a universal classical behaviour at long times, thus universal statistical properties of the energy levels, described in section 3.4. At short times (long energy range), the specificities of the system appear to be related to the periodic orbits of the system, as explained in section 4.

iments. The best example is provided by flat microwave cavities where solving the Maxwell equations is equivalent to calculating the eigenstates of the corresponding two-dimensional Schrödinger billiard [18]. The advantage is that a measure of the "wavefunction" is possible.

2.   TIME SCALES - ENERGY SCALES

For a correct understanding of the connections between the quantum and the classical properties of a chaotic system, it is crucial to know the relevant time scales (and the corresponding energy scales) of the problem.

The shortest time scale is simply the period of the shortest periodic orbit T_{Min}. A slightly longer time scale is given by the time taken for chaos to manifest, that is the inverse of the typical Lyapounov exponent. The larger the sensitivity on initial conditions, the shorter this time scale. These two time scales have of course nothing to do with $\hbar$. The corresponding energy scale, $2\pi\hbar/T_{\text{Min}}$, see Fig. 1, is the largest energy scale of interest in the problem.

There is also a basic quantum time scale. To understand its origin, let us consider a time-independent bound quantum system with Hamiltonian H, in an arbitrary initial state $|\psi(t=0)\rangle$. Its evolution can be expressed using the discrete eigenstates and eigenvalues of the Hamiltonian H

$$H|\phi_i\rangle = E_i|\phi_i\rangle\,, \tag{10}$$

with the following expression:

$$|\psi(t)\rangle = \sum_i c_i \exp\left(-i\frac{E_i t}{\hbar}\right)|\phi_i\rangle, \tag{11}$$

where the constant coefficients c_i are computed from the initial state using:

$$c_i = \langle\phi_i|\psi(0)\rangle. \tag{12}$$

The autocorrelation function of the quantum system is a diagonal element of the time-evolution operator:

$$C(t) = \langle\psi(0)|\psi(t)\rangle = \langle\psi(0)|U(t,0)|\psi(0)\rangle = \sum_i |c_i|^2 \exp\left(-i\frac{E_i t}{\hbar}\right). \tag{13}$$

It is a discrete sum of oscillating terms, and, consequently, a quasi-periodic function of time. This is *extremely different* from a classical autocorrelation function for a chaotic system which is decreasing over the characteristic time scale T_{Min} and does not show any revival at longer times [2].

The Fourier transform of the autocorrelation function is:

$$\tilde{C}(E) = \frac{1}{2\pi\hbar}\int_{-\infty}^{\infty} e^{iEt/\hbar} C(t)\, dt = \sum_i |c_i|^2 \delta(E - E_i), \tag{14}$$

that is a sum of δ-peaks at the positions of the energy levels.

If we now consider the Fourier transform not over the whole range of time from $-\infty$ to $+\infty$, but over a finite time interval, we obtain a smoothed version of the quantum spectrum:

$$\tilde{C}_T(E) = \frac{1}{T}\int_{-T/2}^{T/2} e^{iEt/\hbar} C(t)\, dt = \sum_i |c_i|^2 \frac{\sin\frac{T(E-E_i)}{2\hbar}}{\frac{T(E-E_i)}{2\hbar}}, \tag{15}$$

where all the peaks are smoothed δ-peaks of width $2\pi\hbar/T$. For short T, the different broadened peaks centered at the energy levels E_i overlap, and $\tilde{C}_T(E)$ is a globally smooth function, like its classical counterpart. In such a situation, it is possible (although nothing proves that is is always the case) that the quantum $\tilde{C}_T(E)$ mimics the classical chaotic behaviour. The important point is that, for large T, the different peaks do not overlap and the discrete nature of the energy spectrum **must** appear in $\tilde{C}_T(E)$, whatever the initial state. The typical time needed for resolving individual quantum energy levels is called the Heisenberg time and is simply related to the mean level spacing Δ through:

$$T_{\text{H}} = \frac{2\pi\hbar}{\Delta}. \tag{16}$$

After this time, the quantum system *cannot* mimic the classical chaotic behaviour which has a continuous spectrum. Since T_H depends on $\hbar$, one can understand how quantum tends to classical dynamics as $\hbar$ goes to zero. The mean level spacing is given by the Weyl's rule and scales as $\hbar^d$, with d the number of degrees of freedom (see Eq. (19)). For two- (or higher) dimensional systems, T_H tends to infinity as $\hbar \to 0$, see Fig. 1.

In some sense, after the Heisenberg time, the quantum system "knows" that the energy spectrum is discrete, it has resolved all individual peaks and the future evolution cannot bring any essential new information. As a consequence, the system cannot explore a new part of the phase space, it freezes its evolution, repeating forever the same type of dynamics.

Other time scales may exist in specific systems. For example, in an open Hamiltonian system, the typical time scale for escaping the chaotic region is obviously important. Also, in mixed chaotic-regular systems, different time scales coexist in the different regions of phase space (and at their boundaries) making general statements extremely difficult. For systems coupled to their environments, dissipation and decoherence of the quantum wavefunction is known to play a very important role [3] and these effects may be dominant over chaotic effects. For the sake of simplicity, we will restrict to the case where T_{Min} and T_H are the only relevant time scales. In the semiclassical limit $\hbar \to 0$, the corresponding energy scales $2\pi\hbar/T_{\mathrm{Min}}$ and $2\pi\hbar/T_H = \Delta$ are both small compared to the energy itself. This means that we will always look at relative small changes in the energy, such that the classical dynamics does not substantially changes over the energy range considered. This is of course possible in the semiclassical regime thanks to the large density of states. For low excited states, such a local approach lacks any relevance.

3. STATISTICAL PROPERTIES OF ENERGY LEVELS RANDOM MATRIX THEORY

3.1. LEVEL DYNAMICS

The goal of traditional spectroscopy is to assign quantum numbers to the different energy levels in order to obtain a complete classification of the spectrum. When little is known about the system, it is difficult to identify the good quantum numbers and their physical interpretation, or even to know whether they exist or not. A simple tool is to look at the level dynamics, that is the evolution of the various energy levels as a function of a parameter. As good quantum numbers are associated with conserved quantities, i.e. operators commuting with the Hamiltonian, energy levels with different sets of good quantum numbers are not coupled and thus generically cross each other [19]. On the contrary, if two states are coupled, the energy levels will repel each other, producing an avoided crossing. The width of the avoided crossing, i.e. the minimum energy difference between the two energy curves, is a direct measure of the strength of the coupling.

Thus, looking at the level dynamics gives some qualitative information on the properties of the system. This is illustrated in Fig. 2 which shows the evolution of the energy levels of a hydrogen atom as a function of the magnetic field strength. At low magnetic field, Fig. 2a, there are only level crossings. A given eigenstate can be unambiguously followed in a wide range of field strength, since it crosses (or has very small avoided crossings with) the other energy levels, which proves that there are at least approximate good quantum numbers. At higher magnetic field, Fig. 2b, the sizes of the avoided crossings increase and individual states progressively loose their identities. In other words, the good quantum numbers are destroyed.

A crucial observation is that the transition from crossings (or tiny avoided crossings) to large avoided crossings takes place where the classical dynamics evolves from regular to chaotic. The transition is smooth – with the proportion of large avoided crossings progressively increasing – and there is a large intermediate region where crossings and large avoided crossings coexist. This is in complete agreement with the classical transition from regularity to chaos. From a pure quantum point of view, this phenomenon is extremely difficult to understand, as the matrix representing the Hamiltonian in any basis has the same structure whatever the magnetic field strength. The dramatic effect on the energy level dynamics is a direct manifestation of chaos in the quantum properties of the system.

In the fully chaotic regime, the energy levels and the eigenstates strongly *fluctuate* when the magnetic field is changed. In that sense, the quantum system shows a high sensitivity on a small perturbation, like its classical equivalent. The energy spectrum of a classically chaotic system displays an extreme *intrinsic* complication, which means the death of traditional spectroscopy. Such extremely complex spectra have been observed experimentally in atomic systems in external fields [20], on the eigenmodes of microwave billiards (when a parameter of the billiard shape is varied) [18] and numerically on virtually all chaotic systems [6, 19]. It should be emphasized that level dynamics in the chaotic regime looks extremely similar whatever the system is, as long as its classical dynamics is chaotic. It is probably the simplest and most universal property.

3.2. STATISTICAL ANALYSIS OF THE SPECTRAL FLUCTUATIONS

This qualitative property has been put on a firm ground by the study of the statistical properties of energy levels [6, 19, 21]. The idea is the following: there are too many levels and their evolution is too complicated to deserve a detailed

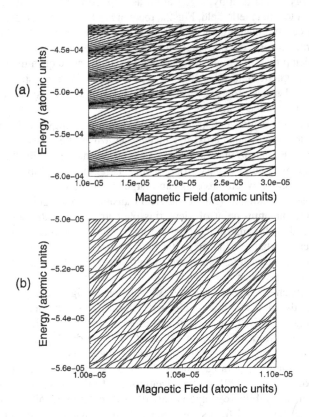

Figure 2.   Map of the energy levels of a hydrogen atom versus magnetic field γ (in atomic units of 2.35×10^5 T) for typical Rydberg states. At low energy (a), the classical dynamics is regular and the energy levels (quasi-) cross. At high energy (b), the classical dynamics is chaotic, the good quantum numbers are lost and the energy levels strongly repel each other. The strong fluctuations in the energy levels are characteristic of a chaotic behaviour.

explanation, level by level. In complete similarity with a gas of interacting particles where the detailed positions of the various particles do not really carry the relevant information which is rather contained in some *statistical* properties, we must use a statistical approach for the description of the energy levels of a chaotic quantum system. In order to compare different systems and characterize the *spectral fluctuations*, we must first define proper quantities. For a complete description, see [6, 19].

Density of states.   The density of states is:

$$d(E) = \sum_i \delta(E - E_i), \tag{17}$$

where the E_i are the energy levels of the system. The cumulative density of states counts the number of energy levels below energy E. It is thus:

$$n(E) = \int_{-\infty}^{E} d(\epsilon) \, d\epsilon = \sum_i \Theta(E - E_i). \tag{18}$$

This is a step function with unit steps at each energy level. When there is a large number of levels, one can define the averaged cumulative density of states $\bar{n}(E)$, a function interpolating $n(E)$ by smoothing the steps. Its derivative is the averaged density of states $\bar{d}(E)$. There are several cases where this is the only relevant quantity for the physics of the system. For example, in a large semiconductor sample, the averaged density of states at the Fermi level is what determines the contribution of electrons to the specific heat at low temperature [22].

   The averaged density of states can be determined in the semiclassical approximation (see section 4) by the Weyl's rule (also known as the Thomas-Fermi approximation):

$$\bar{d}(E) = \frac{1}{(2\pi\hbar)^d} \int d\mathbf{p} \, d\mathbf{q} \, \delta(H(\mathbf{q}, \mathbf{p}) - E). \tag{19}$$

It only depends on the classical Hamilton function H and not on the regular or chaotic nature of the dynamics.

Unfolding the spectrum.   The next step is to eliminate the slow changes in the averaged density of states by defining an unfolded spectrum through the following quantity:

$$\hat{N}(x) = n(\bar{n}^{-1}(x)), \tag{20}$$

which is nothing but the cumulative density of states represented as a function of a rescaled variable such that the "energy levels" now appear equally spaced by one unit. By construction, the rescaled energy levels $x_i = \bar{n}(E_i)$ have a density of states equal to one. It allows to compare spectra obtained for different parameters or even for completely different systems.

Nearest Neighbor Spacing Distribution.   The simplest quantity is the distribution of nearest neighbour spacings, i.e. of energy difference between two consecutive levels $s_i = x_{i+1} - x_i$. This distribution is traditionally denoted $P(s)$. By virtue of the unfolding procedure, the average spacing is one. Its behaviour near $s = 0$ measures the fraction of very small spacings (quasi-degeneracies), hence the degree of level repulsion.

Number Variance.   The use of the nearest neighbor spacing distribution is simple, but not very logical from the statistical physics point of view. Indeed, $P(s)$ involves *all* correlation functions among the energy levels. It is simpler to consider separately the two-point, three-point, etc... correlation functions. The two-point correlation function R_2 depends only on the energy difference if the spectrum is stationary (i.e. statistically invariant by a global translation, which is likely for a large unfolded spectrum). Near 0, it again measures the degree of level repulsion. A more global quantity is the number variance $\Sigma^2(L)$ which is the variance of the number of levels contained in an energy interval of length L. It is a measure of the rigidity of the spectrum, that is, it measures how the spectrum deviates from a uniform spectrum of equally spaced levels. A related quantity is the so-called spectral rigidity $\Delta_3(L)$ which measures how much the cumulative density of states differs from its best linear fit on an energy interval of length L. It is another alternative to the two-point correlation function. Its advantage is that it is very robust against imperfections such as spurious or missing energy levels and can be determined rather safely from a limited number of energy levels. This is of major importance for example in analyzing experimental atomic [23, 24] or nuclear spectra [6].

3.3.   REGULAR REGIME

In the regular regime (see Fig. 2a), consecutive energy levels generally do not interact. Thus, from the statistical point of view, they can be considered as independent random variables. The distribution of spacings is the one of uncorrelated levels, that is a Poisson distribution:

$$P(s) = e^{-s}, \tag{21}$$

which nicely reproduces the numerical results obtained on different systems (see Fig. 3a) and also several experimental results [25, 18]. Note that the maximum of the distribution is near $s = 0$ which shows that quasi-degeneracies are very probable and that level repulsion is absent.

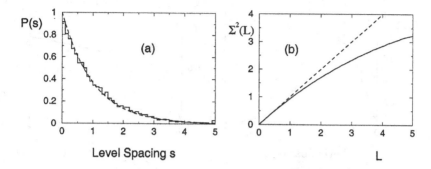

Figure 3. Statistical properties of energy levels for the hydrogen atom in a magnetic field, obtained from numerical diagonalization of the Hamiltonian in the regular regime. (a) Nearest neighbor spacing distribution. The distribution is maximum at 0 and well fitted by a Poisson distribution (dashed line). (b) Number variance. Again, the Poisson prediction (dashed line) works quite well. The saturation at large L is due to the residual effects of periodic orbits and is well understood.

This is a *universal* result which applies generically to regular systems. Other statistical quantities can be described as well. The two-point correlation function is simply $R_2(x) = 1$ and the number variance is:

$$\Sigma^2(L) = L. \tag{22}$$

Figure 3b shows the numerical result for the hydrogen atom in a magnetic field in the regular regime. The agreement with the prediction is good, at least for low L. The saturation at large L can be quantitatively understood using periodic orbit theory (see section 4). In simple words, it is due to long range correlations in the spectrum induced by periodic orbits.

3.4. CHAOTIC REGIME – RANDOM MATRIX THEORY

In the chaotic regime, the strong level repulsion induces a completely different result for the spacing distribution – see Fig. 4a – with practically no small spacing, and also a lack of large spacings.

A simple model is able to predict the statistical properties of energy levels. It assumes a maximum disorder in the system and that – from a statistical point of view – all basis sets are equivalently good (or bad). It therefore models the Hamiltonian by a set of random matrices which couple any basis state to all the other ones. Depending on the symmetry properties of the Hamiltonian (especially with respect to time reversal, see below), different ensembles of random matrices have to be considered. Let us assume for the moment that the system is time-reversal invariant and can be represented by a real symmetric matrix H_{ij} in some basis. The ensemble of random matrices is built by assuming

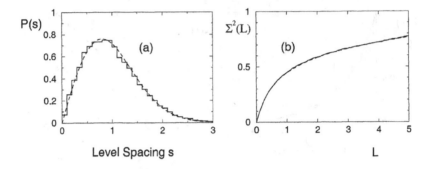

Figure 4.   Same as Fig. 3, but in the classically chaotic regime. (a) The probability of finding almost degenerate levels is very small (level repulsion). The results are well reproduced by the Wigner distribution (dashed line) and Random Matrix Theory. (b) The number variance is much smaller than in the regular case, showing the rigidity of the energy spectrum. The results agree perfectly with the prediction of Random Matrix Theory (dashed line).

that it is globally invariant over any orthogonal transformation (all basis sets are equivalent) and that the various matrix elements are *independent* random variables[1]. With these basic assumptions, one obtains easily that all matrix elements have a Gaussian distribution with zero average and variance:

$$< H_{ij}^2 >= (1 + \delta_{ij})\sigma^2 , \tag{23}$$

where σ is the only remaining free parameter.

These properties define the Gaussian Orthogonal Ensemble (GOE) of random matrices[2].

Knowing the probability density, we can extract the statistical properties of the eigenvalues. It involves the use of either beautiful old-fashioned mathematics [21] or almost incomprehensible supersymmetry techniques [26]. Most formulas are explicit but not very illuminating; they can be found in [6, 21].

The spacing distribution cannot be calculated in closed form, but it turns out to be very close to the Wigner distribution:

$$P(s) = \frac{\pi s}{2} \, e^{-\frac{\pi s^2}{4}} . \tag{24}$$

This distribution, shown in Fig. 4a, agrees extremely well with the numerical results got on the hydrogen atom in a magnetic field. Similar results have

---

[1] The latter hypothesis is not at all crucial. It can be easily relaxed, generating other ensemble of random matrices with similar statistical properties.

[2] An alternate derivation of the GOE is based on information theory. If we look for the probability density which maximizes the entropy $S = - \int P(H) \ln P(H) dH$ with the constraint that the average value of $\text{Tr}(H^2)$ is fixed, one rediscovers immediately the GOE. The idea behind this derivation is that we know basically nothing about the distribution and have to take it as general as possible.

been obtained on dozens of quantum chaotic systems, both numerically and experimentally. Experimental examples are the energy levels of highly excited nuclei [6], rovibrational levels of the NO_2 molecules [16], energy levels of the hydrogen atom in a magnetic field [24] and electromagnetic eigenmodes of microwave cavities [18] and vibrations of elastic blocks (see R. Weaver in this book and references therein).

The transition from a Poisson distribution in the classically regular regime to a Wigner distribution in the chaotic regime gives a **characterization** of quantum chaos, at least for highly excited states.

Other statistical properties have been studied and are found in good agreement with the predictions of Random Matrix Theory [23]. For example, the number variance, shown in Fig.4b, is in perfect agreement with the GOE prediction which, for large L, is

$$\Sigma^2(L) \simeq \frac{1}{\pi^2} \ln 2\pi L. \tag{25}$$

Note that the number variance is much smaller here than in the regular case. The spectrum is *extremely* rigid, as for $L = 10^6$, Σ^2 is only of the order of 3. This means that the typical fluctuation of the number of levels is 1 or 2 additional or missing levels over a range of one million level. In the Poisson model, the typical fluctuation would be $\sqrt{L} = 1000$ levels! This extraordinary large rigidity is due to the strong couplings existing between all the states in the model. If a fluctuation makes the level repulsion abnormally large between two states, they cannot repel too strongly because they are themselves strongly pushed by the other levels. From maximum disorder at the microscopic level, a globally rigid structure is born.

Finding universal properties in the local statistical properties of energy levels for chaotic systems is not a real surprise. As discussed in the preceding section, this range of energy (mean level spacing Δ) corresponds to a long time behaviour ($h/\Delta = T_H \gg T_{Min}$), where chaos is classically fully developed with its universal properties. Universality is also observed in the corresponding quantum dynamics. On the other hand, at shorter times of the order of T_{Min}, non-universal properties exist in the classical behaviour. This implies also a deviation from the predictions of Random Matrix Theory on a large energy scale, as has been numerically and experimentally observed [12, 25, 24].

Random Matrix Theory can also predict the behaviour of quantities beyond the energy levels. For example, it can predict the distribution of the wavefunction amplitude [27], the lifetimes of resonances in open systems [28, 29] or the distributions of transition matrix elements [13].

If time-reversal symmetry (or more generally all anti-unitary symmetry) is broken, the Hamiltonian cannot be written as a real symmetric matrix, but rather as a complex Hermitian matrix. One has to change the ensemble of random matrices to use and define the Gaussian Unitary Ensemble (GUE). The probability distribution is such that both $\mathrm{Re}H_{ij}$ and $\mathrm{Im}H_{ij}$ are Gaussian distributed. This adds *more* level repulsion because two arbitrary states have two chances to be coupled and to repel. The calculations are similar to the GOE case and the predicted distributions agree very well with numerical results [6, 30].

One also has to consider the special case of half-integer spin systems with time-reversal invariance: there, all levels are doubly degenerate (Kramers degeneracy). If some rotational invariance exists, this degeneracy is hidden and the GOE should be used in each rotational series. If the rotational invariance is broken, every level will be exactly doubly degenerate and the Gaussian Symplectic Ensemble (GSE) of random matrices has to be used [6, 19].

It is important to notice the role of symmetries for level statistics. If a good quantum number survives in a system (for example a discrete two-fold symmetry), the states with the same good quantum number will interact, but they will ignore the other states. Thus, even if each series with a fixed quantum number obeys the GOE statistics, the total spectrum will appear as the superposition of several uncorrelated GOE spectra, which has completely different statistical properties. It is very important to be sure that one has a pure sequence of levels before analyzing it. This may be difficult in a real experiment because of stray mixing between series, usually much easier in numerical experiments.

Finally, intermediate regimes have been studied, for example between the regular Poisson and the chaotic GOE regimes. In general, this transition is not universal.

4.   SEMICLASSICAL APPROXIMATION

The previous section has shown the existence of universal fluctuation properties associated with chaos for quantum systems. These properties take place at short energy range, of the order of the mean level spacing, that is for times of the order of the Heisenberg time, much longer that the period of the shortest periodic orbit. This also implies that a detailed analysis of all energy levels and eigenstates does not make sense: no interesting information can be brought to the physics of the chaotic phenomenon, beyond the statistical aspects. On the other hand, this does *not* mean that these energy levels do not carry any information; it is just that this information has to be extracted in a different way. More precisely, as the individual specificity of a chaotic system manifests at relatively short times, before universal chaotic features dominate, it has to be found in the long energy range characteristics of the quantum spectra.

For such a short time scale, as discussed in section 2, a semiclassical approximation might be used. It is the goal of this section to show how this can be implemented and eventually used to make some quantitative predictions on quantum chaotic systems which go beyond simple statistical statements.

The simplest class of systems are the completely integrable systems, where there exist as many independent constants of motion as the number of degree of freedoms. For such systems, there is a standard semiclassical theory – known as EBK quantization rule – which is a simple extension of the well known WKB theory for time-independent one-dimensional systems [31]. It allows, for example, to calculate accurately the energy levels shown in Fig. 2a.

4.1. SEMICLASSICAL PROPAGATOR

For a chaotic system, a direct solution of the time-independent Schrödinger equation seems out of reach. It is easier to calculate a semiclassical approximation of the unitary evolution operator. This is also more convenient if one wants to compare to the classical dynamics, as the regular or chaotic character expresses in the time domain.

The propagator is defined as a matrix element of the evolution operator in the configuration space representation:

$$K(\mathbf{q}', t'; \mathbf{q}, t) = \langle \mathbf{q}' | U(t', t) | \mathbf{q} \rangle. \tag{26}$$

The semiclassical approximation for the propagator relies on a separation of phase and amplitude and the neglect of higher order terms in $\hbar$. One then finds the time-dependant Hamilton-Jacobi equation for the action [1], which can be locally solved along trajectories. The result is known as the Van Vleck propagator [32-34]:

$$K(\mathbf{q}', t'; \mathbf{q}, t) = \sum_{\text{Clas. Traj.}} \left(\frac{1}{2i\pi\hbar} \right)^{d/2} \left| \text{Det} \frac{\partial^2 R(\mathbf{q}', t'; \mathbf{q}, t)}{\partial \mathbf{q}' \partial \mathbf{q}} \right|^{1/2} \times$$
$$\exp\left(i \frac{R(\mathbf{q}', t'; \mathbf{q}, t)}{\hbar} - i \frac{\pi\nu}{2} \right), \tag{27}$$

where the sum is over all the classical trajectories going from $(\mathbf{q}, t)$ to $(\mathbf{q}', t')$. The function $R(\mathbf{q}', t'; \mathbf{q}, t)$ is the classical action, i.e. the integral of the Lagrangian along the trajectory:

$$R(\mathbf{q}', t'; \mathbf{q}, t) = \int_t^{t'} L(\mathbf{q}, \dot{\mathbf{q}}, \tau) \, d\tau. \tag{28}$$

The non-negative integer ν counts the number of caustics – points where the determinant in the equation becomes singular – encountered along the trajectory and is called a Morse index.

A few remarks should be made on this formula:

- The structure of this formula is a phase expressed as a purely classical quantity evaluated along a trajectory, divided by $\hbar$, and a smoothly varying amplitude, in complete analogy with the standard WKB approach [31].

- The fact that the same quantity R appears both in the phase and in the amplitude is not surprising. It ensures the unitarity of the time evolution. In fact, the prefactor $\sqrt{\text{Det}}$ is of purely classical origin. It just represents how a classical phase space density initially localized in $\mathbf{q}$ and uniformly spread in $\mathbf{p}$ evolves according to the Liouville equation, Eq. (6).

- At short time difference $t' - t$, there is only one classical trajectory connecting $(\mathbf{q}, t)$ to $(\mathbf{q}', t')$ (more or less a straight line). The existence of multiple trajectories connecting the starting and ending points is essential for chaotic systems: at long times, the trajectories become very complicated and their number grows exponentially, which renders the use of the semiclassical propagator more and more difficult.

- A completely different derivation of the Van Vleck propagator is possible using the Feynman path integral [35] formulation of quantum mechanics. The propagator can be *exactly* written as a superposition of contributions of all paths connecting the starting and ending points. The phase of each contribution is the integral of the Lagrangian along the path divided by $\hbar$. In the semiclassical limit, the sum over paths can be calculated by the stationary phase approximation. The paths with stationary phase are precisely the classical trajectories, and the prefactor in the stationary phase integration exactly gives the Van Vleck amplitude. This approach explains why the contributions of the different classical trajectories have to be added coherently in the propagator. It also explains why the semiclassical propagator can cross the caustics safely and the origin of the Morse index (prefactor in the stationary phase integration).

4.2.   GREEN'S FUNCTION

The next step is to go from the time domain to the energy domain. The Green's function is:

$$G(E) \equiv \frac{1}{E - H} = \frac{1}{i\hbar} \int_0^\infty U(\tau, 0)\, \exp\left(i\frac{E\tau}{\hbar}\right) d\tau, \qquad (29)$$

where the last equality is valid for E in the upper half complex plane.

In configuration space, this gives:

$$G(\mathbf{q}', \mathbf{q}, E) \equiv \langle \mathbf{q}'|G(E)|\mathbf{q}\rangle = \frac{1}{i\hbar} \int_0^\infty K(\mathbf{q}', \tau; \mathbf{q}, 0) \exp\left(i\frac{E\tau}{\hbar}\right) d\tau. \quad (30)$$

In order to get a semiclassical approximation for the Green's function, one plugs the Van Vleck propagator, Eq. (27), into Eq. (30). The integral over τ can be performed by stationary phase approximation. The stationary phase condition writes:

$$\frac{\partial R(\mathbf{q}', \tau; \mathbf{q}, 0)}{\partial \tau} + E = 0. \tag{31}$$

From the Lagrange-Hamilton equation [1, 34], the partial derivative is minus the Hamilton function. Hence, stationary phase selects trajectories going from $\mathbf{q}$ to $\mathbf{q}'$ with precisely energy E. This allows to write the semiclassical Green's function at energy E as a sum over classical trajectories with energy E, a physically satisfactory result. The phase of each contribution is the sum of R (calculated along the orbit) and $E\tau$, which gives the reduced action S [1]. The detailed calculation of the various prefactors is not very difficult, but rather tedious; see [34] for details. It is convenient to distinguish the coordinate $q_{\parallel}$ chosen along the trajectory and the the coordinates $\mathbf{q}_{\perp}$ transverse to the orbit. One finally obtains the semiclassical Green's function [32, 36, 34]:

$$G(\mathbf{q}', \mathbf{q}, E) = \frac{1}{i\hbar} \sum_{\text{Clas. Traj.}} \frac{\left| \text{Det} \dfrac{\partial^2 S(\mathbf{q}, \mathbf{q}', E)}{\partial \mathbf{q}_{\perp} \partial \mathbf{q}'_{\perp}} \right|^{1/2}}{(2i\pi\hbar)^{(d-1)/2} \sqrt{|\dot{q}_{\parallel} \dot{q}'_{\parallel}|}} \exp\left(\frac{iS(\mathbf{q}, \mathbf{q}', E)}{\hbar} - i\frac{\pi\nu}{2} \right),$$

$$\tag{32}$$

with

$$S(\mathbf{q}, \mathbf{q}', E) = \int_{\mathbf{q}}^{\mathbf{q}'} \mathbf{p}.d\mathbf{q}. \tag{33}$$

Again, the prefactor simply represents the classical evolution of a phase space density with fixed energy, according to the Liouville equation. This semiclassical approximation breaks down for very short trajectories. Indeed, the integral over τ cannot be performed by stationary phase approximation in such a case. A specific short time expansion is possible when $S(\mathbf{q}, \mathbf{q}', E)$ is not much larger than $\hbar$. It basically consists in ignoring the effect of the potential and using the free Green's function [32].

The Green's function by itself is not very illuminating. In order to obtain some information on the energy spectrum and eigenstates, one needs a more global quantity.

4.3.   TRACE FORMULA

The Green's function, Eq. (29), has a singularity at each energy level, like the density of states, Eq. (17). They are actually related by the simple equation:

$$d(E) = -\frac{1}{\pi}\text{Im}\,\text{Tr}G(E) = -\frac{1}{\pi}\text{Im}\int d\mathbf{q}\,G(\mathbf{q}, \mathbf{q}, E)\,. \qquad (34)$$

If one uses the semiclassical Green's function, Eq. (32), the density of states is obtained as a sum over *closed* trajectories, starting and ending at position $\mathbf{q}$. The last integral over position $\mathbf{q}$ can again be performed by stationary phase. The Lagrange equations tell us that the partial derivative of $S(\mathbf{q}, \mathbf{q}', E)$ with respect to $\mathbf{q}'$ is the final momentum $\mathbf{p}'$ while its partial derivative with respect to $\mathbf{q}$ is minus the initial momentum $\mathbf{p}$. Stationary phase thus selects closed orbits where the initial and final momenta are equal, that is **periodic orbits**.

Putting everything together gives the celebrated trace formula (also known as the Gutzwiller trace formula from one of its authors), written here for a two-dimensional system [36, 33, 32, 34]:

$$d(E) = \bar{d}(E) + \sum_{\text{p.p.o. } k,\text{ repetitions } r} \frac{T_k}{\pi\hbar} \frac{\cos\left[r\left(\frac{S_k}{\hbar} - \nu_k\frac{\pi}{2}\right)\right]}{\sqrt{|\det(1 - M_k^r)|}}, \qquad (35)$$

where the sum is performed over all primitive periodic orbits (i.e. periodic orbits which do not retrace the same path several times) and all their repetitions $r > 0$. T_k is the period of the orbit, S_k its action, ν_k its Maslov index and M_k the 2 by 2 monodromy matrix describing the linear change of the transverse coordinates after one period; for details, see [32, 34].

The term $\bar{d}(E)$ whose expression is given in Eq. (19) comes from the "zero-length" trajectories. Indeed, for such trajectories, the semiclassical approximation for the Green's function breaks down and a repaired formula (see previous section) has to be used, which produces this smooth term.

The trace formula deserves several comments:

- The trace formula is a central result in the area of quantum chaos, as it expresses a purely quantum quantity (the density of states) as a function of classical quantities (related to periodic orbits) and the constant $\hbar$.

- It uses only periodic orbits, which proves that they are the skeleton of the chaotic phase space. In that sense, they replace the invariant tori used for regular systems.

- Each periodic orbit contributes to the density of states with an oscillatory contribution. The period of these modulations correspond to a change of the argument of the cosine function $S_k/\hbar$ by 2π. As the derivative of the action with respect to energy is the period T_k of the orbit, the corresponding characteristic energy scale is $2\pi\hbar/T_k$. In the semiclassical regime, this is much larger than the mean level spacing $\Delta = 2\pi\hbar/T_H$. Hence, the trace formula describes the long range correlations in the energy spectrum.

- The trace formula breaks the simple connection between a given energy level and a simple structure in phase space. An energy level is a δ singularity in the density of states while each orbit contributes to a modulation of the density of states with finite amplitude. Thus, to build a δ peak requires a coherent conspiracy of an infinite number of periodic orbits.

- The present formula is restricted to isolated periodic orbits such that the phase space distance to the closest periodic orbit is larger than $\hbar$. For non isolated periodic orbits, the simple stationary phase treatment fails. A specific treatment is required and various similar formula can be written. Especially, for integrable systems, the sum over periodic orbits can be performed analytically using a Poisson sum formula [37]: the result is exactly equivalent to the WKB (in one dimension) and EBK (in multidimensional systems) quantization schemes.

- The formula is valid only at lowest order in the Planck's constant $\hbar$. Including higher orders in the various stationary phase approximations is tedious, but feasible [38].

- At the very beginning, the trace formula relies on the semiclassical Green's function. This requires that the de Broglie wavelength of the quantum particle is smaller than any relevant spatial scale of the classical system. This is wrong in the presence of small scatterers. Hence, the trace formula can in principle be used only for scattering by smooth potentials. In some specific cases of non-smooth potentials, such as billiards, the breakdown of the semiclassical approximation can be repaired and the trace formula can also be used. The case of point scatterers is more tricky and discussed below, see section 4.7.

- If one is not interested in the density of states, but in some other physical quantity, it is often possible to get similar expressions. The general strategy is to express the quantity of interest using the Green's function of the system, then to use the semiclassical Green's function. For example, the photo-ionization cross-section of an hydrogen atom in a magnetic field has been calculated in [39] as a sum over periodic orbits starting and ending at the nucleus.

The practical use of the trace formula is difficult. Indeed, extracting individual energy levels by adding oscillatory contributions requires in principle an infinite number of periodic orbits. In practice, it may be argued that it is enough to sum up all orbits with periods up to the Heisenberg time. Indeed, longer orbits will produce modulations on an energy scale smaller than the mean level spacing, and are thus expected to cancel out and to be irrelevant (only useful to make peaks narrower but not moving theirs positions). In the semiclassical limit, T_H is so much longer than T_{Min} that the proliferation of long orbits makes the procedure unpractical. However, for relatively low excited states, T_H is not much larger than T_{Min} and the trace formula has been successfully used to compute several states [34, 40].

Another possibility is to use an open system with resonances instead of bound states. There, the density of states has only bumps related to the resonances and the use of a finite (and hopefully small) number of periodic orbits may correctly reproduce the quantum properties. This is shown in Fig. 5 for the hydrogen atom in a magnetic field at scaled energy $\epsilon = 0.5$. With few hundred periodic orbits, we are able to reproduce the finest details in the apparently random fluctuations of the photo-ionization cross-section [41]. This is a striking illustration of the strength of semiclassical methods.

4.4.   CONVERGENCE PROPERTIES OF THE TRACE FORMULA

Because of the proliferation of long periodic orbits, it is not clear whether the sum in the trace formula converges or not. There is a competition between the exponential proliferation of periodic orbits and the decrease of the individual amplitudes (long orbits tend to be very unstable hence creating large denominators in the trace formula). On the average, long orbits explore more or less uniformly the whole phase space so that their instability depends only on the period. For example, for a bound 2-dimensional system, one has:

$$ A_k = \frac{T_k}{\sqrt{|\det(1 - M_k)|}} \simeq T_k \, e^{-\frac{1}{2}\lambda T_k} , \tag{36} $$

where λ is the average Lyapounov exponent per unit of time. Similarly, it is known that the number $N(T)$ of orbits with period less than T typically behaves as [42]:

$$ N(T) \simeq \frac{e^{\lambda T}}{\lambda T} . \tag{37} $$

Thus the proliferation of orbits gives an increasing factor of the order of $\exp(\lambda T)$ in the trace formula which overcomes the decrease of the amplitude which scales as $\exp(-\lambda T/2)$ and the sum does not converge [32]. This means that, depending on the order in which the various periodic orbit contributions are added, the result can be anything! However, it is rather clear that the periodic orbits are not independent from each other and that the information that they

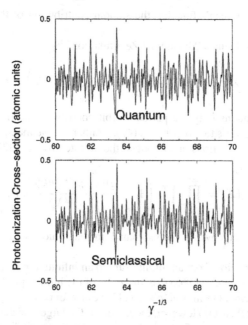

Figure 5.   The photo-ionization cross-section of the hydrogen atom in a magnetic field γ, (plotted versus $\gamma^{-1/3}$), in the chaotic regime. The smooth part of the cross-section is removed, in order to emphasize the apparently erratic fluctuations. Upper panel: the "exact" quantum cross-section calculated numerically. Lower panel: the semiclassical approximation of the cross-section, as calculated using periodic orbit theory. All the fine details – which look like random fluctuations – are well reproduced by periodic orbit theory.

contain is somewhat structured: most of the information contained in the very long orbits can be extracted from shorter periodic orbits.

The idea is thus to use the structure of the classical orbits to make the trace formula more convergent. The first step is to pass from the density of states to the so-called spectral determinant defined by:

$$f(E) = \prod_i (E - E_i), \qquad (38)$$

which has a zero at each energy level. It can be rewritten as:

$$f(E) = \exp\left(-\frac{1}{\pi}\int^E \mathrm{Im}\, \mathrm{Tr}G(\epsilon)\, d\epsilon\right). \qquad (39)$$

This quantity is of course highly divergent but can be regularized through multiplication by a smooth quantity having no zero [32]. This corresponds to removing in the Green's function the smooth contribution of the zero length trajectories.

One can thus define a "dynamical Zeta function" by:

$$Z(E) = \exp\left(-\frac{1}{\pi}\int^{E} \text{Im } \text{Tr}G_{\text{reg}}(\epsilon)\, d\epsilon\right), \qquad (40)$$

where G_{reg} contains only the periodic orbit contributions. By inserting the semiclassical Green's function in Eq. (40), a rather simple manipulation allows to sum over the repetitions and to obtain the infinite product [32]:

$$Z(E) = \prod_{m=0}^{\infty}\prod_{\text{p.p.o.}}\left(1 - \frac{\exp\left(i\frac{S_k}{\hbar} - i\nu_k\frac{\pi}{2}\right)}{|\lambda_k|^{1/2}\lambda_k^m}\right), \qquad (41)$$

where λ_k is the eigenvalue (larger than 1 in magnitude) of the monodromy matrix M_k.

The transformation from an infinite sum to an infinite product does not cure the lack of convergence. Of course, the zeros of the infinite product are not the zeros of its individual terms (which do not have any for real energy, because $|\lambda_k|$ is always larger than 1). However, the larger m, the larger the denominator and the more convergent the infinite product over primitive periodic orbits. Hence, the most significant zeros – the most important ones for the physical properties – will come only from the $m = 0$ term.

Different approaches have been proposed for a practical calculation of the infinite product. For a bound system, the energy spectrum is known to lie on the real axis, which is not evident from Eq. (41). This requires some subtle compensations between the various terms. By imposing this property, it is possible to show that periodic orbits with long periods (longer than the Heisenberg time) do not carry any significant information to the infinite product which can be somehow cut at the Heisenberg (or rather half of it, see [43]). This makes it possible to calculate semiclassically the energy levels of chaotic systems with a fair accuracy. It is however limited to bound systems.

Another approach is possible. When expanding the infinite product, there are some crossed terms between orbits appearing with a positive sign which might cancel approximately with negative terms from more complicated orbits. The idea of the cycle expansion is to group such terms so that maximum cancellation takes place. Suppose that I have two simple orbits labelled 0 and 1 and a more complicated orbit labelled 01 which is roughly orbit 0 followed by orbit 1. If the action (resp. Maslov index) of orbit 01 were exactly the sum of the actions (resp. Maslov indices) of orbits 0 and 1 and its unstability eigenvalue λ the product of the instability eigenvalues of orbits 0 and 1, complete cancellation would take place. As these properties cannot be exact, only partial cancellation takes

State	Periodic Orbit	Cycle Expansion	Exact quantum
1s1s	-3.0984	-2.9248	-2.9037
2s2s	-0.8044	-0.7727	-0.7779
2s3s		-0.5902	-0.5899
4s7s		-0.1426	-0.1426
5s5s	-0.131	-0.129	-0.129

Table 1.1 .   Some energy levels (in atomic units) of the $^1S^e$ series of the helium atom, compared to the simple semiclassical quantization using only the simplest periodic orbit and the more refined "cycle expansion" which includes a set of unstable periodic orbits. The agreement is remarkable, which proves the efficiency of semiclassical methods for this chaotic system (courtesy of D. Wintgen).

place. But, for longer and longer orbits, the cancellation is better and better and the cycle expansion might be convergent (although there is no proof).

This simple idea can be generalized to take into account all the orbits if an efficient coding scheme exists – also known as a good symbolic dynamics – for the periodic orbits. Then, there are some cases like the 3-disks scattering problem [32], where the cycle expansion can be made convergent and can be used to efficiently calculate the quantum properties of the system.

4.5.   AN EXAMPLE : THE HELIUM ATOM

The idea of the cycle expansion has been successfully used by Wintgen and coworkers [44] to calculate some energy levels of the helium atom. Although the system is not fully chaotic, most of the dynamics is [45]. Of special interest is the eZe configuration where the electrons and the nucleus lie on a straight line, with the electrons on opposite sides of the nucleus. In such a configuration, one can find a symbolic dynamics for the periodic orbits: any periodic orbit can be uniquely labelled by the sequence in which the electrons hit the nucleus. The motion transverse to the eZe configuration is stable and can be taken into account. By calculating the zeros of the infinite product, Wintgen et al. have been able to perform a fully semiclassical calculation of several energy levels of the helium atom. Some results are displayed in Table I. For the ground state, it differs by only 0.7% from the exact quantum result. For excited states, it is even better. Thus, these authors have been able to solve a problem open since the beginning of the century, when pioneers of quantum mechanics tried to quantize helium after having successfully quantized the hydrogen atom. However, these pioneers had no idea of the classically chaotic nature of phase space, they were not even thinking of a trace formula – not to speak of cycle expansion. They could not have the key idea that the correspondence between a classical orbit and an energy level is not one-to-one but that an infinite number of periodic orbits is needed to build a chaotic quantum eigenstate. It is only after 70 years of work on classical and semiclassical dynamics that their goal could be met.

4.6.   LINK WITH RANDOM MATRIX THEORY

Although it is clear that the semiclassical approach gives a not-too-bad description of low excited states and of global (i.e. collective) properties of highly excited states, it is not clear whether it is able to reproduce the individual properties of highly excited states. In other words, the question is to know whether the semiclassical approximation may be used up to and beyond the Heisenberg time. The answer is not presently very clear. Especially, there are some strong indications that results of the Random Matrix Theory can be obtained using a semiclassical approach, but we are still lacking a global understanding of the connection.

I now illustrate this point using the so-called form factor. It is related to the autocorrelation function, Eq. (13), averaged over the initial state. It can also be viewed as the average return probability, as discussed elsewhere in this book [46]. The idea is to use for the form factor something like:

$$K(t) \approx \frac{1}{\mathcal{N}} \left| \sum_i \exp\left(-i\frac{E_i t}{\hbar}\right) \right|^2, \tag{42}$$

where the sum extends over all states in an interval of energy centered around E and width $\Delta E \ll E$. The width has to be chosen sufficiently large to contain many states (so that the state-to-state fluctuations are washed out) but also sufficiently small such that the classical dynamics does not significantly change across the energy interval. The prefactor is, just for convenience, chosen as:

$$\mathcal{N} = \frac{\Delta E}{\Delta} = \bar{d}(E)\Delta E. \tag{43}$$

It represents simply the number of levels in the energy interval. At long time, the various oscillating terms in $K(t)$ are out of phase and just add incoherently. This happens when the contributions between two consecutive energy levels are typically dephased by 2π, thus for times longer than the Heisenberg time, Eq. (16), leading to:

$$K(t \gg T_{\mathrm{H}}) = 1. \tag{44}$$

To make the connection with the trace formula, it is useful to re-express $K(t)$ as a function of the density of states. From Eq. (42), $K(t)$ appears as the square modulus of the Fourier transform of the density of states. This quantity is however very singular at time $t = 0$ where all oscillatory terms add coherently, producing a huge δ peak. In order to remove this uninteresting behaviour, we define the form factor using the fluctuating part only of the density of states – removing from $d(E)$ the smooth part $\bar{d}(E)$, see Eq. (19):

$$K(t) = \frac{1}{N} \left| \int_{E-\Delta E/2}^{E+\Delta E/2} [d(E) - \bar{d}(E)] \exp\left(-i\frac{Et}{\hbar}\right) \right|^2 \tag{45}$$

or

$$K(t) = \frac{1}{\Delta} \int \left\langle \left[(d - \bar{d})\left(\mathcal{E} + \frac{x}{2}\right)\right] \left[(d - \bar{d})\left(\mathcal{E} - \frac{x}{2}\right)\right] \right\rangle \exp\left(-i\frac{xt}{\hbar}\right) dx, \tag{46}$$

where $<>$ denotes the local average over the energy $\mathcal{E}$. Thus, $K(t)$ is nothing but the Fourier transform of the autocorrelation function of the spectrum, the latter being – up to a constant factor – the two-point correlation function R_2 introduced above in section 3.2. More precisely, one has:

$$K(t) = 1 + \int (R_2(u) - 1) \exp\left(-2i\pi u \frac{t}{T_H}\right) du. \tag{47}$$

At this stage, $K(t)$ can be obtained exactly from Random Matrix Theory, by substituting the well known form of R_2 [6]. To make the connection with semiclassics, it can also be calculated using the semiclassical approximation (trace formula) for the density of states. One substitutes Eq. (35) in Eq. (46), and expands the classical quantities (such as the action) in the vicinity of the energy E of interest, and finally obtains:

$$K(t) = \frac{\Delta}{2\pi\hbar} \sum_{i,j} A_i A_j \exp\left(\frac{i(S_i - S_j)}{\hbar}\right) \delta\left(t - \frac{T_i + T_j}{2}\right). \tag{48}$$

For very short times – of the order of T_{Min}, the period of the shortest periodic orbit – $K(t)$ displays a series of δ peaks located at the combinations of the various periods. Because of the proliferation of orbits, these peaks rapidly overlap and $K(t)$ becomes a smooth function. It is not straightforward to calculate it, but a simple guess is to assume that, because the actions S_i, S_j change with energy, the non-diagonal term in the double sum will be washed out. With this *diagonal* approximation, the sum can be calculated. Indeed, the asymptotic exponential divergence of the density of periodic orbits, Eq. (37), exactly cancels out the exponential decrease of the amplitudes, Eq. (36), and one gets:

$$K(t) = \frac{\Delta t}{2\pi\hbar} = \frac{t}{T_H}. \tag{49}$$

As this expression diverges at long times – which is clearly unphysical – it cannot be valid for arbitrarily long times. Indeed, around the Heisenberg time, the density of periodic orbits is so large that the diagonal approximation breaks down. The non-diagonal terms (whose number increases faster than the number of diagonal terms) are not completely washed out. These non-diagonal terms conspire to cancel the divergence of the diagonal terms and to finally obtain a constant value, Eq. (44), at long times.

We are thus left with two asymptotic forms of the form factor, Eqs. (49) and (44) for short and long time respectively. The simplest form for $K(t)$ is thus to assume that it grows linearly up to T_H and is then constant and equal to one. It turns out that this is **exactly** the result predicted by Random Matrix Theory for the Gaussian Unitary Ensemble. Thus, the semiclassical approximation certainly partly contains Random Matrix Theory. However, it is not clear whether it contains all of it. Indeed, we have shown that, for a specific quantity, the two approaches lead to the same prediction, but we are not able to derive Random Matrix Theory from semiclassics. Also, our semiclassical calculation is limited to the short-time limit (the long-time limit requires some quantum ingredient) and we do not know how to interpolate between the two limits. The fact that the simplest form (linear, then constant) is the exact result must be considered as a piece of luck.

If we now turn to systems with time-reversal symmetry, the preceding arguments have to be modified. Indeed, the diagonal approximation breaks down for pair of time-reversed orbits. These orbits have exactly the same action, period and stability exponent, so that the corresponding crossed terms in the double sum, Eq. (48), are exactly equal to the diagonal terms. Altogether this doubles the form factor. This is of course not unexpected as $K(t)$ is strongly related to the returning probability. For multiply scattering media, it is well known that coherent back-scattering enhances by a factor 2 the probability of being exactly back-scattered. We here have a manifestation of the same physical effect – constructive interference between a path and its time-reversal – in a chaotic system instead of a multiply scattering medium.

Finally, one gets at short time:

$$K_{\mathrm{GOE}}(t) = \frac{2t}{T_H}.\qquad(50)$$

Once more, this is the prediction of Random Matrix Theory for the Gaussian Orthogonal Ensemble at short times. At long times, $K(t)$ is still unity. It turns out, that for intermediate time (of the order of the Heisenberg time), the Random Matrix result has a complicated form which interpolates between the two asymptotic limits [6].

4.7. THE SEMICLASSICAL APPROXIMATION IN THE PRESENCE OF SMALL SCATTERERS

The semiclassical approximation used in the previous sections is valid only for a sufficiently small de Broglie wavelength. In a scattering system, this fails if the scattering objects are smaller than the de Broglie wavelength. There, the scattering events have to be treated using a quantum approach. In this section, I show how this can be done.

It is important to realize that this is a rather common situation. Let us give few examples:

- Single or multiple scattering of light by microscopic particles like atoms or molecules. A simple approximation is to consider the scattering particles as quasi-point particles or Rayleigh scatterers. The scattering amplitude then takes a universal form.

- In quantum mechanics, one can consider a series of disordered point scatterers. Each point scatterer scatters isotropically. Coherent back-scattering, for example, is known to be observed in this case.

- From the previous example, one can build even a simpler example using only one scatterer. In order to have multiple scattering, we need to add a device which sends back the scattered wave onto the scatterer. This can be for example the walls of a billiard. The billiard itself may be regular or chaotic. In the first case, the wave reflected by the walls will be strongly correlated with the initial wave, even after multiple scattering. In the second case, the billiard itself acts more or less as a random medium which reflects the scattered wave to the scattering particle with a random-like amplitude. Multiple reflection replaces multiple scattering.

- We can build an atomic analog of the previous billiard. Indeed, for a Rydberg atom, the ionic core composed of the nucleus and the inner electrons act as a scattering object on the Rydberg electron. The size of the ionic core is of the order of few Bohr radii, comparable to the de Broglie wavelength of the outer electron. The scattering amplitude is described by a set of phase shifts [47] (related to the quantum defects well known in atomic physics). In the very specific case of the hydrogen atom, there is no inner electron and only the nucleus scatters. In this case, it is not trivial but true that scattering by the pure Coulomb potential can be described semiclassically [48]. For all other atoms, the inner electrons act more or less like a point-scatterer. The combination of the outer external field and the inner Coulomb field is a device which reflects back the scattered wave to the ionic core. The physical picture is that the Rydberg electron propagates semiclassically between successive collisions with the ionic core, but each collision has to be properly treated using the quantum scattering amplitude.

4.8.   THE DIFFRACTIVE GREEN'S FUNCTION

In this section, I show how to build the semiclassical Green's function in the presence of point-scatterers. There are some subtleties associated with the regularization of singularities when point scatterers are used. Technically, this is feasible but not straightforward, and the results are rather natural. Hence, I will here cheat a little and use some sloppy language and mathematics. However, the reader should know that it is possible to write everything rigourously, see [49].

Basically, a point-scatterer can be imagined as being associated with a δ-peak in the potential. Thus, I write the Hamiltonian as:

$$H = H_0 + V, \tag{51}$$

where H_0 is a smooth Hamiltonian – for which the standard semiclassical approximation can be used – and V represents the effect of the scatterers:

$$V(\mathbf{q}) = \sum_i v_i \delta(\mathbf{q} - \mathbf{q}_i), \tag{52}$$

where v_i is the strength of scatterer i located at position $\mathbf{q}_i$.

If G_0 represents the Green's function associated with the Hamiltonian H_0, the full Green's function associated with H can be expanded in a multiple scattering series:

$$G = G_0 + G_0 V G_0 + G_0 V G_0 V G_0 + \dots. \tag{53}$$

In configuration space, the matrix elements of V are just δ-functions which makes it possible to obtain the following simple expansion of the Green's function:

$$G(\mathbf{q}', \mathbf{q}, E) = G_0(\mathbf{q}', \mathbf{q}, E) + \sum_i G_0(\mathbf{q}', \mathbf{q}_i, E) v_i G_0(\mathbf{q}_i, \mathbf{q}, E)$$

$$+ \sum_{i,j} G_0(\mathbf{q}', \mathbf{q}_i, E) v_i G_0(\mathbf{q}_i, \mathbf{q}_j, E) v_j G_0(\mathbf{q}_j, \mathbf{q}, E) + \dots \tag{54}$$

If the medium is sufficiently dilute for 2 consecutive scattering events to be separated by more than a de Broglie wavelength, the semiclassical approximation for the unperturbed Green's function G_0, Eq. (32), can be used. The corresponding expansion for the full Green's function G is known, for obvious reason, as the diffractive Green's function.

If needed, small scatterers which are not point scatterers can be treated as well just by replacing the constant scattering amplitude v_i by the full non-isotropic quantum scattering amplitude, the incoming and outgoing directions being given by the classical trajectories appearing in the semiclassical approximation.

The diffractive Green's function is the fundamental object for calculating the properties of systems with small scatterers.

4.9.   DIFFRACTIVE TRACE FORMULA

In order to calculate e.g. the density of states, one has to compute the trace of the Green's function. This can be performed quite easily using the diffractive Green's function. Indeed, the contributions at the various scattering orders are simply additive. At zeroth order, the contribution is simply the standard Gutzwiller trace formula, Eq. (35). The higher orders can be also estimated. Let us calculate for example the second order. It involves the integral:

$$\int G_0(\mathbf{q}, \mathbf{q}_2, E)\, v_2\, G_0(\mathbf{q}_2, \mathbf{q}_1, E)\, v_1\, G_0(\mathbf{q}_1, \mathbf{q}, E)\, d q\,. \qquad (55)$$

It is very similar to the standard (non diffractive) integral for the trace formula, except that one has to consider now a large series of orbits where the particle propagates classically between successive scattering events. The integral over the initial(=final) point $\mathbf{q}$ can be performed by the stationary phase approximation. For a reason already explained in section 4.3, this selects classical orbits having the same initial and final momentum. This implies that the two contributions $G_0(\mathbf{q}, \mathbf{q}_2, E)$ and $G_0(\mathbf{q}_1, \mathbf{q}, E)$ describe two parts of the same trajectory going from $\mathbf{q}_2$ to $\mathbf{q}_1$ through $\mathbf{q}$. Thus, the trace of the diffractive Green's function select *diffractive periodic orbits* which bounce between the various scatterers. The various prefactors appearing in the stationary phase integral nicely combine to produce the diffractive trace formula which gives the total density of states:

$$d(E) = d_G(E) + d_D(E)\,, \qquad (56)$$

where $d_G(E)$ is the Gutzwiller contribution, Eq. (35), and $d_D(E)$ is the diffractive contribution:

$$d_D(E) = \sum_{\text{Diffractive periodic orbits } p} \frac{T_p}{\pi\hbar} \prod_{\text{scatterers } i} v_i\, G_0(\mathbf{q}_{i+1}, \mathbf{q}_i, E)\,, \qquad (57)$$

where the $\mathbf{q}_i$ denotes the positions of the successive scatterers along a diffractive periodic orbit.

An immediate consequence of the diffractive trace formula is that the density of states is modulated not only by the periodic orbits but also by all the diffractive periodic orbits. In the case of a single scatterer (placed for example in a billiard), the diffractive orbits are composed of combinations of closed orbits all starting and ending at the scatterer, the action and the period of the diffractive orbit being simply the sum of the actions and periods of the elementary closed orbits.

Experimentally, such diffractive orbits have been observed on Rydberg states of non-hydrogenic atoms in magnetic field [48] and the predictions of the diffractive trace formula are in excellent agreement with both the numerical and the experimental results.

Because diffractive orbits are composed by combining elementary parts of orbits, the number of diffractive orbits increase very rapidly, typically exponentially with the action or period of the orbit. Thus, even if the unperturbed Hamiltonian H_0 is completely regular, the number of terms in the diffractive trace formula increases exponentially, exactly like for chaotic systems. This leads to the same convergence properties than the initial trace formula for chaotic systems.

The same trick for improving convergence can be used. Because of the additivity of the normal and diffractive parts of the Green's function, the dynamical zeta function, Eq. (40), for the system in the presence of the scatterers can be written as a product:

$$Z(E) = Z_G(E)Z_D(E), \tag{58}$$

where $Z_G(E)$ is the standard "Gutzwiller" zeta function, Eq. (41), and $Z_D(E)$ is the diffractive zeta function which can be written as:

$$Z_D(E) = \prod_p \left(1 - \prod_i v_i\, G_0(\mathbf{q}_{i+1}, \mathbf{q}_i, E)\right). \tag{59}$$

The zeros of the total zeta function $Z(E)$ give the energy levels of the system in the presence of the scatterers. It should be emphasized that all these expressions are infinite products and that the zeros of the product are not the zeros of the individual terms. Indeed, the zeros of $Z_G(E)$ are the energy levels of the system in the absence of the scatterers and thus are *not* zeros of $Z(E)$. This implies that they are in fact poles of the diffractive part $Z_D(E)$. This shows that there are some subtle relations between the diffractive orbits and the ordinary periodic ones.

Techniques similar to the ones used in section 4.4 can be used to obtain energy levels of the system. Indeed, the organization of the diffractive orbits as direct combinations of simpler orbits is favourable to cycle expansions and related techniques [32]. These could be useful towards understanding of localization phenomena, especially when the density and strength of the scatterers are large.

The proliferation of scattering orbits looks very similar to the proliferation of periodic orbits in chaotic systems. One should however be careful: the physics and the formula look identical, but the similarity is not complete. If the unperturbed systems is already chaotic, it has been shown both using Random Matrix Theory and the semiclassical approach that the spectral properties are essentially not affected by the scatterers [50]. This is not a real surprise: the scatterers do not add more disorder in a already fully chaotic system.

On the other hand, if the initial system is regular, the spectral properties of the perturbed system are close to but definitely different from those predicted by Random Matrix Theory [51]. For example, the spacing distribution $P(s)$ behaves linearly for small S, like the Wigner distribution, Eq. (24), but with a different constant. Also, at large spacing, $P(s)$ decreases exponentially, not like a Gaussian. These intermediate statistical properties are still heavily studied [52].

Conclusion

In these lectures, I hope I could convince the audience that we have some partial answers to the questions raised in the introduction. Chaos manifests itself in the quantum properties of the systems like the energy levels and the eigenstates, in at least two ways:

- On a narrow energy interval - roughly at the level of individual eigenstates - the quantum structures display strong, apparently random, fluctuations and a high sensitivity on any small change of an external parameter. This is the quantum counterpart of the classical sensitivity on initial conditions.

- On a large energy scale, where spectral properties are averaged over several states, the specific features of the studied system become manifest and are mainly related to the periodic orbits of the classical system.

For regular systems, efficient semiclassical methods exist. For chaotic systems, we understand the role of periodic orbits. Yet, we are not often able to compute individual highly excited states of a chaotic system from the knowledge of its classical dynamics. Using periodic orbits, we can compute low resolution spectra. Whether periodic orbit formulas are the end of the game or just an intermediate step towards a more global understanding is unknown.

References

[1] L. Landau and E. Lifchitz E., *Mécanique*, Mir, Moscow (1966).

[2] A.J. Lichtenberg and M.A. Lieberman, *Regular and stochastic motion*, Springer-Verlag, New-York (1983).

[3] W. Zurek W., Physica Scripta, **T76**, 186 (1998), and references therein.

[4] G. Casati and B. Chirikov, *Quantum Chaos*, Cambridge University Press, Cambridge (1995).

[5] M. Hillery, R.F. O'Connel, M.O. Scully and E.P. Wigner, Phys. Rep. **106**, 121 (1984).

[6] O. Bohigas in *Chaos and quantum physics*, edited by M.-J. Giannoni, A. Voros and J. Zinn-Justin, Les Houches Summer School, Session LII, North-Holland, Amsterdam (1991).

[7] U. Smilansky in *Chaos and quantum physics*, edited by M.-J. Giannoni, A. Voros and J. Zinn-Justin, Les Houches Summer School, Session LII, North-Holland, Amsterdam (1991).

[8] H. Friedrich and D. Wintgen, Phys. Rep. **183**, 37 (1989).

[9] C.H. Iu et al, Phys. Rev. Lett. **66**, 145 (1991).

[10] A. Holle et al, Phys. Rev. Lett., **61**, 161 (1988); J. Main et al, Phys. Rev. A **49**, 847 (1994).

[11] T. van der Veldt, W. Vassen and W. Hogervorst, Europhys. Lett. **21**, 903 (1993).

[12] D. Delande in *Chaos and quantum physics*, edited by M.-J. Giannoni, A. Voros and J. Zinn-Justin, Les Houches Summer School, Session LII, North-Holland, Amsterdam (1991).

[13] H. Hasegawa, M. Robnik and G. Wunner, Prog. Theor. Phys., **98**, 198 (1989).

[14] F.L. Moore et al, Phys. Rev. Lett. **75**, 4598 (1995); H. Amman et al, Phys. Rev. Lett. **80**, 4111 (1998).

[15] A. Arimondo, W.D. Phillips and F. Strumia, *Laser Manipulation of Atoms and Ions*, North-Holland, Amsterdam (1992).

[16] R. Georges, A. Delon and R. Jost, J. Chem. Phys. **103**, 1732 (1995).

[17] G. Mueller et al, Phys. Rev. Lett. **75**, 2875 (1995); T.M. Fromhold et al, Phys. Rev. Lett. **72**, 2608 (1994).

[18] H.-J. Stockmann, J. Stein and M. Kollmann in [4]; H.-J. Stockmann et al, Phys. Rev. Lett. **64**, 2215 (1990); H.-D. Graf et al, Phys. Rev. Lett. **69**, 1296 (1992).

[19] F. Haake, *Quantum signatures of chaos*, Springer-Verlag (1991).

[20] C.H. Iu et al, Phys. Rev. Lett. **63**, 1133 (1989).

[21] M.L. Mehta, *Random Matrices*, Academic Press, London, (1991).

[22] E. Akkermans, G. Montambaux, J.-L. Pichard and J. Zinn-Justin, *Mesoscopic Quantum Physics*, Elsevier Science B.V., North-Holland, Amsterdam (1995).

[23] D. Delande and J.C. Gay, Phys. Rev. Lett. **57**, 2006 (1986).

[24] H. Held et al Europ. Lett. **43**, 392 (1998).

[25] G.R. Welch et al, Phys. Rev. Lett. **62**, 893 (1989).

[26] T. Guhr, A. Mueller-Groeling and H.A. Weidenmuller, Phys. Rep. **299**, 189 (1998).

[27] A. Kudrolli, V. Kidambi and S. Sridhar, Phys. Rev. Lett. **75**, 822 (1995).

[28] B. Grémaud, D. Delande and J.C. Gay, Phys. Rev. Lett. **70**, 615 (1993).

[29] T. Ericson, Phys. Rev. Lett. **5**, 430 (1960).

[30] K. Sacha K., J. Zakrzewski and D. Delande, Phys. Rev. Lett. **83**, 2922 (1999).

[31] M.V. Berry and K.E. Mount, Rep. Prog. Phys. **35**, 315 (1972).

[32] P. Cvitanovic *et al*, *Classical and quantum chaos: a cyclist treatise*, http://www.nbi.dk/ChaosBook.

[33] M. Brack M. and R.K. Bhaduri, *Semiclassical Physics*, Addison-Wesley (1997).

[34] M.C. Gutzwiller, *Chaos in Classical and Quantum Mechanics*, Springer, Berlin (1990).

[35] L.S. Schulman, *Techniques and Applications of Path Integration*, Wiley, New-York (1981).

[36] R. Balian and C. Bloch, Ann. Phys. **85**, 514 (1974).

[37] M. Berry in *Chaotic behaviour of deterministic systems*, Les Houches Summer School 1981, G. Iooss, R. Helleman and R. Stora Ed., North-Holland, Amsterdam (1983).

[38] P. Gaspard in [4]; D. Alonso and P. Gaspard, Chaos **3**, 601 (1993).

[39] E. Bogomolny, JETP **69**, 275 (1989).

[40] D. Wintgen, Phys. Rev. Lett. **61**, 1803 (1988); G. Tanner et al, Phys. Rev. Lett. **67**, 2410 (1991).

[41] R. Marcinek and D. Delande, Phys. Rev. A, to be published.

[42] M. Berry in *Chaos and quantum physics*, edited by M.-J. Giannoni, A. Voros and J. Zinn-Justin, Les Houches Summer School, Session LII, North-Holland, Amsterdam (1991).

[43] E. Bogomolny and J. Keating, Phys. Rev. Lett. **77**, 1472 (1996).

[44] D. Wintgen et al, Prog. Theo. Phys. **116**, 121 (1994); G. Tanner et al, Rep. Prog. Phys, in press.

[45] D. Wintgen, K. Richter and G. Tanner, Chaos, **2**, 19 (1992); J.M. Rost et al, J. Phys B **30**, 4663 (1997); B. Grémaud and D. Delande, Europ. Lett. **40**, 363 (1997).

[46] E. Akkermans and G. Montambaux, contribution in this book.

[47] M.J. Seaton, Rep. Prog. Phys. **46**, 167 (1983).

[48] P. Dando, T.S. Monteiro, D. Delande and K.T. Taylor, Phys. Rev. A **54**, 127 (1996).

[49] P. Šeba, Phys. Rev. Lett. **64**, 1855 (1990); S. Albeverio, F. Gesztesy, R. Hoegh-Krohn and H. Holden, *Solvable Models in Quantum Mechanics*, Springer, New-York (1998).

[50] E. Bogomolny, P. Leboeuf and C. Schmit, arXiv:nlin.CD/0003016; M. Sieber, arXiv:nlin.CD/0003019.

[51] T. Jonckheere, B. Grémaud and D. Delande, Phys. Rev. Lett. **81**, 2442 (1998).

[52] E.B. Bogomolny, U. Gerland and C. Schmit, Phys. Rev. E **59**, R1315 (1999).

Chapter 2

RANDOM MATRIX THEORY OF SCATTERING IN CHAOTIC AND DISORDERED MEDIA

J.-L. Pichard

Service de Physique de l'Etat Condensé,
CEA-Saclay, 91191 Gif sur Yvette cedex, France
pichard@drecam.saclay.cea.fr

Abstract   We review the random matrix theory describing elastic scattering through zero-dimensional ballistic cavities (having chaotic classical dynamics) and quasi-one dimensional disordered systems. In zero dimension, general symmetry considerations (flux conservation and time reversal symmetry) are only considered, while the combination law of scatterers put in series is taken into account in quasi-one dimension. Originally developed for calculating the distribution of the electrical conductance of mesoscopic systems, this theory naturally reveals the universal behaviors characterizing elastic scattering of various scalar waves.

Keywords: random matrix theory, scattering theory, quantum chaos, Anderson localization, disordered systems

Introduction

This chapter is a short introductory review of the random matrix descriptions of elastic scattering. Additional information can be found in more exhaustive reviews [1-3]. The more recent reviews are given by Bohigas [4], Beenakker [5] and by Guhr, Müller-Groeling and Weidenmüller [6]. The basic references for random matrix theory are the book of Mehta [7] (see also Porter [8]) and the series of papers published by Dyson [9] in 1962.

1.   GAUSSIAN ENSEMBLES OF HERMITIAN MATRICES

For a statistical description of a matrix ensemble, one has to define a measure of the space of the matrices having the required symmetries. If one is interested

P. Sebbah (ed.), Waves and Imaging through Complex Media, 125–140.
© 2001 *Kluwer Academic Publishers. Printed in the Netherlands.*

by the distribution of a restricted set of parameters suitable for describing the matrices (e. g. the eigenvalues of an hermitian matrix), one has to use the system of coordinates using those parameters. The jacobian of the transformation (e. g. from matrix elements coordinates towards eigenvalue-eigenvector coordinates) yields correlations between those parameters. Those correlations (level repulsions) are at the origin of universal behaviors first observed in complex nuclei, then in small metallic particles, quantum billiards, hydrogen atom in a magnetic field, mesoscopic quantum systems, electro-magnetic cavities... (see contributions of O. Legrand, D. Delande and R. Weaver in this book). The simplest illustration is given by the Gaussian ensembles of Hermitian matrices introduced in this section. Another illustration is given in the following section: the distribution of the radial parameters characterizing a scattering matrix S or a transfer matrix M.

For doing statistics with real numbers, one defines the probability $P(dx)$ to have a real number x inside an infinitesimal interval of length dx: $P(dx) = \rho(x)\mu(dx)$ where $\rho(x)$ is a density and $\mu(dx) = dx$ the measure of an infinitesimal interval of the real axis. Similarly, for doing statistics with matrices X, one defines the measure $\mu(dX)$ of an infinitesimal volume element dX of the matrix space in which X is defined and one gives a density probability $\rho(X)$. The measure $\mu(dX)$ is given by the symmetries of X, while the density $\rho(X)$ may contain physical assumptions (e.g. minimum information density given a few physical constraints).

For instance, let us introduce the Gaussian Orthogonal Ensemble (GOE) of real symmetric matrices H and the probability distribution of their eigenvalues E_i. A real symmetric matrix $H = H^T = H^*$ of size N has $(N^2 + N)/2$ independent entries. The infinitesimal volume element dH has a measure $\mu(dH)$ given by the product of the infinitesimal variations dH_{ij} of the $N(N + 1)/2$ independent entries:

$$\mu(dH) = \prod_{i \leq j}^{N} dH_{ij}. \tag{1}$$

A possible definition of the GOE density probability $\rho(H)$ is given by a maximum entropy criterion. Minimizing [10] the information entropy

$$I(\rho(H)) = -\int \rho(H) \ln \rho(H)\mu(dH), \tag{2}$$

with an imposed expectation value for the trace of H^2 gives

$$\rho(H) \propto \exp\left(-\frac{\mathrm{tr}\, H^2}{2\sigma^2}\right). \tag{3}$$

Since

$$\exp\left(-\frac{\mathrm{tr}\,H^2}{2\sigma^2}\right)\mu(dH) = \prod_{i=1}^{N}\exp\left(-\frac{H_{ii}^2}{2\sigma^2}\right)dH_{ii}\prod_{i<j}^{N}\exp\left(-\frac{H_{ij}^2}{\sigma^2}\right)dH_{ij},$$

the $N(N+1)/2$ independent matrix elements H_{ij} ($i \leq j$) are uncorrelated variables with Gaussian distributions of variance σ^2 and $\sigma^2/2$ for the diagonal and the off-diagonal entries respectively. This ratio between the variances is important since it makes the GOE ensemble invariant under change of basis: $\rho(H)$ is only function of the N eigenvalues E_i of H through $\sum_{i=1}^{N} E_i^2$ and does not depend on the eigenvectors of H. To calculate $P(E_1,\dots,E_N)$, one has to go from the parameterization of H in terms of its matrix elements H_{ij} to the parameterization of H in terms of its eigenvalue-eigenvector coordinates. The Jacobian of this change of coordinates is at the basis of the level repulsion and of the spectral rigidity characteristic of usual random matrix theories.

A real symmetric matrix is diagonalizable by an orthogonal transformation. Let us first consider a $2d$ rotation of angle θ. One has

$$O_2 = \begin{pmatrix} \cos\theta & \sin\theta \\ -\sin\theta & \cos\theta \end{pmatrix} \tag{4}$$

and by differentiation

$$dO_2 = O_2\begin{pmatrix} 0 & d\theta \\ -d\theta & 0 \end{pmatrix}, \tag{5}$$

$$\begin{aligned} O_N^T O_N &= I_N & (O_N + dO_N)^T(O_N + dO_N) &= I_N \\ dO_N &= O_N dA_N & dA_N^T &= -dA_N, \end{aligned}$$

where I_N denotes the unit $N \times N$ matrix. dA_N is a $N \times N$ real antisymmetric matrix and $\mu(dO_N) = \prod_{i<j}^{N} dA_{ij}$.

H is diagonalizable by an orthogonal transformation O: $H = OH_DO^T$ where H_D is a real diagonal matrix of entries E_i and of measure $\mu(dH_D) = \prod_{i=1}^{N} dE_i$. By differentiation, one gets

$$dH = Od\mathcal{H}O^T, \tag{6}$$

$$d\mathcal{H} = dAH_D - H_D dA + dH_D, \tag{7}$$

where we have used $dO = OdA$ and $dA^T = -dA$. The real symmetric matrix dH is related to $d\mathcal{H}$ by an orthogonal transformation. The Jacobian is equal to one and $\mu(dH) = \prod_{i<j}^{N} dH_{ij} = \prod_{i<j}^{N} d\mathcal{H}_{ij}$. The product of the infinitesimal diagonal elements of $d\mathcal{H}$ gives $\mu(dH_D)$, the off-diagonal contribution gives $\prod_{i<j}^{N} |E_i - E_j| dA_{ij}$, and one eventually obtains the measure $\mu(dH)$ in terms

of the measures $\mu(dO)$ and $\mu(dH_D)$

$$\mu(dH) = \prod_{i<j}^{N} |E_i - E_j| \mu(dH_D) \mu(dO).$$ (8)

In terms of the eigenvalue-eigenvector coordinates of H, the GOE distribution becomes:

$$P(dH)\mu(dH) = P(E_1, \dots, E_N)\mu(dH_D)\mu(dO),$$ (9)

where the joint probability distribution of the eigenvalues is identical to the Gibbs factor of a set of N point charges free to move on the real axis of the complex plane with a pairwise logarithmic repulsion and a quadratic confining potential at an inverse temperature $\beta = 1$:

$$P(E_1, \dots, E_N) \propto \exp\left[-\beta \sum_{i<j}^{N} \ln|E_i - E_j| + \sum_{i=1}^{N} \frac{E_i^2}{2\sigma^2} \right].$$ (10)

The pairwise repulsion coming from the Jacobian makes unlikely level degeneracies and explains why dramatically non-random an energy-level series really is. This is the Coulomb gas analogy usual in Random Matrix Theory. To appreciate how this random matrix approach is adapted to include symmetry breaking effects, let us assume that the matrix H is the Hamiltonian of an electron moving in a chaotic cavity. Applying a magnetic field removes time reversal symmetry, H becomes hermitian ($H = H^\dagger$) and $\mu(dH) = \prod_{i=1}^{N} dH_{ii}^1 \prod_{i<j} dH_{ij}^1 dH_{ij}^2$, taking into account the infinitesimal variations of the real and imaginary parts of its matrix elements ($H_{ij} = H_{ij}^1 + iH_{ij}^2$). H is now diagonalizable by a unitary transformation U and $dU = U da$ where da is an infinitesimal anti-hermitian matrix ($da = -da^\dagger$), and $\mu(dU) = \prod_{i=1}^{N} da_{ii}^2 \prod_{i<j}^{N} da_{ij}^1 da_{ij}^2$. One obtains for hermitian matrices $dH = U d\mathcal{H} U^\dagger$ where $d\mathcal{H} = da H_D + dH_D - H_D da$. The Jacobian of a unitary transformation being equal to one, one eventually finds $\mu(dH) = \prod_{i<j} |E_i - E_j|^2 \mu(dH_D)\mu(dU)/(\prod_i da_{ii}^2)$, the square coming from the fact that the non diagonal contribution of $d\mathcal{H}$ is now complex. Breaking time reversal symmetry, one keeps the Coulomb gas analogy, with a temperature divided by a factor two ($\beta = 1 \to 2$). For electrons of spin $1/2$, one can also break spin rotation symmetry (SRS) by spin orbit scattering, an effect which preserves time reversal symmetry (TRS). The matrix elements of H are no longer real ($\beta = 1$) or complex ($\beta = 2$), but quaternion real ($\beta = 4$) and the level distribution is still given by the Coulomb gas analogy with a temperature divided by a factor 4: $\beta = 1 \to 4$.

The main feature of those three Gaussian ensembles of random matrices is that $\rho(H)$ does not couple eigenvalue and eigenvector variables. These ensembles are invariant under canonical transformations: orthogonal transformations

($\beta = 1$) when the system invariant under TRS and SRS symmetries, unitary transformations ($\beta = 2$) in the absence of TRS and symplectic transformations ($\beta = 4$) with TRS and without SRS. The eigenvectors are totally random, the measure of the matrices O or U are given by the Haar measures over the orthogonal or unitary groups respectively, and the integration over the eigenvectors is trivial. This is the totally random character of the eigenvectors which makes the energy levels correlated (pairwise logarithmic repulsion) and subject to universal symmetry breaking effects (e.g. $\beta = 1, 4 \rightarrow 2$ when TRS is broken).

2. RADIAL PARAMETERIZATION OF SCATTERING MATRICES S AND MEASURES

Similar random matrix theories can be adapted to matrices having different symmetries and can give the joint probability distribution of a subset of variables which can be used for their parameterization. Another example is provided by the unitary matrices describing complex elastic scatterers. Let us consider a perfect waveguide characterized by N quantized modes propagating to the right and N time reversed modes propagating to the left. Let us introduce in the middle of this waveguide a complex elastic scatterer described by its $2N \times 2N$ scattering matrix S. This matrix describes the various transmission and reflection amplitudes to the right or to the left, has to be unitary for conserving the flux amplitudes ($SS^\dagger = S^\dagger S = I_{2N}$), and must be symmetric ($S = S^T$) if one has time reversal invariant scattering:

$$S = \begin{pmatrix} r & t' \\ t & r' \end{pmatrix}. \tag{11}$$

The $N \times N$ matrices t and t' describe transmission amplitudes of the incoming fluxes to the right and left directions respectively, while r and r' describe reflections. Let us consider time reversal invariant scattering where S is symmetric and can be decomposed as

$$S = U^T.U, \tag{12}$$

$$S + dS = U^T(I_{2N} + idM)U, \tag{13}$$

where U and dM are respectively a unitary and an infinitesimal real symmetric matrices. This decomposition is not unique, one can multiply U by an arbitrary orthogonal transformation O ($U \rightarrow UO$) but the measure $\mu(dS) = \prod_{i \leq j} dM_{ij}$ is uniquely defined since the Jacobian of the transformation $dM \rightarrow dM = OdMO^T$ is equal to one. $\mu(dS)$ was expressed by Dyson [9] in terms of the eigenvectors and eigenvalues of S and their associated measures. The original motivation in Dyson's work was not at all to study a scattering problem, but to use the eigenvalue distribution of S for describing energy-level statistics (the eigenvalues of S being confined on the unit circle of the complex plane, one does not need the somewhat artificial GOE quadratic confining potential). For a scattering problem, the eigenvalue-eigenvector parameterization of S is not adapted and we introduce a more convenient one using 2 unitary $N \times N$ matrices u_1 and u_2 and a diagonal $N \times N$ matrix Λ with N real positive diagonal entries $\lambda_1, \ldots, \lambda_N$. Denoting $T = (1+\Lambda)^{-1}$ and $R = \Lambda(1+\Lambda)^{-1}$, S can be written as:

$$S = \begin{pmatrix} u_1 & 0 \\ 0 & u_2 \end{pmatrix} \begin{pmatrix} -\sqrt{R} & \sqrt{T} \\ \sqrt{T} & \sqrt{R} \end{pmatrix} \begin{pmatrix} u_1^T & 0 \\ 0 & u_2^T \end{pmatrix}. \tag{14}$$

In this parameterization, the transmission and reflection matrices become

$$t = u_2\sqrt{T}u_1^T \quad t' = u_1\sqrt{T}u_2^T \tag{15}$$
$$r = -u_1\sqrt{R}u_1^T \quad r' = u_2\sqrt{R}u_2^T \tag{16}$$

and T and R contain the eigenvalues of $tt^\dagger$ and $rr^\dagger$ (transmission and reflection eigenvalues). One can note that the transfer matrix M which gives the flux amplitudes of the right side of the scatterer in terms of the flux amplitudes of its left side can be written using the same parameterization as

$$M = \begin{pmatrix} u_2 & 0 \\ 0 & u_2^* \end{pmatrix} \begin{pmatrix} \sqrt{I_N + \Lambda} & \sqrt{\Lambda} \\ \sqrt{\Lambda} & \sqrt{I_N + \Lambda} \end{pmatrix} \begin{pmatrix} u_1^T & 0 \\ 0 & u_1^\dagger \end{pmatrix}. \tag{17}$$

M is pseudo unitary (flux conservation)

$$M \begin{pmatrix} I_N & 0 \\ 0 & -I_N \end{pmatrix} M^\dagger = M^\dagger \begin{pmatrix} I_N & 0 \\ 0 & -I_N \end{pmatrix} M = \begin{pmatrix} I_N & 0 \\ 0 & -I_N \end{pmatrix} \tag{18}$$

and when one has TRS ($\beta = 1$), M must also satisfy the requirement:

$$M^* = \begin{pmatrix} 0 & I_N \\ I_N & 0 \end{pmatrix} M \begin{pmatrix} 0 & I_N \\ I_N & 0 \end{pmatrix}. \tag{19}$$

M has the advantage to be multiplicative when one puts the scatterers in series. For the derivation of this parameterization of S, see [11], and use the relation between the matrices M and S. To understand the interest of this parameterization, we introduce the variables x_i from $\lambda_i = \sinh^2 x_i$. One can see that M is decomposed in the product of a unitary transformation, followed by N hyperbolic rotations of angle x_i:

$$\begin{pmatrix} \sqrt{1+\lambda_i} & \sqrt{\lambda_i} \\ \sqrt{\lambda_i} & \sqrt{1+\lambda_i} \end{pmatrix} \rightarrow \begin{pmatrix} \cosh x_i & \sinh x_i \\ \sinh x_i & \cosh x_i \end{pmatrix},$$

before a second unitary transformation. The N parameters λ_i are called the radial parameters of M (or S).

Diagonalizing the matrix containing those parameters in the new parameterization of S by the orthogonal matrix O

$$O = \frac{1}{\sqrt{2}} \begin{pmatrix} (I_N - \sqrt{R})^{1/2} & (I_N + \sqrt{R})^{1/2} \\ (I_N - \sqrt{R})^{1/2} & -(I_N + \sqrt{R})^{1/2} \end{pmatrix}, \tag{20}$$

and denoting

$$U = \begin{pmatrix} u_1 & 0 \\ 0 & u_2 \end{pmatrix} \tag{21}$$

and

$$I = \begin{pmatrix} iI_N & 0 \\ 0 & -iI_N \end{pmatrix}, \tag{22}$$

one can write $S = YY^T$ where the unitary matrix $Y = UOI$. Defining the infinitesimal anti-hermitian and real antisymmetric matrices dA and dB from $dU = UdA$ and $dO = OdB$, using $dS = iY \, dM Y^T$, one obtains:

$$idM = dC + dC^T, \tag{23}$$
$$dC = I^*(dB + O^T dAO)I, \tag{24}$$

which allows us to write the Jacobian matrix of the change of coordinates $dM_{ij} \rightarrow (dA_{ij}, dB_{ij})$. This matrix has a simple block diagonal form and its determinant gives the measure $\mu(dS)$ in terms of the measures $\mu(d\Lambda) = \prod_{i=1}^{N} d\lambda_i$ and $\mu(dU) = \prod_{i=1}^{2} \mu(du_i)$. We have sketched the derivation when S is unitary symmetric ($\beta = 1$), but the extension to the three possible symmetry classes is straightforward and gives for $\mu_\beta(dS)$ the general form [12]

$$\mu_\beta(dS) = P_\beta(\lambda_1, \dots, \lambda_N)\mu(d\Lambda)\mu(dU) \tag{25}$$
$$P_\beta(\lambda_1, \dots, \lambda_N) = \exp -\beta H_\beta(\lambda_1, \dots, \lambda_N) \tag{26}$$
$$H_\beta(\lambda_1, \dots, \lambda_N) = -\sum_{i<j}^{N} \ln|\lambda_i - \lambda_j| + \sum_{i=1}^{N} V_\beta(\lambda_i) \tag{27}$$
$$V_\beta(\lambda) = (N + \frac{2-\beta}{2\beta}) \ln(1 + \lambda). \tag{28}$$

If the scattering is not time reversal symmetric ($\beta = 2$), S is no longer symmetric and one needs two additional unitary matrices u_3 and u_4 for parameterizing S and $\mu(dU) = \prod_{i=1}^{4} \mu(du_i)$. If one considers scattering of spin $1/2$ particles by a TRS scatterer which removes SRS (spin-orbit scattering), $\beta = 4$.

One can similarly show that the measure for the transfer matrices [2] is given in terms of the radial parameters λ_i by:

$$\mu_\beta(dM) = \prod_{i<j}^{N} |\lambda_i - \lambda_j|^\beta \mu(d\Lambda) \prod_{i=1}^{2,4} \mu(du_i). \tag{29}$$

3.   ZERO DIMENSIONAL CHAOTIC SCATTERING

Let us assume that the scatterer represented by the matrix S is a ballistic cavity of irregular shape having chaotic classical dynamics, as sketched in Fig. 1. We have in mind long scattering trajectories corresponding to particles reflected

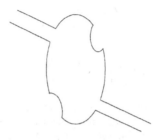

Figure 1.   Scheme of a cavity giving rise to zero-dimensional chaotic scattering.

many times inside the cavity before being transmitted or reflected. Let us assume that the shape of the cavity is slightly modified, or that the wavelength of the incoming fluxes varies. The scattering will be deeply re-organized and a statistical description of the fluctuations of the scattering amplitudes becomes necessary. To this end, we need to define a statistical ensemble of scattering matrices S and we will assume that the scatterer will visit this ensemble when one varies a tunable parameter (shape of the cavity, wavelength, applied magnetic field reorganizing the quantum interferences if we consider electron elastic scattering). The simplest ensemble is the one where all the scattering processes are equiprobable, which does not contain any information about the system excepted its basic symmetries. Those ensembles of minimum information entropy for S are the circular ensembles [9] introduced by Dyson in 1962, for which the probability to find S inside a volume element dS is

$$P(dS) = \frac{1}{V_\beta}\mu_\beta(dS), \qquad (30)$$

where V_β is a normalization constant. One obtains for the radial parameters λ_i a Coulomb gas analogy very similar to the GOE-GUE-GSE Coulomb gas analogies for the energy level of a random Hamiltonian, excepted two noticeable differences: (i) $\lambda_i \in \mathbb{R}^+$ while $E_i \in \mathbb{R}$, the Coulomb gas is free to move only on the positive part of the real axis in the complex plane (ii) the confining potential $V_\beta(\lambda)$ is implied by the symmetries of S (in contrast to the quadratic potential given by a certain choice of $\rho(H)$) and depends on the symmetry parameter β.

Let us see the implication for the total transmission probability $\mathcal{T} = trtt^\dagger = \sum_{i=1}^{N} T_i = \sum_{i=1}^{N}(1 + \lambda_i)^{-1}$. When S gives Fermi wave scattering by a mesoscopic scatterer coupling to electron reservoirs, $\mathcal{T}$ gives its electrical conductance in units of $2e^2/h$ (Landauer formula).

If $N = 1$ (single mode waveguide), the probability distribution of $\mathcal{T}$ exhibits strong symmetry breaking effects (see Fig. 2):

$$P(\mathcal{T}) = \frac{\beta}{2}\mathcal{T}^{\beta/2-1}. \tag{31}$$

If one measures the electrical conductance of a chaotic cavity (quantum bil-

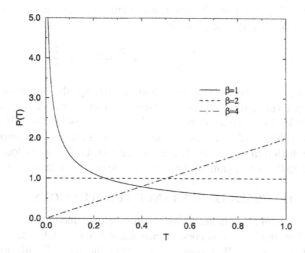

Figure 2. Symmetry breaking effect on the distribution $P(\mathcal{T})$ for a chaotic cavity coupled to leads via single mode contacts ($N = 1$).

liard) coupled to two electron reservoirs by single mode leads, the conductance is likely to be close to 0 if one has TRS and SRS, to have a value uniformly distributed between 0 and $2e^2/h$ if one applies a small magnetic field (no TRS), and a value close to $2e^2/h$ if the electron reflection on the walls of the cavity is accompanied by spin-orbit scattering (no SRS). The distribution $P(\mathcal{T})$ can be explored by changing the Fermi energy with a metallic gate, by small deformations of the cavity or by applying a magnetic field.

If $N \gg 1$ the symmetry breaking effects are much smaller. A first one is the suppression, when there is no TRS, of a small "weak localization" correction to the average transmission $< \mathcal{T} >$, the second effect of removing TRS is to halve the universal variance of $\mathcal{T}$.

3.1.   WEAK LOCALIZATION CORRECTIONS

The large N limit of the density $\rho(\lambda)$

$$\rho_\beta(\lambda) = \sum_{i=1}^{N} \int_{R^+} \cdots \int_{R^+} \prod_{i=1}^{N} d\lambda_i P_\beta(\lambda_1, \ldots, \lambda_N) \delta(\lambda - \lambda_i) \qquad (32)$$

can be calculated using an equation derived by Dyson [13]

$$\int_{R^+} \frac{\rho_\beta(\lambda')d\lambda'}{\lambda - \lambda'} + \frac{\beta - 2}{2\beta} \frac{d\ln\rho_\beta(\lambda)}{d\lambda} = \frac{dV_\beta(\lambda)}{d\lambda}. \qquad (33)$$

Taking the $V_\beta(\lambda)$ characterizing the three circular ensembles for S, one gets for the ensemble averaged total transmission:

$$< \mathcal{T} >= \int_{R^+} \frac{\rho_\beta(\lambda)}{1 + \lambda} = \frac{N}{2} + \frac{\beta - 2}{4\beta} + O(\frac{1}{N}). \qquad (34)$$

The term $\propto N$ is obvious: having chaotic scattering, the probability to be reflected equals the probability to be transmitted. However there is a small correction of order 1 which reduces by $-1/4\,\mathcal{T}$ when there is TRS and SRS, which disappears without TRS and which enhances $\mathcal{T}$ by a factor $1/8$ without SRS. This is the analog in a quantum billiard of the well-known weak-localization corrections to the Boltzmann-Drude conductance of a disordered system.

3.2.   UNIVERSAL CONDUCTANCE FLUCTUATIONS

One of the main phenomena which is naturally explained by random matrix theory [14] are the "universal conductance fluctuations" (UCF) characterizing mesoscopic conductors. When one varies the "conductance" $\mathcal{T}$ with an external parameter (Fermi wavelength, magnetic field ...) one generates fluctuations of magnitude independent of $< \mathcal{T} >$. To calculate $< \delta\mathcal{T}^2 >$ one needs to know the density-density correlation function of the λ-parameters. In the limit $N \to \infty$, one can simply calculate [15] this variance. Exploiting the Coulomb gas analogy, one can write

$$K_\beta(\lambda, \lambda') =< \sum_{ij} \delta(\lambda - \lambda_i)\delta(\lambda - \lambda_j) > -\rho_\beta(\lambda)\rho_\beta(\lambda') \qquad (35)$$

as a functional derivative:

$$K_\beta(\lambda, \lambda') = -\frac{1}{\beta} \frac{\partial\rho_\beta(\lambda)}{\partial V_\beta(\lambda')}. \qquad (36)$$

When $N \to \infty$, $V_\beta(\lambda) \to \int_{R^+} \ln|\lambda - \lambda'|\rho_\beta(\lambda) + const$ (this amounts to neglect the term responsible for the weak localization correction in Eq. (33). In the large

N-limit, $V_\beta(\lambda)$ becomes a linear functional of $\rho(\lambda')$. This implies an important result: $K_\beta(\lambda, \lambda')$ depends only on the nature of the pairwise repulsion between the radial parameters and becomes independent of the confining potential $V_\beta(\lambda)$. The evaluation of the functional derivative giving $K_\beta(\lambda, \lambda')$ is straightforward:

$$\lim_{N \to \infty} K_\beta(\lambda, \lambda') = -\frac{1}{\pi^2\beta} \ln \frac{\sqrt{\lambda} - \sqrt{\lambda'}}{\sqrt{\lambda} + \sqrt{\lambda'}} \tag{37}$$

to eventually give

$$< \delta^2(\mathcal{A}) > \to \frac{-1}{\beta\pi^2} \int_{R+} \int_{R+} d\lambda d\lambda' \ln \frac{\sqrt{\lambda} - \sqrt{\lambda'}}{\sqrt{\lambda} + \sqrt{\lambda'}} \frac{da(\lambda)}{d\lambda} \frac{da(\lambda')}{d\lambda'} \tag{38}$$

for the variance $< \delta^2 A >$ of a linear statistics $A = \sum_{i=1}^N a(\lambda_i)$ of the λ-parameter. Taking $a(\lambda) = (1 + \lambda)^{-1}$, one gets

$$< \delta^2 \mathcal{T} > = \frac{2}{16\beta} . \tag{39}$$

The variance of $\mathcal{T}$ is just a number which depends on β but not on $< \mathcal{T} >$.

When N is finite, weak localization corrections and variances can be exactly calculated using a method introduced by Mehta, Gaudin and Dyson. Using a set of orthogonal polynomials ($\beta = 2$) or shew-orthogonal polynomials ($\beta = 1, 4$), one can perform the integration of $P_\beta(\lambda_1, \dots, \lambda_N)$ over an arbitrary set of λ-parameters in order to have $\rho_\beta(\lambda)$, $K_\beta(\lambda, \lambda')$ and higher order correlation functions. Going to the variables $T_i = (1 + \lambda_i)$ the polynomials required [12] to perform the integrals for $\beta = 2$ are the Legendre polynomials.

4. MANY CHANNEL DISORDERED WAVEGUIDE

So far, we have done the simplest random matrix exercise where all the scatterers are taken with a uniform distribution. Those circular ensembles are suitable to describe scattering in zero-dimensional chaotic cavities, as it has been numerically checked. Another exactly solvable case is provided by a $1d$ series of scatterers. Let us consider a quasi-$1d$ disordered wire (or waveguide) with N channels, as sketched in Fig. 3. Typically, if the (Fermi) wavelength is λ_F, and L_t^{d-1} the transverse section, one has $N = (L_t/\lambda_F)^{d-1}$. Let us consider a "building block" of length δL and of transfer matrix $M_{\delta L}$. A suitable statistical ensemble for $M_{\delta L}$ is given by a maximum entropy ensemble where $< \mathcal{T} >$ is imposed. If δL is larger than the elastic mean free path l, a natural requirement is to impose Ohm's law (the conductance $< g > \propto < \mathcal{T} > \propto Nl(\delta L)^{-1}$). This gives for the building block $P(M_{\delta L}) \propto \exp -(A < \mathcal{T} >)\mu(dM)$, where A is the Lagrange multiplier associated to the imposed constraint. This ensemble preserves the logarithmic pairwise interaction between the λ-parameter and

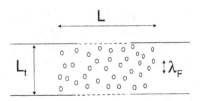

Figure 3.   Scheme of a quasi-1d disordered waveguide of length $L >> L_t$ with $N = (L_t/\lambda_F)^{d-1}$ modes.

the distribution of the auxiliary unitary matrices u_i derived for the circular ensembles, but gives rise to a different potential $V_\beta(\lambda)$. Considering a 1d series of such building blocks, of length L/l, and exploiting the multiplicative combination law of M ($M_{L+\delta L} = M_L.M_{\delta L}$), one can derive a Fokker-Planck equation [5, 16, 17] for $P_\beta(\lambda_1, \dots, \lambda_N, L/l)$:

$$\frac{\partial P_\beta(\lambda_1, \dots, \lambda_N)}{\partial L} = D\Delta_\lambda P_\beta(\lambda_1, \dots, \lambda_N) \tag{40}$$

$$D = \frac{2}{(\beta N + 2 - \beta)l} \tag{41}$$

$$\Delta_\lambda = \sum_{i=1}^{N} \frac{\partial}{\partial \lambda_i} \lambda_i(1 + \lambda_i) J \frac{\partial}{\partial \lambda_i} \frac{1}{J} \tag{42}$$

$$J = \prod_{i<j}^{N} |\lambda_i - \lambda_j|^\beta. \tag{43}$$

In this statistical description of M (or S), the radial parameters and the matrices u_i are statistically uncorrelated. The u_i remain distributed with Haar measure over the $N \times N$ unitary group. Increasing the number L/l of blocks put in series only changes the statistics of the radial parameters as given by the above Fokker-Planck equation, This is the limitation of this "isotropic" model which allows us to solve it entirely and to describe quasi-1d localization. But the transverse system dimensions appear only through the parameter N. A strip of purely 1d transverse section and a bar with a 2d section are treated the same way: within the 0d approximation for the transverse dynamics. When $N = 1$, $J = 1$ and the resulting equation was originally derived [18] in 1959 in a work entitled: "waveguides with random inhomogeneities and Brownian motion in the Lobachevsky plane". The nature of the Fokker-Planck equation was indeed correctly identified as a "Heat equation in a space of negative curvature". Looking in the mathematical literature, one can realize that Δ_λ is the radial part of the Laplace-Beltrami operator in the space of the transfer matrices M. Increasing

L yields for M an "isotropic" (invariant under the unitary transformation u_i) Brownian motion in the transfer matrix space which is only characterized by the diffusion constant D.

4.1. QUASI-1D LOCALIZATION

When L increases and exceeds the localization length ξ, the system becomes an Anderson insulator and it is more convenient to use the variables $x_i = L/\xi_i$ ($\lambda_i = \sinh^2 x_i$), where the lengths ξ_i characterize the exponential decays of the transmission channels of the quasi-1d scatterer ($T_i \approx \exp{-(2L/\xi_i)}$). When $L \to \infty$ the variables $1/\xi_i$ are given by the Lyapounov exponents of the multiplicative transfer matrix M, giving N decay lengths $\xi_N \ll \xi_{N-1} \ll \xi_1$, the largest of them defining [19] the localization length ξ of the quasi-1d system. The Fokker-Planck equation becomes:

$$\frac{\partial P_\beta(x_1,\dots,x_N)}{\partial L} = D \sum_{i=1}^{N} \frac{\partial}{\partial x_i} \left(P_\beta + \beta P_\beta \frac{\partial \Omega}{\partial x_i} \right) \tag{44}$$

$$\Omega = \sum_{i<j}^{N} \ln|\sinh x_i^2 - \sinh^2 x_j| - \frac{1}{\beta} \sum_{i=1}^{N} \ln|\sinh x_i|, \tag{45}$$

where the different channels are coupled via Ω. However, when $L \to \infty$, $|\sinh^2 x_i - \sinh^2 x_j| \approx \exp 2x_i$ when $x_i \gg x_j$, the channels become decoupled and $\Omega \to -(2/\beta)\sum_{i=1}^{N}(1 + \beta N - \beta)x_i$. The Fokker-Planck equation becomes solvable and gives [1]:

$$P_\beta(x_1,\dots,x_N,L) = (\frac{\gamma l}{2\pi L})^{\frac{N}{2}} \prod_{i=1}^{N} \exp\left(-\frac{\gamma l}{2L}(x_i - \frac{L}{\xi_i})^2 \right), \tag{46}$$

where $\xi_i = (\gamma l)/(1 + \beta N - \beta)$ and $\gamma = \beta N + 2 - \beta$. One gets two important results:

(i) a universal symmetry breaking effect for the localization length ξ when $\beta = 1 \to 2$:

$$\xi = (\beta N + 2 - \beta)l; \tag{47}$$

(ii) a normal distribution for $\ln \mathcal{T}$ where $-2 < \ln \mathcal{T} >=< \delta^2 \ln \mathcal{T} >$. Those symmetry breaking effects have been observed in magneto-transport measurements performed in disordered wires [20-22], where the conductance $g \propto \exp{-(2L/\xi)}$ and $\xi = (N+1)l \to 2Nl$ when $\beta = 1 \to 2$. However this theory neglecting electron-electron interactions, it would be important to check those universal symmetry breaking effects in the localized regime using other waves (light, sound ...) (see the contribution of A.Z. Genack and A. A. Chabanov and the contribution of R. Weaver in this book).

4.2.   MAPPING ONTO A CALOGERO-SUTHERLAND MODEL OF INTERACTING FERMIONS

When $L \leq \xi$, the system is a disordered conductor where a number $N_{eff} < N$ of channels are still opened, giving $< \mathcal{T} > \approx Nl/L$ (Ohm's law) if one ignores the (small) weak-localization corrections. Approximations can be used also in this limit ($L << \xi$) to calculate the quasi-1d weak-localization corrections $\delta \mathcal{T} = (\beta - 2)/(3\beta)$ and the UCF variance $2/(15\beta)$. One can notice that those values valid for the quasi-1d wire are close, but not identical to those derived from the circular ensembles. The small difference between the UCF variances tells us (see the arguments given for having $K(\lambda, \lambda')$ in the large N-limit) that the radial parameters cannot have [15] exactly the pairwise logarithmic repulsion in quasi-1d. This can be understood when $\beta = 2$ where one can solve the Fokker-Planck equation for any values of N, using a transformation originally introduced by Sutherland to solve Dyson's Brownian motion model. The distribution $P_\beta(x_1, \dots, x_N, L)$ is related [5, 23] to a wave function $\Psi(x_1, \dots, x_N, L)$ by the transformation

$$P = \Psi \exp - \left(\frac{\beta \Omega}{2} \right) \tag{48}$$

and the Fokker-Planck equation for P becomes a Schrödinger equation for Ψ in imaginary time.

$$-l \frac{\partial \Psi}{\partial L} = H\Psi, \tag{49}$$

where the Hamiltonian is given by

$$H = \frac{-1}{2\gamma} \sum_{i=1}^{N} \left(\frac{\partial^2}{\partial x_i^2} + \frac{1}{\sinh^2 2x_i} \right) + (\beta - 2)U(x_i, x_j) \tag{50}$$

$$U(x_i, x_j) = \frac{\beta}{2\gamma} \sum_{i<j}^{N} \frac{\sinh^2 2x_j + \sinh^2 2x_i}{(\cosh 2x_j - \cosh^2 2x_i)^2}. \tag{51}$$

For $\beta = 2$, the "particles" do not interact and the equation can be solved to give [23] a small change in the Coulomb gas analogy for the radial parameters:

$$- \ln |\lambda_i - \lambda_j| \rightarrow -\frac{1}{2} \ln |\lambda_i - \lambda_j| - \frac{1}{2} \ln |\text{arsinh}^2 \sqrt{\lambda_i} - \text{arsinh}^2 \sqrt{\lambda_j}|$$

yielding the change $< \delta^2 \mathcal{T} > = \frac{2}{16\beta} \rightarrow \frac{2}{15\beta}$.

For $\beta \neq 2$, one has not yet found how to solve the Schrödinger equation. Let us note that the Calogero-Sutherland Hamiltonian derived from the original Dyson's Brownian motion model is now solved [24] when $\beta/2 = p/q$ where p and q are integer.

Summary

We have seen how to derive the probability distributions of S and how to extract the distribution of $\mathcal{T}$ using very little information (symmetries, combination law in quasi-1d). This can be simply done as far as the radial parameters are decoupled from the auxiliary matrices u_i, which limits the method to 0d and quasi-1d. A more difficult task remaining to be achieved is to go beyond this limit, and to describe elastic scattering in 2d and 3d, including possible Anderson localization. One can ask to what extend a real chaotic cavity or a quasi-1d disordered waveguide is accurately described by those random matrix theories. Numerical calculations [25] of the distribution of $\mathcal{T}$ in (suitably) designed chaotic cavities (suitably) connected to two many-channel ballistic waveguides confirm the random matrix results. For quasi-1d disordered wires, a field theory approach has been derived [26] assuming local diffusive dynamics. One obtains a non-linear σ-model which gives using supersymmetry $< \mathcal{T}(\mathcal{L}) >$ and $< \delta^2 \mathcal{T}(L) >$ in the large N-limit. The results turn out to coincide [27, 28] with those given by the Fokker-Planck equation in this limit. The finite N behaviors which are calculable by the Fokker-Planck equation are still out of reach using the σ-model, as well as the moments of $\ln \mathcal{T}$ which are the statistically meaningful observables (related to a normal distribution) in the localized limit.

Acknowledgments

My own research on this random matrix theory adapted to scattering was published in a series of papers done with C.W.J. Beenakker, Y. Imry, R. A. Jalabert, P. A. Mello, K. Muttalib, K. M. Frahm, K. Slevin, A. D. Stone and N. Zanon, collaborations which are gratefully acknowledged.

References

[1] J.-L. Pichard, in *Quantum coherence in Mesoscopic systems*, edited by B. Kramer, NATO ASI Series B254 (plenum New York) p369 (1991).

[2] A. D. Stone, P. A. Mello, K. A. Muttalib and J.-L. Pichard, in *Mesoscopic Phenomena in Solids*. edited by B. L. Altshuler, P. A. Lee and R. A. Webb, (North Holland, Amsterdam) p 369 (1991).

[3] P. A. Mello, in *Mesoscopic Quantum Physics*, edited by E. Akkermans, G. Montambaux, J.-L. Pichard and J. Zinn-Justin, (North Holland, Amsterdam), p 435, (1995).

[4] O. Bohigas, in *Chaos and Quantum Physics*, edited by M.-J. Giannoni, A. Voros and J. Zinn-Justin, (North Holland, Amsterdam), p 87, (1990).

[5] C. W. J. Beenakker, Rev. Mod. Phys. **69**, p731, (1997).

[6] T. Guhr, A. Múller-Groeling and H. A. Weidenmüller, Phys. Rep. **299**, p 189, (1998).

[7] M. L. Mehta, *Random Matrices* (Academic, New-York), (1991).

[8] C. E. Porter, *Statistical Theories of Spectra: Fluctuations*, (Academic, New-York) (1965).

[9] F. J. Dyson, J. Math. Phys. **3**, p140, 157, 1191 and 1199, (1962).

[10] R. Balian, Nuevo Cimento **57**, 183, (1968).

[11] J.-L. Pichard and P. A. Mello, J. Phys. I, **1**, 493, (1991).

[12] R. A. Jalabert, J.-L. Pichard and C. W. J. Beenakker, Europhys. Lett, **27**, 255, (1994); H. U. Baranger and P. A. Mello, Phys. Rev. Lett. **73**, 142, (1994); R. A. Jalabert and J.-L. Pichard, J. Phys. I France **5**, 287, (1995).

[13] F. J. Dyson, J. Math. Phys. **13**, 90, (1972).

[14] K. A. Muttalib, J.-L. Pichard and A. D. Stone, Phys. Rev. Lett. **59**, 2475, (1987).

[15] C. W. J. Beenakker, Phys. Rev. Lett. **49**, 2205, (1994).

[16] O. N. Dorokhov, JETP Lett **36**, 318, (1982).

[17] P. A. Mello, P. Pereyra and N. Kumar, Ann. Phys. (N.Y.) **181**, 290, (1988).

[18] M. E. Gertsenshstein and V. B. Vasil'ev, Theor. Probab. Appl., **4**, 391, (1959).

[19] J.-L. Pichard and G. Sarma, J. Phys. C: Solid State Phys. **14**, L127, (1981).

[20] J.-L. Pichard, M. Sanquer, K. Slevin and P. Debray, Phys. Rev. Lett. **65**, 1812, (1990).

[21] W. Poirier, D. Mailly and M. Sanquer, Phys. Rev. **B** 59, 10856, (1999).

[22] Y. Khavin, M. E. Gershenson and A. L. Bogdanov, Phys. Rev. **B** 58, 8009, (1998).

[23] C. W. J. Beenakker and B. Rejaei, Phys. Rev. Lett. **71**, 3689, (1993).

[24] D. Serban, F. Lesage and V. Pasquier, Nucl. Phys. **B** 466, 499, (1996).

[25] H. U. Baranger and P. A. Mello, cond-mat/9812225 and references therein.

[26] K. B. Efetov and A. I. Larkin, Sov. Phys. JETP 58, 444, (1983).

[27] K. M. Frahm, Phys. Rev. Lett. **74**, 4706, (1995).

[28] K. M. Frahm and J.-L. Pichard, J. Phys. I France **5**, 847, (1995).

Chapter 3

WAVE CHAOS IN ELASTODYNAMICS

R. L. Weaver

Department of Theoretical and Applied Mechanics
University of Illinois
104 So Wright Street, Urbana, IL 61801
r-weaver@uiuc.edu

Abstract Recent years have demonstrated that ultrasonic waves in solids permit investigations of wave chaos, reverberation and multiple scattering that complement those of microwaves, optics and electronics. The large quality factors that can be achieved with elastic waves, their relatively slow wave speed, and their good impedance mismatch with the laboratory, lead to certain advantages for the experimentalist and new perspectives for the theorist. In particular these features allow transients to be observed, allow good isolation from leads and antennas, and convenient time and length scales in the experiment. A short course on continuum elastodynamics in solids is followed by a review of the laboratory work on spectral and transport properties of diffuse ultrasonic waves in solids. These include demonstrations of energy equipartition, of eigenvalue correlations and spectral rigidity, enhanced backscatter in enclosures, wave diffusion, and Anderson localization.

Keywords: elastic waves, chaos, localization, random matrix theory, continuum elasticity

Introduction

The past two decades, and in particular the past few years, have witnessed a marked increase in the use of elastic waves to investigate residual coherence and mesoscopic phenomena. Laboratory study of ultrasonics in solids provides a convenient venue in which to investigate wave chaos, reverberation, and multiple scattering. The phenomena of incoherence, interference, and the perennial wave/ray duality, manifest themselves with elastic waves in solids as they do with microwaves, optics and electrons. But there are particularities of elastic waves also, differences that lend themselves to advantages and disadvantages

P. Sebbah (ed.), Waves and Imaging through Complex Media, 141–186.
© *2001 Kluwer Academic Publishers. Printed in the Netherlands.*

for the experimentalist, and to different perspectives for the theorist. This talk is intended to inform as to the advantages and disadvantages of tests using elastic waves, and to communicate some of the insights that have arisen from this work. The paramount differences lie, not in the classical, non quantum, nature of these waves, but in ready access to the time domain, transients are observable, and in the ability to isolate the sample from leads and antennas. As with other classical waves we also encounter a degree of absorption and a lack of incoherent scattering. All these points will be illustrated in what follows. The talk is divided into two parts. The first part consists of a brief course on continuum elastodynamics in solids. The basic theory, phenomenology, and applications of linear elastic waves are presented in a manner intended to be readily assimilated by a modern physicist or mathematician. The second part reviews the laboratory work on spectral and transport properties of diffuse ultrasonic waves in solids, with particular emphasis on Anderson localization in disorder and on the spectral and response statistics of reverberant enclosures.

1.   CONTINUUM ELASTODYNAMICS AND ULTRASONICS IN SOLIDS

The basic laws of linear elastic waves were in place in the first half of the 19th century. Navier, and Cauchy [1], established the role of the notions of strain and stress, and the Hooke-like constitutive law relating them. Poisson [2] demonstrated that these laws implied, in an isotropic medium, two uncoupled waves, each with its own wave speed. By way of historical context it is noteworthy that this led MacCullagh [3] in 1839 to demonstrate that further assumptions on the properties of the medium led to a wave equation which agreed with observations in optics, thus establishing the luminiferous ether as a mechanical model for light.

Modern courses in mathematical linear continuum elastodynamics are usually aimed at the ultrasonic practitioner, structural acoustician, or seismologist. Typical texts and references are those of Graff [4], Miklowitz [5], Achenbach [6], Truell *et al.* [7], Aki [8] or Auld [9]. The Miklowitz and Achenbach books emphasize fundamentals, and exact solutions for simple geometries, half spaces, plates, rods, and full spaces, the classic problem being "Lamb's problem," for the dynamic response of an isotropic elastic half space to a point load on the surface. Graff's book is similar, but with greater emphasis on approximate theories for plates and beams and rods. There is some small attention to scattering from isolated heterogeneities, e.g. spheres, in these books. The theory of the scattering of elastic waves is summarized in monographs and review articles. See for example, Pao and Mow [10] for scattering from separable surfaces, Achenbach and Gautesen [11] for application of Keller's geometrical theory of diffraction to elastic wave ray theory, and Waterman [12] for more direct numerical treatment of scattering from elastic inclusions and voids. Some discussion of the extensive literature on the theory of mean responses (averages of single powers of the Green's function) in multiple scattering elastic media may be found in Karal and Keller [13], in Stanke and Kino [14], and in Varadan *et al.* [15]. Diagrammatic, and related, theories for mean square responses in multiply scattering elastic media are found in Weaver [16], in Turner and Weaver [17] and in Ryzhik *et al.* [18]. This list is intended to reflect the main stream of elastic wave theory, and is not exhaustive. None of these books or papers address the diffuse field residual coherence issues that bring this series of lectures together. The much more limited elastic wave literature on such topics is reviewed in the next section.

Basic equations. A derivation of the wave equations of continuum linear elastodynamics begins with kinematic descriptions of the relation between displacement fields and strains, constitutive relations between strains and stress, and a statement of momentum balance. The chief dependent variable is a differentiable vector displacement field **u**. It is a function of independent variables position **x** and time t

$$\mathbf{u} = \mathbf{u}(\mathbf{x}, t).$$

Within the restriction to linear elastic waves we need not concern ourselves with the distinction between position **x** as an Eulerian or Lagrangian variable, that is, with the issue of whether **x** represents a fixed coordinate system or a coordinate system attached to material points. The distinction is subtle, but critical when seeking rational and consistent derivations in nonlinear elasticity [19].

Within the usual continuum mechanics assumptions, the action of material points on each other is local, ie., associated with a (tensor) stress field

$$\underline{\underline{\sigma}} \quad \text{or} \quad \sigma_{ij}(\mathbf{x}, t),$$

whose divergence gives the net internal force acting on a material point. Momentum balance then reads

$$\underline{\nabla} \cdot \underline{\underline{\sigma}} + \underline{f} = \rho \underline{\ddot{u}}$$
$$\text{or} \tag{1}$$
$$\sigma_{ij,j} + f_i = \rho u_{i,tt},$$

where f is a volumetric density of external body force and ρ is material mass density. A comma amongst the subscripts indicates a partial derivative: $()_{,j} = \partial_j() = \partial()/\partial x_j$ and $()_{,t} = \partial_t() = \partial()/\partial t$.

These equations are supplemented with an essentially phenomenological statement of the constitutive relationship between the stress and the displacement field. If the stress is to be independent of rigid body displacements, then it can depend only on spatial gradients of **u**. If it is to depend only locally on the field **u** then we presume that it depends only on the first spatial derivatives of **u**. If it is also to be independent of rigid body rotations, then it must depend only on the symmetric part of the displacement gradient. On linearizing, this means it depends only on (linearized) strain:

$$\varepsilon_{ij}(\mathbf{x}, t) = \frac{1}{2}[\partial_i u_j + \partial_j u_i] \,. \tag{2}$$

Locality demands that the stress depend on the strain at the same place. If we also demand that the stress depend on the strain at the same time, and not on the strain rate, then the most general linear constitutive relationship giving stress σ in terms of strain ε is a simple proportionality

$$\underline{\underline{\sigma}} = c \cdot \underline{\nabla}\, \underline{u}$$
$$\text{or} \tag{3}$$
$$\sigma_{ij} = c_{ijkl}\varepsilon_{kl} \,,$$

where the fourth rank tensor c is a material quantity called the stiffness tensor or Hooke's tensor. The symmetry of stress (and strain) imply that c is symmetric in the first two indices (and the last two indices). From the existence of a strain energy (or alternatively from path independence of work done) we can also conclude that c is symmetric between the first two and the last two indices: $c_{ijkl} = c_{klij}$.

If the material is isotropic, then its stiffness tensor c must be of the form

$$c_{ijkl} = \lambda\delta_{ij}\delta_{kl} + \mu\{\delta_{ik}\delta_{jl} + \delta_{il}\delta_{jk}\} \,, \tag{4}$$

where λ, μ are the so called "Lamé Moduli." Here μ is the familiar shear modulus; the equally familiar bulk modulus relating dilation to pressure is $\kappa = \lambda + 2\mu/3$. The Young's modulus E is $\mu(3\lambda + 2\mu)/(\lambda + \mu)$. Poisson's ratio $\nu = \lambda/2(\lambda + \mu)$ is the conventional measure of the ratio of moduli; all isotropic linearly elastic media are nondimensionalized in terms of their Poisson ratio $-1 < \nu < 1/2$. Most common materials have ν in the vicinity of 0.30. In anisotropic media (e.g. composite materials, single crystals, polycrystals with texture) Hooke's tensor is much more complicated.

The above equations may be combined into a single governing partial differential equation of motion

$$[c_{ijkl}(\mathbf{x})u_{k,l}(\mathbf{x}, t)]_{,j} + f_i(\mathbf{x}, t) = \rho(\mathbf{x})u_{i,tt}(\mathbf{x}, t) \,, \tag{5}$$

consisting of three simultaneous coupled Partial Differential Equations, for the three fields u_i. The equations may also be derived from least action

$$A[\mathbf{u}] = -\frac{1}{2}\int dt \int_V d^3x \{u_{i,j}(\mathbf{x}, t)c_{ijkl}(\mathbf{x})u_{k,l}(\mathbf{x}, t) - \rho(\mathbf{x})u_{,t}^2(\mathbf{x}, t)\} \,. \tag{6}$$

The principle of least action generates natural boundary conditions, to be applied at surface points lacking geometric constraints,

$$T_j(\mathbf{x}) = n_i(\mathbf{x})\sigma_{ij}(\mathbf{x}, t) = 0, \quad \mathbf{x} \text{ on } \partial V, \tag{7}$$

implying that the traction $\mathbf{T}$ (equal to the surface normal times the stress) must vanish on the surface. This boundary condition is the most common, as it represents conditions of an unconstrained, free, surface. The other obvious boundary condition is that of a rigid surface, such that u itself vanishes. This is not easy to effect in the laboratory and is therefore not as often studied. Other boundary conditions commonly suggest themselves also. These include the rigid-smooth boundary conditions in which one or two components of surface displacement are set to zero, the resulting natural boundary condition then requires the remaining component(s) of traction to vanish.

The above partial differential equation simplifies greatly if stiffness and density are independent of position, i.e. if the material is homogeneous,

$$c_{ijkl}u_{k,lj}(\mathbf{x}, t) + f_i(\mathbf{x}, t) = \rho\, u_{i,tt}(\mathbf{x}, t). \tag{8}$$

This may be compared to the more familiar scalar wave equation:

$$\text{Constant} \cdot \nabla^2 \Psi(\mathbf{x}, t) + f(\mathbf{x}, t) = \Psi_{,tt}(\mathbf{x}, t). \tag{9}$$

In homogeneous isotropic media the equations simplify further. In this case they are sometimes written in vector form

$$(\lambda + \mu)\nabla(\nabla \cdot \mathbf{u}) + \mu\nabla^2\mathbf{u} + \mathbf{f} = \rho\, \partial^2\mathbf{u}/\partial t^2$$
$$\text{or} \tag{10}$$
$$(\lambda + 2\mu)\nabla(\nabla \cdot \mathbf{u}) + \mu\nabla \times \nabla \times \mathbf{u} + \mathbf{f} = \rho\, \partial^2\mathbf{u}/\partial t^2.$$

The coupling of the three components of u remains. In this form, however, the coupling may be removed by the Helmholtz decomposition

$$\mathbf{u} = \nabla\Phi + \nabla \times \mathbf{H}, \tag{11}$$

in terms of scalar and vector displacement potential fields Φ and $\mathbf{H}$. The first term above is termed the dilatational part of u; the second term the solenoidal, or equivoluminal. This decomposition is always possible, but is not unique, nor always useful (for example it profits us little in anisotropic media). It is also global, (the potentials depend on u nonlocally). Nevertheless the decomposition is hallowed by many years of use. It also allows a quick derivation of uncoupled wave equations. Dropping the source term f, making the gauge choice $\nabla \cdot \mathbf{H} = 0$, and substituting Eq. (11) into Eq. (10), we find,

$$(\lambda + 2\mu)\nabla^2\Phi = \rho\, \partial^2\Phi/\partial t^2 \; ; \quad \mu\nabla^2\mathbf{H} = \rho\, \partial^2\mathbf{H}/\partial t^2, \tag{12}$$

in which it is seen that the displacement potential Φ, and the (Cartesian) components of $\mathbf{H}$, are governed by uncoupled scalar wave equations. The former is associated with a wave speed $c_{\text{dilatational}} = [(\lambda + 2\mu)/\rho]^{1/2}$; the latter with a wave speed $c_{\text{equivoluminal}} = [\mu/\rho]^{1/2}$.

Plane wave solutions.   The same result is obtained by seeking plane wave solutions of Eq. (10). We try

$$\mathbf{u} = \mathbf{U}(ct - \hat{\mathbf{n}} \cdot \mathbf{x}), \tag{13}$$

which is a plane wave of constant displacement profile, traveling in the $\hat{\mathbf{n}}$ direction at a speed c. On substituting in Eq. (10) we find

$$(\lambda + \mu)\hat{\mathbf{n}}(\hat{\mathbf{n}} \cdot \mathbf{U}'') + \mu(\hat{\mathbf{n}} \cdot \hat{\mathbf{n}})\mathbf{U}'' = \rho c^2 \mathbf{U}'', \tag{14}$$

where $''$ represents two derivatives with respect to argument. If the displacement $\mathbf{U}$ is in the $\hat{\mathbf{n}}$ direction (i.e. if wave is longitudinal), then Eq. (14) is satisfied for $c = [(\lambda + 2\mu)/\rho]^{1/2}$. For this reason the quantity $[(\lambda + 2\mu)/\rho]^{1/2}$ is also called the longitudinal wave speed. It is typically of the order of 5 or 6 mm/μsec in most materials, so the convenient frequency of 1 MHz corresponds to a 5 or 6 mm wavelength. If $\mathbf{U}$ is perpendicular to the $\hat{\mathbf{n}}$ direction (i.e. if the wave is transverse) then Eq. (14) is satisfied for $c = [\mu/\rho]^{1/2}$. For this reason the quantity $[\mu/\rho]^{1/2}$ is also called the transverse wave speed. It is typically of the order of 3 mm/μsec, so 1 MHz corresponds to a wavelength of about 3 mm. Nomenclature in regard to the distinction between the two wave types is not consistent. Amongst the terminology employed for the faster of the two are, as described above, 'dilational,' and 'longitudinal.' Thus the quantity $[(\lambda + 2\mu)/\rho]^{1/2}$ is often called c_d or c_L. The latter term is technically appropriate only for plane waves. This wave is also called the primary wave (c_p) as it is the first to arrive after a transient excitation, e.g. an earthquake. The subscript p is mnemonic because it can also be construed as standing for pressure or push. The slower of the waves is sometimes called c_e for equivoluminal, or c_s for solenoidal or secondary or shear or shake. It is more commonly called transverse, c_T. The plethora of terminology is sometimes abandoned in favor of the refreshingly austere c_1 and c_2 (Table 3.1).

Longitudinal (c_L)	Transverse (c_T)
Dilatational (c_d)	Distortional or Equivoluminal (c_e) or Solenoidal (c_s)
Primary (c_p)	Secondary (c_s)
Pressure c_p	Shear c_s (also Shear-horizontal SH, and Shear-vertical SV)
push c_p	shake c_s
c_1	c_2

Table 3.1 .   Nomenclature

That any direction $\hat{n}$ is associated with plane waves of different speeds does not depend on the assumption of isotropy. In the more general case (8) also we can seek solutions in the form of plane waves

$$u_i = U_i \ h(\hat{n} \cdot \mathbf{x} - ct). \tag{15}$$

Substitution into Eq. (8) leads to the Christoffel equations [9]

$$[c_{ijkl}n_jn_l - \rho c^2 \delta_{ik}]U_k = 0, \tag{16}$$

showing that ρc^2 is an eigenvalue of an $\hat{n}$ dependent symmetric 3×3 matrix. Thus for each plane wave direction $\hat{n}$ there are three wave speeds, each associated with one of a set of three mutually perpendicular polarization vectors $\mathbf{U}$. These vectors are not necessarily transverse or longitudinal with respect to the wave direction $\hat{n}$. The waves are therefore often called pseudo-longitudinal and pseudo-transverse. That elastic wave speed in an anisotropic medium varies with direction, and does so in a moderately complex manner (more complicated for example than it does in electromagnetics) leads to some fascinating and non-intuitive behaviors. Phonon focussing, beam steering, and internal diffraction are amongst these [20].

Plane waves at free surface.   The Helmholtz decomposition decouples the vector wave equation in the interior of an isotropic homogenous medium. But the dilational and equivoluminal waves that are uncoupled in the interior are in general coupled at surfaces. A plane longitudinal or transverse wave incident upon a plane surface will in general reflect with a degree of mode conversion. An exception to this occurs in the case of rigid-smooth boundary conditions, but the overwhelmingly more common case of a traction-free surface leads to a significant amount of mode conversion. Another exception occurs if a plane shear wave is incident upon the surface with its material displacement vector polarized in the plane of the surface, (termed SH or horizontally polarized shear wave by the seismologists for whom the surface is usually horizontal.) Symmetry readily establishes that such a wave must reflect without mode conversion (Fig. 1a). Figures 1b and 1c illustrate the incidence of plane longitudinal waves, or plane shear waves with displacement having a vertical component (SV), incident upon a homogeneous surface. The reflected waves consist in general of outgoing L and SV waves. Snell's law applies and the angles of reflection are related in a simple manner to the angles of incidence and to the ratios of the wave speeds. The amplitudes of the reflected waves are determined by invocation of the boundary conditions. That two waves are to be expected is readily confirmed after realizing that there are two nontrivial conditions to be satisfied at the surface, for example the vanishing of the vertical and horizontal components of the surface traction. Of parenthetical interest perhaps is the case of total internal reflection of SV waves incident at angles from the normal greater

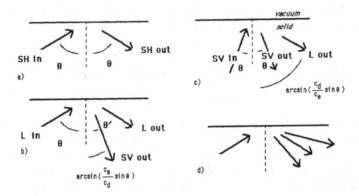

Figure 1.   Plane waves incident upon a homogeneous planar surface will in general reflect with mode conversion.

than a critical angle (typically about 30 degrees) in which the mode converted longitudinal wave evanesces near the surface and carries no energy flux into the interior. In the more complicated case of anisotropic media there are in general three non trivial conditions to be satisfied and three outgoing plane waves (Fig. 1d).

The reflection coefficients for isotropic media, including mode conversion coefficients, are given in many texts [4-6, 9] in closed form, in terms of the material Poisson ratio and the angle of incidence. Plots are also available. The reflection matrix may be put into the form of an S matrix, and the familiar properties of time reversal invariance (symmetry) and energy conservation (unitarity) demonstrated [21] with consequences for detailed balance and equipartition of a diffuse wave field [21-23]. It is worth emphasizing that the mode conversion coefficients are not particularly small. A significant fraction of the energy incident in one form will be converted to the other form. As a consequence, a wave field of an age such that at least a few reflections must have occurred can be expected to be a thorough mixture of shear and longitudinal waves. A diffuse wave field, or an isolated high-order mode of a finite elastic body, is expected to be equipartitioned between the wave types.[22, 23]

Rayleigh Surface waves.   With the right boundary conditions (and traction-free conditions on an isotropic body are amongst the right ones but rigid-smooth and rigid are not) it is also possible to find solutions which propagate along the surface, and evanesce into the interior. The best known of these (and only such solution in the case of an isotropic medium and a traction free surface) is the Rayleigh surface wave, with $c_R \approx 0.93\ c_T$. The precise value of c_R/c_T depends on Poisson ratio. This solution of the elastic wave equations was discovered (Rayleigh in 1887 [24]) long after the governing equations were accepted. The wave can be thought of as a superposition of evanescent SV and L waves in continuous mode conversion into each other. Material motion in a Rayleigh surface wave is (on the surface) a retrograde elliptic motion. At greater depth the motion is in the opposite sense, see Fig. 2. The displacement field is composed of both **H** and Φ fields, which for the time-harmonic case diminish exponentially with depth.

$$\Phi \sim exp(-\alpha z) \ \text{ with } \ \alpha = (2\pi/\lambda)[1 - c_R^2/c_L^2]^{1/2}$$
$$\mathbf{H} \sim exp(-\beta z) \ \text{ with } \ \beta = (2\pi/\lambda)[1 - c_R^2/c_T^2]^{1/2} \, .$$

The **H** field decays with depth more slowly than does the Φ field, with an e-folding distance of about half a wavelength. Surface waves are sensitive to material properties only within about this distance of the surface.

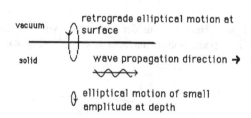

Figure 2. The depth of penetration of a Rayleigh surface wave is of the order of a wavelength, thus they respond to the properties of the material only within about a wavelength of the surface.

Surface waves are generated by mode conversion from bulk waves incident on flawed surfaces or corners (Fig. 3). They are not generated by bulk waves incident on homogeneous surfaces. They are also readily generated by dynamic loads, but only those acting within about a half wavelength of the surface. Owing to their concentration of energy near the surface these waves are often the more destructive parts of an earthquake. They are also the more slowly propagating, a fact that allows the detection of a P wave to serve as an early warning (some 10's of seconds typically) of an earthquake.

Waveguides, e.g. a Plate. As with microwaves in pipes, acoustics in ducts, and optics in fibers, elastic waves can propagate in waveguides. Chief amongst the elastic waveguides of interest are plates and rods, in part due to their ubiquity in structural applications. The simplest elastic waveguide is an infinite isotropic

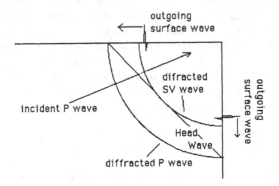

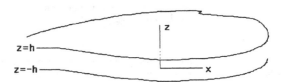

Figure 3.   A plane bulk wave incident on a corner or a flaw near the surface will specularly reflect and mode convert to other bulk waves. It will also diffract, and mode convert to surface waves [25]. The specularly reflected parts are not indicated in this figure.

homogeneous plate. Here we consider such a plate with traction-free boundary conditions on its top and bottom surfaces at $z = \pm h$. (Fig. 4)

Figure 4.   The simplest elastic waveguide is a homogeneous plate. The guided waves in such a structure are termed Lamb waves.

One may choose to analyze elastic waves in such a structure as composed of obliquely propagating plane waves which multiply reflect at the top and bottom surfaces. If they are SH waves they do so without mode conversion. The behavior of SH waves is relatively simple, the mathematics being equivalent to that of scalar waves in a plane waveguide. If we picture an initial obliquely travelling P or SV ray, however, we determine that there will be multiple reflections and mode conversions at plate surfaces. After a few such reflections an initial single ray has generated a large number of other rays. A very complicated situation arises very quickly.

Except at the shortest of distances, it is better to consider instead a guided-mode picture in which fields are sought in the form:

$$\Phi = A \cos \alpha z \exp\{i\omega t - ikx\}$$
$$\mathbf{H} = \hat{\jmath} B \sin \beta z \exp\{i\omega t - ikx\}$$
$$\text{(symmetric waves)}$$

(17)

or

$$\Phi = C \sin \alpha z \exp\{i\omega t - ikx\}$$
$$\mathbf{H} = \hat{\jmath} D \cos \beta z \exp\{i\omega t - ikx\}$$
$$\text{(anti-symmetric waves)},$$

(18)

where the vertical wavenumbers are $\alpha = [\omega^2/c_L^2 - k^2]^{1/2}$ and $\beta = [\omega^2/c_T^2 - k^2]^{1/2}$. It may be noted that the curl of $\mathbf{H}$, and the gradient of Φ, (and therefore the corresponding displacement vector $\mathbf{u}$) lie in the x-z plane. The up-down reflection symmetry of the structure assures the decoupling of the symmetric from the antisymmetric fields. Vanishing traction at the free surfaces requires a specific ratio between A and B (or C and D) and also requires satisfaction of a dispersion relation:

$$\frac{\tan \alpha h}{\tan \beta h} = -[\frac{4k^2 \alpha \beta}{(k^2 - \beta^2)^2}] \quad \text{anti-symmetric waves},$$
$$\frac{\tan \alpha h}{\tan \beta h} = -[\frac{(k^2 - \beta^2)^2}{4k^2 \alpha \beta}] \quad \text{symmetric waves}.$$

(19)

These are the Rayleigh-Lamb dispersion relations. They are transcendental relationships giving k as an implicit function of ω, or vice-versa. There are an infinite number of solutions ω for fixed k; the solution is multi-branched. Each solution corresponds to a specific k and ω value, and a specific ratio of A to B (or C to D), ie, to a specific displacement field called a "Rayleigh Lamb" mode. Typical numerical solutions to the above relations are given in a plot like that of Fig. 5. As with guided microwaves and optics, each branch of the dispersion relation is dispersive. These Rayleigh-Lamb modes of propagation in a plate are studied extensively in ultrasonics. It is not difficult to show that a plane Lamb wave incident upon a plate edge or other heterogeneity will in general mode convert with all other Lamb waves, and with the SH waves. Unless the plate edges have rigid-smooth boundary conditions, the Rayleigh-Lamb modes of an infinite plate are not eigenmodes of vibration of a thick finite plate.

In the limit that the plate is thin compared to a wavelength all but the three lowest branches fail to propagate. In the lower left corner of Fig. 5 one can identify the three remaining waves as familiar structural-acoustic waves.

$$\omega = c_{plate}k; \quad \omega = c_e k; \quad \omega = Bhk^2.$$

(20)

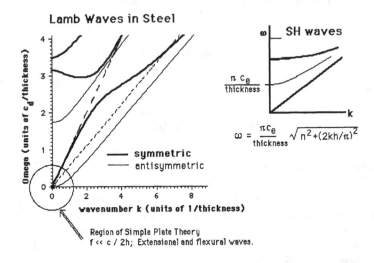

Figure 5.   The solution of Eq. (19) for the case of Poisson ratio = 0.28. The dashed lines indicate the relations $\omega = kc_e$ and $\omega = kc_d$. At low frequencies the relations asymptote to $\omega = kc_e$ (for the SH wave), to $\omega = kc_{plate}$, with $c_{plate} \approx 0.9\, c_d$ (for the symmetric extensional wave at low frequency) and to $\omega = Bhk^2$ for the antisymmetric flexural wave. $B = c_{Bar}[3(1 - \nu^2)]^{-1/2}$, with c_{Bar} being the "bar wave speed" = $\sqrt{E/\rho}$ typically 5 mm/μsec.

One of them has $\omega = c_{plate}k$ (c_{plate} typically $\approx 0.90\, c_L$) and is sometimes called an extensional wave. Material motion in this wave is mostly longitudinal, but there is small amount of motion normal to the plate surface associated with the Poisson effect in which longitudinal extensions are accompanied by lateral contractions. There is another nondispersive wave in this limit, one with $\omega = c_e k$. This is an SH wave with material motion entirely within the plane of the plate. There is a flexural wave as well, highly dispersive, with ω proportional to k^2 ; its motion is largely normal to the plate surface. The higher branches are unimportant (except perhaps as they evanesce near edges and flaws and sources) as long as frequencies are low enough that they cannot propagate. The first cutoff for the higher modes occurs at $\omega = \pi c_e/2h$, corresponding to 1 MHz in a $2h$ =1.5 mm thick plate. Significant deviations from the simple dispersion relations (20) occur, however, at frequencies less than this. More extensive discussion of structural waves may be found in Graff [4]. Other wave guides, e.g. multi layered plates, or rods, also have multi-branched dispersion relations. At low frequencies they also become familiar structural waves. In a rod at long wavelength, for example, there are four propagating structural waves: two flexural waves, one extensional wave, and one torsional wave.

Self-adjointness and modes.   In a finite elastic body, and under a variety of boundary conditions derivable from a least action principle, the elastic wave equation has a complete set of orthogonal vector valued eigenfunctions $\mathbf{u}(\mathbf{x})$. As in the Helmholtz equation for a scalar wave with Dirichlet or Neumann boundary conditions, and as for E&M waves with appropriate boundary conditions, our governing equations admit normal modes. The vector-field valued linear operator L which acts on a vector field $\mathbf{u}$ in a finite body that satisfies either of the typical boundary conditions (or many others) is *self-adjoint*

$$(L\,u)_i = [c_{ijkl}(\mathbf{x})u_{k,l}(\mathbf{x})]_{,j}\, /\, \rho(\mathbf{x})\,, \tag{21}$$

with respect to the inner-product

$$< \mathbf{u}\mid \mathbf{v} > = \int_V u_j(\mathbf{x})v_j(\mathbf{x})\rho(\mathbf{x})d^3\mathbf{x}\,. \tag{22}$$

This allows one to conclude that the eigenvalue problem

$$[c_{ijkl}(\mathbf{x})u_{k,l}(\mathbf{x})]_{,j} = \lambda\rho(\mathbf{x})u_i(\mathbf{x})\,, \tag{23}$$

has a countably (in a finite body) infinite spectrum, with complete orthogonal eigenfunctions, and eigenvalues λ that may be identified with the natural frequencies of vibration of the body

$$\omega_r^2 = -\lambda\,.$$

The eigenfunctions $\mathbf{u}^r$ are real vector-valued functions of $\mathbf{x}$. It is clear that there is no particular reason for them to be purely longitudinal, or transverse, or surface. On the contrary, they are generally going to be mixtures of all types of waves.

The scalar wave equation separates in a number of coordinate systems; correspondingly the eigenmodes of the Helmholtz equation are known in a number of simple geometries. For the far more complex case of a finite three dimensional elastic body with traction free surfaces, the only case in which the elastic eigenmodes are known in closed form is the case of an isotropic elastic sphere [26]. There is a small industry engaged in the numerical calculation, and laboratory measurement, of the eigenmodes and eigenfrequencies of other elastic objects, with the purpose of characterizing materials and objects by their resonant ultrasonic spectra [27].

Detection and Generation of ultrasound in solids.   The standard choice for the generation and detection of ultrasound is contact piezoelectrics. Most commercial transducers are of this kind. The last few decades have seen advances in synthetic piezoelectric materials; modern commercial transducers are in consequence robust and sensitive and easy to use. Standard commercial transducer designs can be, however, poor choices for investigations of wave chaos and residual coherence. They are usually built to have large aperture to wavelength ratios, so as to have narrow beam patterns that may or may not be an experimental complication. In order to more easily interpret signals in terms of material displacements, point piezoelectric devices are sometimes preferred [28-30]. The large contact area of conventional transducers has an added disadvantage as the transducer becomes a more invasive device, one that modifies the structure and adds losses to the system. Liquid couplants (water or gel or light oil in the case of a normal displacement sensor, honey or other very viscous material in the case of a sensor of in-plane motions) add considerable dissipation. The experimental work that will be cited here is mostly restricted to measurements using dry-coupled point piezoelectric detectors.

Contact piezoelectric devices are not easily scanned in space. Often scanning is done in immersion in a water bath [31], an approach that is usually counter-indicated in diffuse field work, as the lifetime of a diffuse wave field against leakage to a water bath is short.

Ultrasonics in solids is also studied using non contact laser techniques. As a source, a Q-switched YAG (optical pulse widths of the order of a few nanoseconds, beam diameters of about a mm) will generate ultrasound by thermoelastic mechanisms or by ablation. At low fluences (< 5 mJ) the source mechanism is thermoelastic; the heating of the surface leads to a sudden thermal expansion that emits elastic waves [32]. At larger fluences the surface boils and the consequent recoil momentum applies an impulsive normal force to the sample. The latter mechanism is far more efficient but damages the surface. The chief virtues of the technique, relative to piezoelectrics, are that it is scannable and moderately well characterized. There are fluctuations (in laser power) from pulse to pulse that make the technique not entirely reproducible. The technique generates transients only, not harmonic signals. Lock-in techniques are therefore not available for improving signal to noise ratios.

Laser-interferometers provide a non contact method for detection. Their virtues (in comparison with piezoelectrics) include that they are well-calibratable; signals can be understood in terms of surface displacements, in nanometers. They are also non-invasive; there are no losses associated with detection by interferometer. They are easily scanned in space. Their chief faults are that they are much less sensitive than piezoelectrics, noise floors of the order of a fraction of a nanometer per $\sqrt{\text{MHz}}$ of bandwidth are not uncommon. By way of comparison an early study by the author [33] quoted numbers using water-coupled point piezoelectrics equivalent to a noise floor of 10^{-3} nanometer per $\sqrt{\text{MHz}}$. Another difficulty is cost; off-the-shelf laser ultrasonic systems are many tens of thousands of US dollars, the majority of which is related to the detector.

The difficulties involved in using lasers, and a continuing demand for non-contact methods, has led to a certain amount of work in air-coupled ultrasonics. The enormous impedance mis-match between air and typical solids, and between typical piezoelectric materials and air, and the fact that a signal must pass each of these interfaces twice, results in very weak signals. This has been mitigated to some extent by recent advances in device design, but insertion losses remain high.

Electromagnetic-Acoustic Transducers (EMAT's) offer an alternative non contact method. A time varying magnetic field in the vicinity of a conductor provides a mechanical force on the material by means of the Lorentz force on the eddy currents. This generates SH waves near the conductor's surface. The effect is reciprocal and can be used for detection as well. Like other non-contact methods it suffers from poor sensitivity in comparison with contact piezoelectrics. Some materials (e.g. Nickel) have a significant magnetostrictive coefficient by which strain and magnetic field are coupled. The effect is not commonly used.

Capacitive sensors are also used. These detect normal surface displacement by detecting the variations in the charge on a capacitor, one face of which is the sample surface. They are not, in the sense needed here, "non-contact," as the device is usually supported by contact (at positions away from the sensor itself) with the specimen. That contact, while not disturbing the interpretation of the signal as proportional to mechanical displacement, can disturb the sample and lead to extra dissipation. They are much less sensitive than contact piezoelectrics.

In all cases frequencies are such that modern digitizers, with digitization speeds routinely exceeding 100's of MSamples/second, can repetition-average (if the source is reproducible), thereby enhancing signal to noise ratios.

There are a few additional methods for the generation of ultrasound in solids that are worth noting. One of the simplest and least expensive is the breaking of a mechanical pencil lead [34]; the step surface load that results when the lead fractures has a rise time of the order of a microsecond and is of amplitude comparable to that of pulsed contact piezoelectrics; it is somewhat reproducible. Slightly higher amplitude, and with a faster rise time, is the fracture of a thin glass capillary [35], by compressing it with a razor blade. A continuous wide-band noise source has been shown to be obtained from a tiny turbulent jet of helium gas [36]. The impact of a ball-bearing (diameter $\sim$1 mm) is another low-tech method. These and other non-piezoelectric source mechanisms are reviewed in [37].

Beyond Linear Elastodynamics, real material behavior.   Laboratory specimens are not perfectly elastic, homogenous, or isotropic. It is necessary to understand deviations from these idealizations. The more important, or more obvious, deviations include those due to absorption, nonlinearity, microstructure, and the modulation of properties by other fields.

Absorption is a chronic complication. It limits the range of time over which a transient signal can be studied; it limits the sharpness of resonances, and diminishes signal to noise ratios. Those effects of absorption are, however, essentially only nuisances. More problematic is the, as yet not thoroughly investigated, effect of absorption on the more intriguing predictions of multiple scattering and wave chaos theory, e.g. level statistics, mode amplitude statistics, and localization.

Losses to air are usually negligible. Acoustic impedance mismatches between air and typical solids are large. Between air and aluminum for example, it is $\sim 4 \times 10^4$. Thus a typical ray will reflect from an aluminum/air interface a number of times of order 40,000 before escaping into the air. Loss to air is important only for the most reverberant materials. For water, on the other hand, the impedance match is much closer. Diffuse elastic waves in a solid in immersion will escape after a number of reflections of the order of 10.

Losses are in practice usually dominated by some variety of internal friction; modelers typically invoke thermo-elastic mechanisms or dislocation drag or molecular relaxation. In pure materials and in single crystals, loss mechanisms have been studied at great length for many decades. There is a series of "International Conferences on Internal friction and ultrasonic absorption in solids" in which many such investigations are reported. In engineering materials containing internal heterogeneities (cracks, precipitates, crystallites) a variety of incompletely understood loss mechanisms can operate. The author's investigations, while not primarily aimed at understanding absorption, have concluded that losses in his aluminum samples are increased by machining, diminished by polishing and cleaning, unaffected by heat treatments, increased by surface contaminations, unaffected by temperature between -30 and +30° C, and approximately proportional to the ratio of Surface Area to Volume. He finds quality factors Q in the range from 10^4 to 10^5. In iron and copper alloys he finds much lower Q's. In alkali glasses the Q's are also low, $\sim 10^3$. In plastics they are typically less than 100. Ultrasonic Q's in single crystals of quartz [38] can be greater than 10^6, at cryogenic temperatures Q's higher than 10^9 have been reported [39].

Elastic wave losses are phenomenologically modeled by linear viscoelasticity, that is, by modifying the continuum constitutive relationship giving the stress in terms of the history of the local strain.

$$\sigma_{kl}(t, \mathbf{x}) = \int_{-\infty}^{t} C_{klij}(t - \tau, \mathbf{x})\varepsilon_{ij}(\tau, \mathbf{x})d\tau . \qquad (24)$$

The time-dependent fourth rank tensor C is a dynamic stiffness. On Fourier transformation of the above one finds

$$\tilde{\sigma}_{kl}(\omega, \mathbf{x}) = [\mathrm{Re}\tilde{C}_{klij}(\omega, \mathbf{x})]\tilde{\varepsilon}_{ij}(\omega, \mathbf{x}) + i\,[\mathrm{Im}\tilde{C}_{klij}(\omega, \mathbf{x})]\tilde{\varepsilon}_{ij}(\omega, \mathbf{x}) , \qquad (25)$$

showing that the proportionality between strain and stress is retained, but that it becomes complex and (weakly) frequency dependent. In practice, for band-limited processes, one usually neglects the frequency dependence and merely replaces the usual Hooke's tensor with a complex quantity. The imaginary part represents the dissipation. The reduced elastodynamic wave equation then becomes

$$[\mathrm{Re}\tilde{C}_{ijkl}(\mathbf{x})u_{k,l}(\mathbf{x}) + i\,\mathrm{Im}\tilde{C}_{ijkl}(\mathbf{x})u_{k,l}(\mathbf{x})]_{,j} = -\omega^2\rho(\mathbf{x})u_i(\mathbf{x}) . \qquad (26)$$

The formalism is entirely analogous to that invoked in nuclear reaction physics or in billiards with open channels [40] in which the Hamiltonian gains an imaginary (and weakly energy dependent) part that makes the system non-self-adjoint.

Losses into the transducers often must be quantified; we have found them to be minimizable by use of small dry contacts. Losses into supports are similar. We typically use sharp pins, in turn supported by a layer of foamed plastic.

At many times we have found ourselves speculating on the possible relevance of material nonlinearities, in particular when a laboratory result is unexpected. In principle there are nonlinear corrections to the constitutive relation Eq. (3) between stress and strain. There are also potential kinematic nonlinearities (in most materials these are of less importance) related to the convection of momentum by material displacement. In pure materials typical third order elastic constants (the second derivative of stress with respect to strain) are of magnitude similar to that of the second order elastic constants. This implies that the effect of nonlinearity is insignificant except after a number of cycles of the order of the inverse of the rms strain. Commonly used strain levels in the experiments to be quoted are less than 10^{-6}, sometimes much less. Thus one would not anticipate nonlinear behavior. On the other hand, in materials with what is sometimes called "mesoscopic nonlinearity," e.g. rock, and other micro-cracked bodies, nonlinearity can be much more significant, and is sometimes investigated with a view to Nondestructive evaluation of materials [41]. In view of the consequent uncertainty as to the relevance of nonlinearity, and in view of the ease with which tests of its importance can be conducted, we usually confirm its unimportance by conducting tests at a variety of input amplitudes. We have yet to detect any effect of that variation in our studies in aluminum and other simple materials, except of course for a worsening of signal to noise ratios as the input pulse amplitude is diminished.

Laboratory specimens also have, in principle, complex microstructure that is not in accord with the usual presumption of a homogeneous interior. Typical metals are polycrystalline, with grain sizes 1 to 50 micron. This implies that frequencies below 10 MHz are virtually unscattered, and a ray-picture is valid. After long times, however, or at shorter wavelengths, or for large-grained materials, geometrical optics can be misleading; rays scatter. Another consequence of the polycrystalline nature of the material is that preferred orientations of grains (created perhaps by cooling stresses or rolling) can induce an anisotropy that is usually weak, but nevertheless complicates a simple ray-picture. Glasses, or perfect crystals, do not have a scattering microstructure. Composite materials can have strongly scattering microstructure, but usually have significant internal friction as well.

There is at least one more fashion in which real ultrasonics differs from the idealized model. Material properties, most importantly stiffness, can vary with the application of an external field. The temperature dependence of wave speeds is well known. Unlike the case in scalar acoustics in which there is only one wave speed and temperature fluctuations merely rescale time, in an elastic body in which shear and longitudinal waves have different dependencies on temperature, a new temperature can be equivalent to a new sample [38, 42]. The Copenhagen group [30, 38] has speculated that variations in elastic stiffness of quartz with applied electric field can play a similar role, and may even allow diabolic points in a spectrum to be explored by simultaneous variations of two applied electric fields.

Applications.   This is not the place for a thorough review of the many technical applications of ultrasonics in solids, but a brief list is not unwarranted. The best known is flaw detection, in which pulse-echo techniques are used to identify, and sometimes characterize, cracks and inclusions in structures. This is especially challenging in materials with scattering microstructures, in materials with significant anisotropy and in specimens with rough or curved surfaces. The characterization of materials, by means of measurements of wave speed, attenuation and backscatter, and the characterization of interfaces by measurements of reflections and transmissions is another important engineering application. Analysis of Acoustic Emission signals, from brittle crack growth or other small scale stress release mechanisms (domain wall motion, and twinning for example) has long been explored in attempts to understand the sources. The best known example of this is in seismology, where interpretation of received waveforms in terms of source mechanisms is severely complicated by the complex medium in which the waves propagate. Recent years have also seen a number of developments in which small sensors take advantage of dependencies of ultrasound on the environment [43].

We have found that ultrasound in certain high Q solids (aluminum and quartz mostly) lends itself well to laboratory studies of wave chaos and multiple scattering. At frequencies of the order of MHz signals are directly accessible in the time domain. Waveform digitization at these speeds, even of long duration signals, has become inexpensive in recent years. At these wavelengths it is not difficult to construct samples that are large compared to wavelength, and yet have scatterers with size comparable to wavelengths. The next section reviews the recent history of ultrasonics in solids as applied in investigations of wave chaos and localization.

2. DIFFUSE FIELDS

2.1. SPECTRAL PROPERTIES OF THE HIGHER MODES OF GENERIC FINITE BODIES

Diffuse fields, level densities and equipartition. It was appreciated at an early stage that a diffuse ultrasonic field should experience a degree of equipartition. Davis Egle, in 1980 [22] argued that the large ratio between typical P to S and S to P mode conversion coefficients at a free surface should lead to a multiply reflected ultrasonic field being quickly dominated by S waves. Weaver pointed out [23] that this result is independent of the details of the boundary conditions and may be obtained by a comparison of the densities of the (pseudo) modes of longitudinal and shear type. Standard mode counting procedures give what may be termed the first term of a Weyl-series for elastic waves in an isotropic body. The modal density is proportional to Volume, and to the square of frequency.

$$\frac{dN}{d\omega} = \frac{\omega^2 V}{2\pi^2}[2/c_T^3 + 1/c_L^3] + \cdots.$$

This shows that shear waves constitute about 93% (!) of the energy of a diffuse field. The longitudinal waves are almost irrelevant. Similar arguments show that the number of modes associated with Rayleigh surface waves is

$$\frac{dN_{\text{surface waves}}}{d\omega} = \frac{\omega S}{2\pi c_R^2}.$$

Here S is the free surface area of the solid.

More detailed calculations [44] have shown that approximately 70% of the mean square normal displacement at the free surface of a thick body with an equipartitioned diffuse field is in the form of Rayleigh waves. Thus most of what we detect in such samples is Rayleigh waves. In-plane surface displacements, on the other hand, are dominated by the SH waves. It must be noted that these Rayleigh waves do not form their own, independent, diffuse field, but have their origins in mode conversion at corners and flaws. Mode conversion between surface and bulk waves is generally rapid.

The next term in the Weyl series includes the Rayleigh waves, but also includes other effects, as the surface modifies the local density of states of the bulk waves as well. It is proportional to surface area and has been calculated [45]

$$
\begin{aligned}
\frac{dN}{d\omega} = D(\omega) = \frac{t_{Heisenberg}}{2\pi} &= \frac{V}{2\pi^2}[\omega^2/c_L^3 + 2\omega^2/c_T^3] \\
+ \frac{S\omega}{8\pi c_L^2}[\frac{2 - 3(c_L/c_T)^2 + 3(c_L/c_T)^4}{(c_L/c_T)^2 - 1}] &+ O(\frac{L}{c}),
\end{aligned}
\tag{27}
$$

but would be different if the boundary conditions were different. The third term in the series has not been calculated. These considerations imply that typical actual normal modes of such bodies are in the bulk about 93% Shear and about 7% longitudinal. On the surface, their normal displacements are mostly Rayleigh. The strength of typical mode conversion coefficients suggests that there will be little fluctuation away from these numbers.

Equipartition amongst ultrasonic elastic waves has been investigated in a thick plate [46], where theory predicts interesting structure as a function of frequency. An impulsive (step) point force (effected with broken pencil leads) was applied normally to the surface of a finite flat plate. Figure 6 describes the measurement. The signal contained useful components up to half a MHz. The response was found to reverberate over long times. The absorption time was tens of msec. This may be compared to the transit time (L/c) of 100 μsec. These frequencies and time scales were such that waveforms could be captured by a digital waveform recorder. In contrast to the microwave and optical and electronic cases, time-domain waveforms are easily accessible.

At frequencies of 500 kHz, in an aluminum plate of 12 mm thickness, and 30 cm width, there are several guided Rayleigh-Lamb modes that can propagate. The theory of the generation of the various Lamb waves by a surface step load is straightforward, albeit non-trivial [47]. The theory of reflections and mode conversions of the Lamb waves at plate edges is virtually unexplored. Nevertheless the theory for the mean power spectrum for the equipartitioned long-time-scale response is simple [48]. It is not difficult to understand one of the key results of that theory: At those distinct frequencies at which guided Lamb waves have their cutoffs we may expect interesting discontinuities in power density, related to discontinuities in modal density. Figure 7 superposes theory and measurements and shows that the essential predictions are met. Equipartition is taking place.

It is worth emphasizing that the equipartition taking place here is rather different from that of conventional statistical mechanics where a phase-destroying

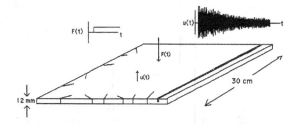

Figure 6.   Sketch of the experimental system used to confirm certain predictions of equipartition for an elastic plate. A step load F(t) was applied at one point, and the displacement signal power spectral density determined at anther point. A groove along one side served to break the up/down symmetry and mix the symmetric and antisymmetric Lamb waves; slits on other sides served to diffusely scatter the waves.

thermal bath mixes the modes and leads to exchange of energy amongst modes of different natural frequencies. Here the equipartition is only amongst (pseudo) modes with frequencies within a short range of each other, a range comparable to the inverse of the transit time L/c. This distinction is discussed further elsewhere [48, 49].

Level Correlations.   Elastic wave systems are complicated. Nevertheless, their level densities, as seen above, seem to conform to simple theoretical expectations. Do the higher-order eigenstatistics do so also? In 1989 we investigated the level-spacing and level-rigidity statistics of aluminum blocks [50]. Spectra were obtained by Fourier transforming transient waveforms, with frequency content of a few hundred kHz, and durations of 10's of msec. Signals were detected using oil-coupled pin transducers as detectors. Signals were generated by the impact of small ball bearings. Again the time scales were such that signals could be recorded in the time domain and then Fourier transformed. In each block a number, about 150, of resonant peaks were sufficiently distinct to allow assessment of eigenfrequency density and eigenfrequency correlations. The observed density was found to fit the two-term Weyl series (27). Of greater interest was a comparison of eigenfrequency repulsion and spectral rigidity with the predictions of Random Matrix Theory. We speculated that the complex dynamics of these bodies, with mode conversion amongst bulk waves at the surfaces and edges, and amongst the bulk and surface waves at the corners, would lead to a dynamics sufficiently complicated that it appeared statistically equivalent to that of a matrix from the GOE. Three different bodies were studied. Blocks 'A' and 'B' (see Fig. 8) had several saw cuts that served to break parity symmetries and whose tips and corners would generate extra mode

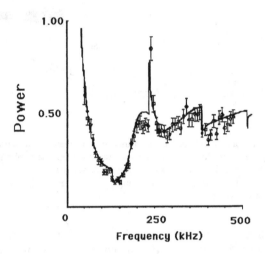

Figure 7.   The power spectral density observed in vertical surface displacement on a plate
of finite thickness shows discontinuities and singularities related to the discontinuities in modal
density at the various cutoffs of the Lamb modes. Solid line is theory [48]. Data points are
measurements [46].

conversion scattering. It was conjectured that the spectrum should be a GOE
spectrum. Block 'C' retained one reflection symmetry so it was conjectured
that the spectrum should be a superposition of two GOE spectra of equal mean
density.

   Results were found to be in accord with the predictions of these speculations.
Level repulsion and spectral rigidity were in excellent agreement with the predictions of Random Matrix Theory. It was subsequently pointed out [51] that
these objects are nominally pseudo-integrable, and that the ray dynamics is not
obviously chaotic. That is certainly a correct description of an idealization of
the objects as rectangular solids with planar slits. It is not necessarily a correct description of the actual bodies in which the slits have finite thickness and
curved ends. In any case it has since then become clear that pseudo-integrable
bodies with sufficient number of diffuse point scatterers (e.g. corners, slit tips)
have apparent GOE level statistics over considerable range.

   Similar work has been conducted by the Copenhagen group [30]. In Quartz
at a Q of 10^5, and using two sapphire styli bonded to piezoelectric chips as
dry-coupled point transducers, and swept CW insonification, they identified
about 1400 resonant frequencies between 600 and 900 kHz. As the size of a

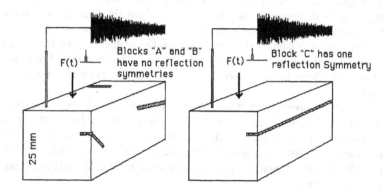

Figure 8. Sketch of aluminum bodies used for the first investigation of elastic wave eigen-frequency correlations. Impulse loads F(t) applied by the impact of small ball bearings led to transient signals with ring down times of order 10 msec, and frequencies up to 300 kHz.

removed octant in the corner was increased, a gradual transition was observed corresponding to the gradual loss of the (sole) reflection symmetry of the original block. For a range of octants of small size (radius 0.0 to 1.7 mm) level repulsion increased with octant size as the flip symmetry was slowly broken. Spectral rigidity increased also but remained that of a pseudo-integrable system (linear at long range.) This is expected based on a ray picture in which the elastic rays mode convert and diffract in a complicated fashion at the plane faces and corners, but are not chaotic. For a large defocussing octant of 10 mm radius (and offset so that its center is slightly within the block's original volume) the spectrum showed the full rigidity of the GOE.

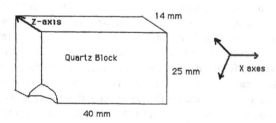

Figure 9. Sketch of the quartz block used for a later study [30] of elastic wave level correlations.

More recently the group has investigated the "dynamics" of the spectrum as temperature is varied [38] and found agreement with RMT predictions in regard to level curvature. They have suggested that simultaneous variations of two parameters (e.g. quasi-dc electric fields and/or temperature) might allow investigation of "diabolical Lasagna," and scrutiny of diabolical points where the sheets of the Lasagna meet at double-cones. In particular it should be possible to study the Berry phases incurred as external parameters circumnavigate such points.

All elastic wave systems investigated to date have shown good time reversal invariance; there has been no sign, in level correlations or backscatter, of any time reversal noninvariance. Estimates have suggested that it would be difficult to observe GUE statistics in an elastic wave system, as it is Coriolus force that provides the proper analog for electrons in a magnetic field, and the necessary spin speeds are large.

Eigenfunction statistics.   That the level statistics of elastodynamics systems are in conformance with the Bohigas conjecture seems now to be well established. Indeed, level statistics often manifest repulsion and (short range) rigidity very like that of the GOE even in nominally pseudo-integrable systems. This has had some impact on practical acoustics, with implications in reverberation rooms for power variances [52, 53], and for response correlations [54]. An experimentally more complicated issue arises when one asks for the statistics of the eigenfunctions.

The theoretical understanding is clear. Random Matrix Theory says that the Wigner hypothesis applies, that modes are random superpositions of standing plane waves of all directions. This in turn implies that modes are real Gaussian random processes in space with correlations given by Bessel functions. The occurrence of scars in which an occasional mode is clearly associated with a small volume of phase space coincident with a marginally unstable periodic orbit is a non-universal behavior and an exception to this otherwise reliable guideline. Numerical results on 2-d billiards, 2-dimensional scans of microwave cavities, and 2-dimensional scans of gravity waves in shallow water tanks [55] agree that most modes are well approximated as such random superpositions of standing plane waves. In agreement with RMT, mode amplitudes in these systems are found to be Gaussian random numbers with definite spatial correlations.

Elastic wave systems are not easy to scan spatially, especially in 3-dimensions, though laser interferometric detection and laser thermo-elastic excitation [32] can, and are, scanned across surfaces. For this reason it is power variances that provide the simplest means for the investigation of eigenfunction statistics in elastodynamics. Power variances have been studied in reverberation room acoustics for many years, and are not yet fully understood. These rooms are used to evaluate total acoustic power produced by various sources, as the rooms provide an integration of the output power over all outgoing directions. But if the source acts at a single frequency, the response of the room will fluctuate, depending on whether or not that frequency coincides with a mode of the room, or on whether or not the source and receiver coincide with nodes of that mode. An old problem in reverberation room acoustics has been the prediction of that fluctuation; it is not yet solved. The issue is mathematically similar to a standard problem in nuclear physics: the understanding of cross section statistics [40]. To the extent that the nuclear Hamiltonian is a GOE system coupled to a set of decay channels, and to the extent that an isolated reverberation room is a GOE system with some set of decay mechanisms, the problems are identical. The acoustic problem is actually a little bit richer, as the acoustic case allows measurements of S matrix phase as well as cross sections, and allows independent variation of the decay mechanisms (e.g. by adding oil), and allows probes with negligible associated dissipation.

That reverberation room response does fluctuate, and that the degree of fluctuation is a function of modal overlap, is clear. Figure 10 shows the power transmission coefficient, $|G(\omega, x_{receiver}, x_{source})|^2$, essentially the square of the Green's function, as a function of frequency over a short frequency interval in which the resonances are distinct. This data was taken in a small aluminum body with defocussing and symmetry breaking elements. Modal overlap $M = 2\pi\bar{\gamma}\partial N/\partial\omega$, where $\bar{\gamma}$ is the mean resonance width, ie. the mean imaginary part of an eigenfrequency, is small. Here M is about 0.1. Figure 11 on the other hand, shows the same spectrum but over a range of frequencies for which the modal overlap is much greater; the resonances are not distinct. Here M is about 10 and the power transmission coefficient fluctuates to a substantially lesser degree.

For various reasons this matter has some importance. We have speculated that measures of the fluctuations could provide assessments of modal density, a quantity that is difficult to measure when resonances are not distinct. Reverberation room acousticians remain concerned with the question of understanding and predicting the fluctuations. A method called Statistical Energy Analysis [56] is often appealed to for estimates of average power levels in complex vibrating structures. Fluctuations away from that average have limited the applicability of SEA. Some of these fluctuations may be due to non-universal features of the structures, features that were oversimplified by the practitioner of SEA, but some are undoubtedly simple consequences of the randomness of the structure and cannot be eliminated except by a computationally burdensome detailed modeling. For all these reasons it is important to attempt a better understanding of the fluctuations.

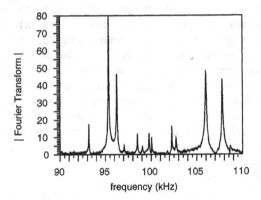

Figure 10.   The absolute value of the Fourier transform of a signal (using dry coupled piezo-electric sources and receivers) in a small aluminum block (Fig. 8) in the range around 100 kHz. We note distinct peaks at the natural frequencies of the specimen, and a variance amongst the peak amplitudes. The peaks are well separated, as the modal overlap at 100 kHz is small.

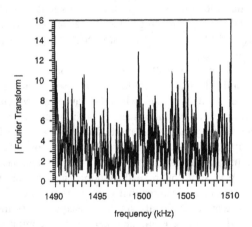

Figure 11.   The absolute value of the Fourier transform of the signal, in the range around 1500 kHz. Distinct peaks at the natural frequencies of the specimen are not discernable, as the modal overlap at 1500 kHz is large.

Following Davy [57] the fluctuations in a frequency band from ω_1 to ω_2 are quantified by means of a relative variance

$$relvar = (\omega_2 - \omega_1) \int_{\omega_1}^{\omega_2} |G(\omega)|^4 \, d\omega \Big/ \left(\int_{\omega_1}^{\omega_2} |G(\omega)|^2 \, d\omega \right)^2 - 1. \quad (28)$$

If G were a complex Gaussian random process, (as it is at high modal overlap) $relvar$ would be unity. If G were a sum of uncorrelated resonances (modes $u^r(\mathbf{x})$) of identical width, γ,

$$G(\mathbf{x}, \mathbf{y}, \omega) = \Sigma_r \frac{u^r(\mathbf{x}) u^r(\mathbf{y})}{\omega^2 - \omega_r^2 - 2i\omega\gamma}, \quad (29)$$

then $relvar$ is $1 + K^2/M$ [57], where K depends on eigenmode statistics $K = <u^4>/<u^2>^2$. RMT, or more specifically the real Gaussian random nature of the eigenmodes, suggests that $K = 3$. An assumption that the modes are real but have level correlations like that of the GOE modifies this prediction. One obtains $relvar = K^2/M - 1$ at $M \ll 1$, and $1 + (K^2 - 3)/M$ at $M \gg 1$ [52].

We have recently begun a project attempting to measure, understand, and predict $relvar$ [53]. We study aluminum blocks constructed with defocussing surfaces and symmetry-breaking cuts. They have level statistics which are in accord with RMT, at least at low frequencies where the mean level spacings are great enough, and the widths are small enough, that the modes are distinct. At greater modal overlap M, where the level statistics are not directly discernable, we can nevertheless investigate the statistics of G itself.

A significant aspect of dissipation in these bodies is that the free decay is not necessarily a simple exponential. Figure 12 shows one case in which the power-spectral density of the signal, essentially the band-limited ultrasonic energy at the receiver, is plotted versus time. For the low frequency band illustrated here the energy decays in a non-exponential fashion. To within the precision of the measurements the decay fits well to a simple three-parameter (E_o, σ, and n) model

$$E(t) = E_o(1 + \sigma t/n)^{-n}. \quad (30)$$

This model follows from a presumption that the signal is an incoherent super-position of exponential decays $\exp -2\gamma t$ from a distribution $p(\gamma)$ of decay rates γ.

$$E(t) \sim \int_0^\infty p(\gamma) \exp(-2\gamma t) d\gamma, \quad \text{with} \quad p(\gamma) d\gamma \sim \gamma^{n-1} \exp(\frac{-n\gamma}{\bar{\gamma}}) d\gamma.$$

$$\quad (31)$$

This chi-square form for $p(\gamma)$ in turn follows from a (naive) model in which it is supposed that the decay rate γ_r of mode number r is given by a first order perturbative estimate

$$\gamma_r = \frac{1}{2\omega} < \mathbf{u}^r(\mathbf{x}) \mid \mathrm{Im}\tilde{C}(\mathbf{x})/\rho \mid \mathbf{u}^r(\mathbf{x}) > = \frac{1}{2\omega} \int_V u^r_{j,i} \, \mathrm{Im}\tilde{C}_{jikl}(\mathbf{x}) \, u^r_{k,l} \, d^3\mathbf{x} .$$

$$(32)$$

If the loss modulus $\mathrm{Im}\tilde{C}$ is distributed smoothly in space, on a scale of several wavelengths, then each mode has about the same γ and the width statistics are uninteresting. If, however, the loss mechanisms are associated with a discrete set $(2n)$ of equipotent isolated points of the body then taking the $\mathbf{u}^r$'s as Gaussian real processes (as in RMT) gives Eq. (31). While this model is admittedly naive, it is striking how well the data fits its prediction. The distributions of widths is related to the smoothness (in space) of the loss mechanisms, and furthermore the distribution of widths (actually, its Laplace transform) is directly observable: $E(t)$.

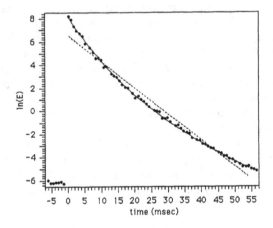

Figure 12.   The logarithm of the observed spectral energy density in a block from Fig. 8 in a frequency band near 195 kHz is plotted versus time. The residual noise level is apparent from the data at negative times. The signal stays well above noise for 60 msec. The decay is fit to two different theories. The conventional linear decay (dashed line) is manifestly contradicted by the data. The three parameter fit to a curved decay (solid line) fits the data quite well.

We have recently made a theoretical estimate for *relvar* (in which the frequency averages in Eq. (28) were replaced by ensemble averages) based on the above distribution of modal widths, based on GOE level correlations, and based on the GOE assertion: $K = 3$. The result has been compared with measurements (see Fig. 13). There is a persistent discrepancy. The observed fluctuations of $|G|^2$ are significantly less than expected, by a factor around 1.5. There are a number of plausible explanations, most of which have been considered and rejected. Perhaps the most viable at this time is a hypothesis that the modes of this dissipative system, almost certainly complex, do not have $K = 3$. In a non-dissipative system the modes are real, and are stochastic superpositions of plane waves, thereby being real Gaussian processes with $K = 3$. In a strongly dissipative system, the complex modes are presumably stochastic superpositions of travelling waves, and are therefore complex Gaussian processes with $K = 2$. In a somewhat dissipative system such as this one, it might well be supposed that the complex modes are more real than imaginary, and have a K somewhere between 2 and 3. This argument has the right sign and magnitude to explain the discrepancy; further measurements are supporting it [58].

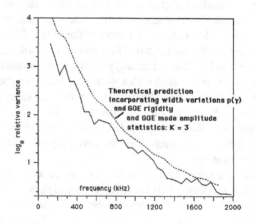

Figure 13. Measurements of *relvar* consistently fall below predictions based on simple RMT arguments.

2.2. TRANSPORT IN MULTIPLE SCATTERING MEDIA, DIFFUSION AND LOCALIZATION

The subject of multiple diffuse scattering is conveniently divided into what might be termed the classical limit of high conductivity in which residual coherence plays no role, and the strong scattering, low conductivity, limit in which localization may be relevant. The former domain is receiving substantial attention in the ultrasonics literature, and is reviewed in this section. The latter domain has received less attention, what there is is reviewed in the next section.

Weak Scattering and Diffusion.   Ultrasonic propagation in random solids (for example composite materials, or polycrystals) at wavelengths comparable to or longer than the size of typical heterogeneities is often studied with a view towards nondestructive characterization of the microstructure. At low frequencies, such that the wavelengths are long compared to the length scales of the heterogeneity, the medium is an effectively homogeneous continuum. Measurements of ultrasonic wave speeds, attenuations, and polarizations provide means for material characterization [59], though microstructural features are not resolved. In such media the chief source of coherent wave attenuation is dissipative, and not correlated with the microstructures of interest. At moderate frequencies attenuation is augmented by diffuse scattering [60] out of an acoustic beam, leading to the possibility of microstructure characterization by means of the frequency-dependence of acoustic velocities and attenuations.

Typical polycrystalline metals and ceramics have grain sizes of the order of 1 to 100 microns. In most materials the crystallites have significant elastic anisotropy (with moduli varying by factors of the order of 2 with orientation.) Thus elastic waves with wavelengths of the order of, or less than, grain diameters can be expected to be scattered strongly by the grain boundaries. For grain radii of 25 microns, and in steel with a shear wave speed of about 3 mm/msec, $ka=1$ corresponds to 20 MHz. Predictions for average waves in polycrystals, i.e. for wave speeds, dispersion, and attenuation, have developed over many decades. Stanke and Kino [14] provide a good review of the literature, and a rational perturbative calculation of mean wave properties to leading order in the crystallite anisotropy.

At higher frequencies where there is significant dependence on microstructure, there is less literature. This is due to the very high mean field attenuations characteristic of this frequency range; coherent propagation is too weak for measurement except in special circumstances [61]. Diffuse multiply scattered fields, however, are not adversely impacted by the strong attenuation of the coherent propagation. Diffuse field studies have successfully characterized sources, fields and media by means of measurements of the evolving wave energy [62-64].

Within the high conductivity regime, ultrasonic diffuse waves in multiply scattering solid media may be further classified in two groups. One concerns theoretical and experimental work [64-66] in which the received signal is modeled as consisting entirely of singly scattered contributions. The other models the received signal as thoroughly multiply scattered, i.e., as a fully developed diffuse field [16, 63]. The former case presumably applies at times sufficiently soon after a transient source has acted. The latter case presumably applies at times sufficiently long after that source has acted. A clear distinction between the two regimes usually requires that the studies be carried out in the time domain. But it also demands that we understand the transition between the regimes. This demand has not yet been met.

Because of the correlation between fracture toughness and grain size in solids, there has long been an interest in ultrasonic characterization of microstructures. A considerable effort has been devoted to using ultrasonic attenuation (of mean plane wave fields) to infer grain size [65]. Diffusely scattered ultrasound, i.e. mean square fields, were first studied in solids in the context of microstructural characterization by means of the (assumed singly-scattered) backscattered field [64, 65]. Figure 14 illustrates the technique.

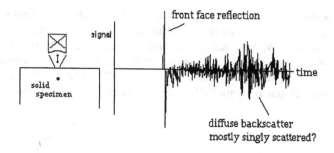

Figure 14.   A typical grain noise backscatter study will insonify through water with a pulse from a focussed immersion transducer. The reflected signal, after the front surface echo, has the appearance of noise, but consists of coherent echoes from randomly positioned crystallites. There is a peak in the backscattered grain noise at a time corresponding to the depth of focus. At sufficiently long wavelengths one can also detect a back surface echo (not pictured).

Multiply scattered ultrasound in polycrystals was studied soon after [16, 62, 63]. The regime between the single scattering and multiple scattering limits is readily accessible in the lab, but little is understood yet as to how best to model it. Theory is based on radiative transfer [17, 18] but due to that theory's large number of parameters, experiments in the transition regime remain unmodeled. Indeed it is sometimes not clear if a short-time backscattering measurement has indeed been confined to the single scattering limit [66], or to what extent another measurement at later times is truly in the multiply scattering limit [63].

Nevertheless, measurements at longer times have been found to fit [62, 63] well to the diffusion model. Transient ultrasound, at frequencies of ten MHz or more in various polycrystalline metals, give rise to signals whose mean squares fit well to the solution of a dissipative diffusion equation (34), at least at times sufficiently long after the action of the source. The fits show that the higher frequencies, and the larger grain sizes, correspond as expected to slower diffusion. In contrast to similar tests in optics or microwaves, access to the time domain and the excellent isolation from leads permit a clear separation between the effects of dissipation and diffusion. The former merely imposes an exponential temporal decay $\exp(-\sigma t)$ on the transport. Fits of this kind allow distinctions between absorption and scattering that cannot be made using coherent attenuation [65], and which are difficult to make when using CW diffuse fields because of lack of time resolution.

Similar issues have arisen in seismology. After the low frequency (and therefore less scattered) part of a seismic wave has passed a seismic recording station, there is often evident a slowly decaying high frequency noise-like signal. It is called the coda, and is presumably related to multiply scattered seismic waves. It is not yet well understood [67].

Localization. In order to explore the very low conductance limit we have also attempted studies of transport in materials with a more strongly scattering microstructure. We have not yet identified a convenient polycrystal for this limit, but have studied [33] an aluminum plate with an artificial microstructure induced by machining many scalloped slits (Fig. 15) and have studied a random open-celled aluminum foam [68]. In the aluminum plate (Fig. 15) broad-band

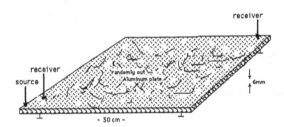

Figure 15. Sketch of the aluminum plate used to study transport in the strong scattering, low conductance, regime. Receivers were placed at several points along the diagonal leading away from the source in one corner.

transient ultrasound is introduced at one point, often a corner, and detected at various other points within the structure. The transport is effectively two dimensional and confined to the finite area. The typical distance between scatterers is about 3 mm and the plate thickness is 6 mm. As in the work in polycrystals, the time domain is readily accessible and losses into transducers and supports are unimportant.

Neither mean free path ℓ nor wavenumber k is a good parameter in this study. The wave field in the plate is a complicated superposition of bulk longitudinal and shear waves and surface Rayleigh waves, all being subjected to mode conversion at the free surfaces and edges. The plate is thick enough to support multiple Lamb waves, which mode convert to each other at the scatterers. There is no one wavelength to identify. Therefore it is difficult to unambiguously identify a mean free path and a wavelength in order to construct the Ioffe-Regel parameter $k\ell$. $\ell = 3$ mm is one estimate, based on the typical spacing between scattering surfaces. $k = 0.4$ mm^{-1} at f = 200 kHz is also plausible, based on a shear wave speed of 3 mm/µsec. While the estimate is crude, it does allow a speculation that localization may be significant for frequencies near 200 kHz. At much higher frequencies wavelengths are clearly too short; at much lower frequencies the plate is best modeled as a homogeneous effective medium with no scattering.

An alternative criterion, equivalent to Ioffe-Regel, is provided by the product of (bare) diffusivity D_o and areal modal density $\frac{\partial^2 N}{\partial\omega\partial A}$. This criterion allows localization length to be estimated in terms of local properties

$$\Lambda \sim \ell \exp(2\pi^2 D_o \frac{\partial^2 N}{\partial\omega\partial A}) \,. \tag{33}$$

Modal density and bare diffusivity can be measured with less ambiguity than k and ℓ.

Figures 16 and 17 show some of the results of tests in this structure. At shorter wavelengths, e.g. Fig. 16, localization is not anticipated. The mean spectral energy in the vicinity of the source is seen to fall rapidly for several msec, and then asymptote onto a final slower, exponential decay. The energy density at a point in the corner opposite to the source rises over a time period of several msec, and then asymptotes onto the same exponential decay. At times greater than 13 msec, the energy is homogeneously distributed across the structure, while it is dissipating at a spatially uniform rate determined by the absorption. The time scale for equilibration is several msec, a period which may be compared with a nominal transit time scale of 300 mm / (3 mm/µsec) = 100 µsec. The behavior is successfully fit to the solution of a dissipative diffusion equation

$$D \nabla^2 E(\mathbf{x}, t) - \sigma E(\mathbf{x}, t) = \delta^2(\mathbf{x})\delta(t) \,, \tag{34}$$

in a square domain with Neumann boundary conditions. Deviations from precise accord may be ascribed to a distribution of values of D across the finite frequency bins. Values of diffusivity D are extracted from that fit. The values vary slightly with frequency, but are consistent with prior estimates $D \sim c\ell/2$. The behavior appears fully classical.

At lower frequencies, however, the energy density does not behave as Eq. (34) would predict. Figure 17 shows the measured energy densities at 200 kHz and an attempt to fit all five profiles (from different receiver positions) to the solution

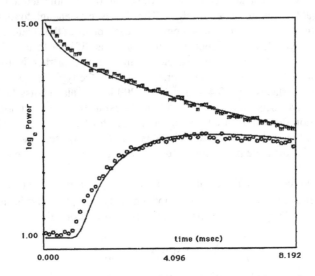

Figure 16.   The power spectral density in a frequency band around 630 kHz is studied near the source in one corner (m) and in the opposite corner (o). The solid lines shows the behavior of a solution to Eq. (34) with parameters chosen to fit the observed evolution. By $t = 13$ msec, the power spectral densities in the two corners become indistinguishable.

of Eq. (34) for a single value of D and σ. As was anticipated by the Ioffe Regel criterion, and by the estimate (33) using measured values of D_o and $\frac{\partial^2 N}{\partial \omega \partial A}$, the behavior is not classically diffusive.

It is not consistent with expectations for the time-domain behavior of localized waves either. The profiles in Fig. 17 show a field that, after 8 msec, is 100 times more intense near the source than it is in the opposite corner of the plate. But it also shows that transport continues at later times, albeit slowly. After 50 msec, the degree of localization has been reduced to a factor of less than 5. It is as if the energy is diffusing, but with a time or length scale varying diffusivity. Extensive numerical simulations [69], with and without absorption, have never shown this kind of behavior. For example, Fig. 18 shows a numerical simulations of the transient response of an Anderson model 2-d lattice. In particular it shows that the transport over large distances ceases at moderate times and a spatial gradient is maintained thereafter. Introducing absorption into the simulation leaves the spatial gradients unchanged. Speculations [33] as to the cause of the residual transport seen in the lab but not in the simulations have so far not been confirmed. Candidates include hopping-induced conductance by means of nonlinear interactions between the ultrasound and ambient structural vibra-

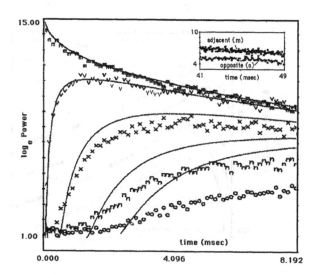

Figure 17.   The power spectral density in a frequency band around 260 kHz is studied near the source in one corner (m) and in the opposite corner (o). The data from three other positions, equi-spaced across the diagonal, are also shown. The solid lines show the behavior of a solution to Eq. (34) with parameters chosen to fit the evolution of the upper data. By $t = 16$ msec, the evolution of the power spectral density has virtually ceased, with a steady state energy density gradient in place across the structure. At that time the energy density in the plate center is approximately one third that near the source, and the energy density in the opposite corner is only 3% of that near the source. At later times, shown in the insert, weak transport continues to take place. At 50 msec the energy density in the opposite corner is 20% that near the source.

tions (analogous to inelastic scattering of electrons), and spatial variations in absorption. Intriguingly, similar behavior has recently been seen (see below) in transport between coupled elastic reverberation rooms in the localized regime.

The effect of dissipation in localizing classical wave systems has received a good deal of attention, and some contention. Absorption of classical waves is not equivalent to inelastic scattering of electrons. Confusions have arisen in part perhaps due to dual meanings of the word "inelastic." An electron that is scattered inelastically is scattered to a neighboring state, it remains in the system; its contribution to the current is augmented by transition between localized states. A classical wave that is absorbed is removed from the system. That the effect of absorption is probably trivial is most easily appreciated by considering an absorption operator $\text{Im}\tilde{C}$, that is diagonal in the basis provided by the modes of the undamped system. This is certainly not the case in practice, but it does provide a simple model for purposes of argument. The natural frequency

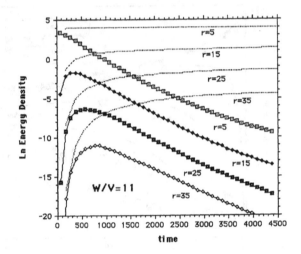

Figure 18.   Numerical simulation of Anderson localization in the time domain shows that the transport time scales are comparable to those of the bare diffusivity, that even at distances of several localization lengths, the steady state late time asymptote is achieved in a moderate amount of time. Transport is not exponentially slow, merely exponentially weak. It also shows that dissipation leads to exponential decay with time, but does not affect the (relative) distribution in space.

of each mode gains an imaginary part when the absorption is introduced; the shape of each mode remains unmodified. Localization is therefore unaffected. The dynamics is affected only inasmuch as each mode then decays as well as oscillates. If each natural frequency has the same imaginary part, then the net change in energy evolution is equivalent to an additional exponentially diminishing factor $\exp(-\sigma t)$. Spatial gradients, if any, are maintained. That this argument captures the essential effects of damping, even for damping operators that are not diagonal in the basis provided by the undamped modes, is supported by numerical simulations (Fig. 18) [69].

Localization of mechanical waves is much better known in one dimensional systems common in structural applications, e.g. Turbines, Trusses, Submarines, railroad track [70]. It is not difficult to observe, in the lab or in numerical simulation. It is thought that localization in nominally periodic, but inevitably flawed, one-dimensional structures may be relevant for understanding fatigue in turbines, acoustics of submarine hulls, large flexible space stations, and other vibrations in lightly damped built structures. A development which is particularly amusing is the occurrence of localization in perfectly periodic nonlinear structures [71] for which a localized nonlinear disturbance in one unit cell can oscillate at a frequency outside the pass band of the adjacent linear (because low amplitude) cells. The large amplitude localized motions act like an impurity in an otherwise perfect structure.

The Anderson transition for elastic waves in multi dimensions is thus not yet seen with complete lack of ambiguity, even for 2-d systems. The anomalous behavior seen in [33] is not yet understood. However, weak localization, and the related "Quantum Echo" [72] have been observed. Recent experiments [29, 73] have shown an enhanced backscatter due to weak localization of reverberant elastic waves in a planar stadium billiard [29], and due to weak localization and quantum echo in a three dimensional reverberant elastic wave billiard [73].

The Paris VII group has studied the elastic wave response of a silicon wafer at frequencies below the cutoff of the higher Lamb waves [29]. Piezoelectric normal point loads chiefly generate flexural waves. Spatially scanned laser interferometric sensors measure normal displacement, for which the flexural waves are the chief contributor. If the wafer preserves its up/down mirror symmetry, then there is no mode conversion from the antisymmetric flexural waves to or from the symmetric extensional and SH waves. Thus the flexural waves form an uncoupled diffuse wave field that is similar to that of the standard model of scalar waves in a 2-d membrane [74]. The anisotropy of the silicon is not significant, nor are the difficult-to model complex reflection coefficients. The group has shown that the broad-band mean square displacement amplitude is enhanced in the vicinity of the source, being twice as great at the source as it is at distant places, and diminishing with distance in a manner consistent with the expected Bessel function. This result follows from familiar arguments involving rays interfering constructively with their time-reversed counterparts.

Weaver and Lobkis [73] recently showed the same effect in the three dimensional elastic body pictured in Fig. 19. Here a diffuse mixture of longitudinal waves, shear waves and surface waves forms a field very different from that of the conventional scalar wave billiard models. The weak localization argument remains valid however: Mean square displacement in the vicinity of the source should be twice what it is elsewhere. The dry-coupled piezoelectric method used did not lend itself to spatial scanning, so the signals from two transducers a fixed distance apart were compared as a function of frequency. The expected Bessel-function like decay of enhancement was observed. The correlation range was found to correspond with that of a diffuse mixture of Rayleigh waves, as is expected from the dominance of the surface field by those waves.

At times long after the action of the transient source, long compared to the break time, the ray picture of weak localization is not trustworthy. RMT predictions, based on an assumed universal GOE structure to the dynamics, indicate that the enhancement factor should grow from the initial two to an

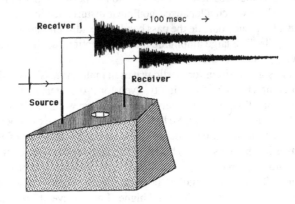

Figure 19.   The aluminum block, a sort of 3-d elastic wave Sinai billiard, used to investigate enhanced backscatter and modal echo. It has non-parallel walls and a circular cylindrical hole. The size of the object (~15 cm) is much greater than the wavelengths of interest. There are about 3000 modes below 100 kHz. The signal in receiver 1 (which is also the source) is compared with that in receiver 2. A reciprocity principle is used to correct for differences in sensitivity of the two transducers.

eventual three [72]. This was termed a quantum echo by Prigodin *et al.*. Figure

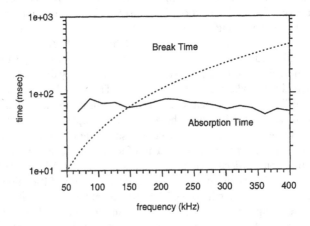

Figure 20.   The object of Fig. 19 was studied for a range values of the ratio of break time to absorption time.

21 shows the measured enhancement at the position of the source, as a function

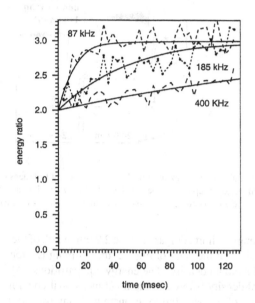

Figure 21. The ratio of the energy, at the source transducer and at a generic other position, as a function of time and frequency. The solid lines show the GOE prediction for this ratio, a value of two at early times, corresponding to weak Anderson localization, followed by a value of three at times near the break time.

of time, for three different frequencies, corresponding to three different break times. The agreement with RMT is very good. It is especially striking that the agreement survives the significant amounts of absorption in the system (Fig. 20). Thus absorption does not appear to have affected the quantum echo.

Anderson localization of elastic waves has also been observed in an interesting structure that might be termed 0-dimensional. Coupled reverberation rooms have long been studied in room acoustics. Consider the coupling of two reverberant bodies, each with many states. If the coupling is sufficiently weak, perturbation theory shows that the modes of the separate bodies will be unmixed, and that therefore sources acting in one body will lead to energy that is largely confined to that body. It also shows that the critical parameter is the ratio of the mean level spacing to the initial conductivity, or leaking rate, of energy between the bodies. The theory of the energy flow in such a system has been outlined recently, together with some numerical simulations [75]. A laboratory version, pictured in Fig. 22, was studied also.

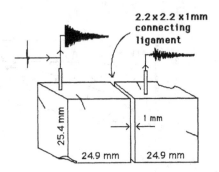

Figure 22.   The aluminum block used for laboratory study of 0-dimensional Anderson localization between weekly coupled reverberant elastic bodies. A small ligament of dimensions 2.2
× 2.2 × 1.0 mm near the center of the gap connects the two sides acoustically.

At wavelengths short compared to the 2.2 mm width of the ligament connecting the two sides of the body of Fig. 22, the coupling is strong compared to the
inverse of the modal density. The energy equipartitions. At long wavelengths
where modal density is low and the nondimensional coupling is weak, typical
pseudo modes of one side fail to resonate with any pseudo modes of the other
side. They do not share energy with the other side. The actual modes of the
whole structure are therefore localized.

The theory [75] makes an intriguing prediction that the energy flow into
the second room, weak though it be in the localized regime, overshoots its
asymptotic value, by as much as 17%. This was observed in some of the
simulations but not in the experiments (Fig. 23). Of additional interest was
a detail in the approach to the asymptotic energy level. Figure 23 shows the
approach, on a time scale of 10 msec, to the predicted asymptote in which
the energy in the second room is a small fraction, 25% to 50% of that of the
first room. The transport continues however, as if some other mechanism is
slowly transferring energy between the rooms. The slow residual transport is
reminiscent of that anomalous behavior seen ten years ago [33] amongst the
elastic waves in the disordered plate. They may have similar origins.

Summary

This review has attempted to establish that elastic waves, though their basic
physics is substantially different from that of electron waves and light and microwaves, nevertheless also manifest the intriguing phenomena of mesoscopic
physics, wave chaos, and localization that are observed with other waves. Like

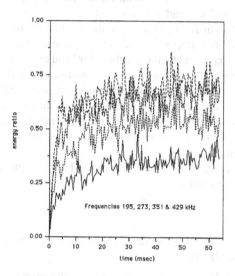

Figure 23. The energy in the side opposite to that in which the source acts is a fraction of the energy close to the source. The behavior is in qualitative agreement with predictions, in that the longer wavelengths localize more. The observed behavior differs from that predicted in that a weak transport continues at late times, and in that the localization is weaker than expected.

them, elastic waves show coherence and incoherence, have modes, scatter, interfere constructively and destructively, and are subject to the familiar dichotomy between conceptualizations based on a ray (particle) picture and a wave picture. The quality factors Q achievable in some solids are sufficient to make ultrasonic investigations of mesoscopic phenomena feasible.

There are two chief differences, however, between studies using ultrasonics and those using other kinds of classical waves. The first of these is due to the low speed of elastic waves in solids. This permits laboratory sized specimens to be studied at moderate frequencies. In consequence, transient phenomena are readily observable. The second is due to the large impedance mis-match between typical solids and air; thus losses to the environment are small, and would be smaller if conducted in a vacuum; specimens can be isolated. The mismatch is also large between typical solids and certain sensors. Thus specimens can be isolated even from the measuring apparatus. This is especially true when laser ultrasonics is used.

The work has observed a number of interesting mesoscopic phenomena. These include diffusion and multiple scattering; spectral statistics: level repulsion, spectral rigidity and parametric level correlations; Anderson localization (in 0, 1 and 2 dimensions); and weak localization, enhanced backscatter and the "Quantum echo" in 2 and 3 dimensions.

The effect of dissipation on transport and localization has been found to be largely trivial. When transport is studied in the time-domain, dissipation manifests as a simple $\exp(-\sigma t)$ factor. On the other hand dissipative but reverberant bodies show weaker fluctuations in power statistics than had been anticipated, behavior that may be related to complex modes, in turn due to the dissipation. The high Q's, the convenient time scales and specimen sizes, the opportunity to investigate the effects of dissipation, the relevance to ultrasonic materials characterization, acoustics and seismology, all suggest that this kind of work is likely to continue.

The author's research in diffuse ultrasonics has been supported by the National Science Foundation.

References

[1] Augustin Cauchy, *Exercise de Mathematiques*, Imprimerie du Bachelier, Paris, **3**, 160 (1828).

[2] S. D. Poisson, *Mémoire sur l'equilibre et le mouvement des corps elastiques*, Mém. de l'Acad. Royale des Sciences, **8**, 357, 623 (1829).

[3] J. MacCullagh, Trans. Royal Irish Acad. **21**, 17 (1848).

[4] K. F. Graff, *Wave Motion in Elastic Solids* (Dover, NY, 1975).

[5] J. Miklowitz, *Elastic waves and waveguides* (North Holland, 1978).

[6] J. D. Achenbach, *Wave propagation in elastic solids* (Elsevier, 1973).

[7] R. Truell, C. Elbaum, and B. B. Chick, *Ultrasonic methods in solid state physics* (New York, Academic Press, 1969).

[8] K. Aki, *Quantitative seismology: theory and methods* (W. H. Freeman, San Francisco, 1980).

[9] B. A. Auld, *Acoustic waves and fields in solids* (Wiley, NY, 1973).

[10] Y.-H. Pao and C. C. Mow, *Diffraction of elastic waves and dynamic stress concentrations* (Crane, Russak and Co., NY, 1973).

[11] J. D. Achenbach and A. K. Gautesen, J. Acoust. Soc. Am. **61**, 413 (1977).

[12] P. C. Waterman, J. Acoust. Soc. Am. **60**, 413 (1976).

[13] F. C. Karal and J. B. Keller, J. Math. Phys. **5**, 537 (1964).

[14] F. Stanke and G. Kino, J. Acoust. Soc. Am. **75**, 665 (1984).

[15] V. K. Varadan, Y. Ma and V. V. Varadan, J. Acoust. Soc. Am. **77**, 375 (1985).

[16] R. L. Weaver, J. Mechs. Phys. of Solids **38**, 55-86 (1990).

[17] J. A. Turner and R. L. Weaver, J. Acoust. Soc. Am. **96**, 3654-74 (1994).

[18] L. V. Ryzhik, G. Papanicolaou, and J. B. Keller Wave Motion **24**, 327-370 (1996).

[19] S. S. Antman, *Nonlinear Problems of Elasticity* (Springer-Verlag New York, 1995). Ivan Stephen Sokolnikoff *Mathematical theory of elasticity*, 2nd ed. (New York, McGraw-Hill, 1956).

[20] J. P. Wolfe, Physics Today, **48**, 34 (1995).

[21] R. L. Weaver, J. Sound and Vibr., **94**, 319-335 (1984).

[22] D. Egle, J. Acoust. Soc. Am. **70**, 476 (1981).

[23] R. L. Weaver, J. Acoust. Soc. Am. **71**, 1608-1609 (1982).

[24] Lord Rayleigh, Proceedings London Mathematical Society, **17**, 4 (1877).

[25] J. Miklowitz, ASME J. Applied Mechanics **49**, 797-815 (1982); V. K. Kinra and B. Q. Vu, Mech. Res. Comm. **10**, 193-8 (1983); B. V. Budaev and D. B. Bogy, Wave Motion **22**, 239-57 (1995).

[26] P. M. Morse and H. Feshbach, *Methods of Theoretical Physics*, p. 1872ff (McGraw Hill, NY, 1953).

[27] A. Migliori, J. L. Sarrao, *Resonant Ultrasound Spectroscopy,* (John Wiley, New York, 1997); W. M. Visscher, A. Migliori, T. Bell, R. Reinert, J. Acoust. Soc. Am. **90**, 2154-62 (1991).

[28] W. Sachse and K. Y. Kim, 311-320, in *Review of Quantitative Nondestructive Evaluation*, **6A**, eds. D. O. Thompson and D. E. Chimenti, (Plenum Press, NY, 1986).

[29] J. de Rosny, A. Tourin and M. Fink, Phys. Rev. Lett. **84**, 1693 (2000);

[30] C. Ellegaard, T.Guhr, K. Lindemann, J. Nygård and M. Oxborrow, Phys. Rev. Lett. **77**, 4918-4921 (1996).

[31] M. Hauser, R. L. Weaver and J. P. Wolfe, Phys. Rev. Lett. **68**, 2604 (1992); R. L. Weaver, M. Hauser and J. P. Wolfe, Zeitschrift fur Physik **B 90**, 27-46 (1993).

[32] L. R. F. Rose, J. Acoust. Soc. Am. **75**, 723 (1984); J. P. Monchalin, A. Moreau, *et al.*, *various papers on laser ultrasound in Nondestructive characterization of materials* **VIII**, R. E. Green, Jr. ed., (Plenum Press, New York, 1998).

[33] R. L. Weaver, Wave Motion **12**, 129-142 (1990).

[34] N. N. Hsu, J. A. Simmons and S. C. Hardy, Materials Evaluation **35**, 100 (1977).

[35] F. R. Breckinridge, C. E. Tschiegg and M. Greenspan, J. Acoust. Soc. Am. **57**, 626 (1975).

[36] S. L. McBride and T. S. Hutchinson, Can. J. Phys. **54**, 1824 (1976).

[37] N. N. Hsu and F. R. Breckinridge, Materials Evaluation, **39**, 60 (1980).

[38] P. Bertelsen, C. Ellegaard, M Oxborrow *et al.*, Phys. Rev. Lett., **83**, 2171 (1999).

[39] A. G. Smagin, Cryogenics **15**, 483-485 (1972).

[40] J. J. M. Verbaarschot, H. A. Weidenmuller and M. R. Zirnbauer, Physics Reports **129**, 367-438 (1985); S. Albeverio, F. Haake, P. Kurasov, M. Kus and P. Seba, J. Math. Phys. **37**, 4888-4903 (1996).

[41] R. A. Guyer and P. A. Johnson, Physics Today **48**, 30 (1999).

[42] R. L. Weaver and O. I. Lobkis, Ultrasonics **38**, 491-4 (2000).

[43] R. S. Muller, R. T. Howe, S. D. Senturia, R. L. Smith, and R. M. White, eds *Microsensors* IEEE, (New York, 1991)

[44] R. L. Weaver, J. Acoust. Soc. Am., **78**, 131-136 (1985).

[45] M. Dupuis R. Mazo, and L. Onsager J. Chem. Phys. **33**, 1452-62 (1960).

[46] R. L. Weaver, J. Acoust. Soc. Am., **79**, 919-923 (1986).

[47] R. L. Weaver and Y.-H. Pao, ASME J. Appl. Mechs. **49**, 821-836 (1982).

[48] R. L. Weaver, J. Sound Vibr. **94**, 319-335 (1984).

[49] R. L. Weaver, J. Acoust. Soc. Am., **80**, 1539-1541 (1986).

[50] R. L. Weaver, J. Acoust. Soc. Am., **85**, 1005-1013 (1989); D. Delande, D. Sornette and R. L. Weaver, J. Acoust. Soc. Am. **96**, 1873-1880 (1994).

[51] O. Bohigas, O. Legrand, C. Schmit and D.Sornette, J. Acoust. Soc. Am. **89**, 1456-8 (1991).

[52] R. L. Weaver, J. Sound Vibr. **130**, 487-491 (1989).

[53] O. I. Lobkis, R. L. Weaver, and I. Rozhkov, "Power Variances and Decay Curvature in a Reverberant System", in press, J. Sound Vibr. (2000).

[54] O. Legrand, F. Mortessagne, P. Sebbah and C. Vanneste, Actes du 4eme Congres Francais d'Acoustique, CFA 97, p.315, Marseille, France, 14-18 Avril 1997.

[55] M. Rollwage, K. Ebeling, and D. Guicking, Acustica **58**, 149-161 (1985).

[56] R. H. Lyon and R. G. DeJong, *Theory and application of statistical energy analysis,* (Butterworths-Heimann, Boston, MA, 1995).

[57] J. L. Davy J. Sound Vibr. **107**, 361-373 (1986); also J. Sound Vibr. **115**, 145-161(1987); also J. Sound Vibr. **77**, 455-479 (1981).

[58] O. I. Lobkis, R. L. Weaver, in press, J. Acoust. Soc. Am. (2000).

[59] e.g. M. Stautberg Greenwood, J. L. Mai and M. S. Good, J. Acoust. Soc. Am. **94**, 908-916 (1993); J. M. Hovem and G. D. Ingram, J. Acoust. Soc. Am. **66**, 1807-1812 (1979); C. M. Sayers and R. L. Grenfell, Ultrasonics **31**, 147-153 (1993); M. A. Biot, J. Acoust. Soc. Am. **28**, 168-178 (1956) and J. Acoust. Soc. Am. **28**, 179-191 (1956); D. Wu, Z. W. Qian, and D.

Shao, J. Sound Vibr **162**, 529-535 (1993); R. C. Courtney and L. Mayer, J. Acoust. Soc. Am. **93**, 3193-3200 (1993).

[60] e.g. L. Schwartz and T. J. Plona, J. Appl. Phys. **55**, 3971-3977 (1984); L. Tsang, J.A. Kong, and T. Habashy, J. Acoust. Soc. Am. **71**, 552-558 (1982).

[61] J. Liu, L. Ye, D. Weitz and P. Sheng, Phys. Rev. Lett. **65**, 2602-2605 (1990).

[62] C.B. Guo, P. Holler and K. Goebbels, Acustica **59**, 112-120 (1985).

[63] R. L. Weaver, W. Sachse, K. Green, and Y. Zhang, Proceedings of Ultrasonics International 91, (Butterworth-Heinemann, Oxford, U.K., 1991) pp. 507-510; R. L. Weaver, in *Non-Destructive Testing and Evaluation in Manufacturing and Construction*, H. L. M. dos Reis, ed. Hemisphere 425-434 (1990)

[64] F. J. Margetan, T. A. Gray and R. B. Thompson, Review of Progress in Quantitative Nondestructive Evaluation, **10**, 1721-1728 (1991).

[65] K. Goebbels, Phil. Trans. R. Soc. London **A320**, 161-169 (1986) ; K. Goebbels, Chapter 4, p. 87-158 in *Research Techniques in NDT* **IV** RS Sharpe, ed. (1980).

[66] M. D. Russell, S. P. Neal, in *Review of Progress in Quantitative Nondestructive Evaluation* **16** (Plenum Press, NY, 1997).

[67] K. Aki and B. Chouet, J. Geophys. Research **80**, 3322-42 (1975); M. Campillo, L. Margerin and N. M. Shapiro, in *Diffuse waves in complex media* J.-P. Fouque ed., 383-404 (Kluwer, Dordrecht, 1999).

[68] R. L. Weaver, Ultrasonics, **36**, 435-442 (1998).

[69] R. L. Weaver, Phys Rev B **49**, 5881-5895 (1994.).

[70] C. H. Hodges, J. Sound Vibr. **82**, 411 (1982); *Localization and the Effects of Irregularities in Structures*, special issue of Applied Mechanics Reviews **49**, 111-120 (1996); C. Pierre, J. Sound Vibr. **139**, 11 (1990); D. Li and H. Benaroya, Appl. Mech. Revs. **45**, (11), 447-460 (1992).

[71] A. F. Vakakis and C. Cetinkaya, SIAM Journal of Applied Mathematics, **53**, 265-282, (1993). O. Gendelman and A. F. Vakakis, Chaos Solitons and Fractals, **11**, 1535-42 (2000).

[72] V. N. Prigodin, B. L. Altshuler, K. B. Efetov and S. Iida, Phys Rev Lett **72**, 546-549 (1994); R. H. Lyon, J. Acoust. Soc. Am. **45**, 546 (1969); R. L. Weaver and J. Burkhardt, J. Acoust. Soc. Am., **96**, 3186-3190 (1994).

[73] R. L. Weaver, O. I. Lobkis, Phys. Rev. Lett. **84**, 4942 (2000).

[74] The spectral statistics of flexural waves have been investigated in billiard-shaped plates. [O. Legrand, C Schmit, and D Sornette, Europhysics Lett **18**, 101 (1992); E Bogomolny and E Hughes, Phys Rev E **57**, 5404 (1998).] These waves do not mode convert to extensional and SH waves at plate edges

and flaws, unless something breaks the up/down mirror symmetry. Thus they constitute an independent diffuse field. Their boundary conditions are substantially different from those of the Helmholtz equation, and these waves therefore provide an independent confirmation of many of the notions of wave chaos theory. Their substantial loss rate to air and to thermoelastic relaxations have limited their usefulness.

[75] R. L. Weaver and O. I. Lobkis , J. Sound Vibr. **231**, 1111-34 (2000).

Chapter 4

TIME-REVERSED ACOUSTICS
AND CHAOTIC SCATTERING

M. Fink and J. de Rosny

Laboratoire Ondes et Acoustique
Ecole Supérieure de Physique et de Chimie Industrielle de la Ville de Paris
Université Denis Diderot, UMR CNRS 7587
10 Rue Vauquelin, 75005 Paris, France
mathias.fink@zeus.loa.espci.fr

Abstract   The objective of this paper is to show that time-reversal invariance can be exploited in acoustics to accurately control wave propagation through random media as well as through chaotic reverberating cavities. To illustrate these concepts, several experiments are presented. They show that chaotic dynamics reduce the number of transducers needed to insure an accurate time-reversal experiment. Multiple scattering in random media and multiple reflection in chaotic cavities enhance resolution in time-reversal acoustics by making the effective size of the time reversal mirror (TRM) much larger than its physical size. Comparisons with phase-conjugated experiments show that this effect is typical of broadband time-reversed acoustics and is not observed in monochromatic phase conjugation. Self averaging properties of broadband time reversal experiments are emphasized.

Keywords: ultrasound, acoustic, coherent backscattering, time reversal, focusing, phase conjugation, chaotic cavity.

Introduction

The objective of this paper is to show that time-reversal invariance can be exploited in acoustics to accurately control wave propagation through random propagating media as well as through chaotic reverberating cavities. Surprisingly, this study proves the feasibility of time-reversal in wave systems with chaotic ray dynamics. Paradoxically, chaotic dynamics is not only harmless but also even useful, as it guarantees ergodicity and thus reduces the number of transducers needed to insure an accurate time-reversal experiment.

187

P. Sebbah (ed.), Waves and Imaging through Complex Media, 187–210.
© 2001 *Kluwer Academic Publishers. Printed in the Netherlands.*

The acoustic wave equation in a non-dissipative heterogeneous medium is invariant under a time reversal operation. Indeed, it contains only a second-order time-derivative operator. Therefore, for every burst of sound $p(\mathbf{r}, t)$ diverging from a source – and possibly reflected, refracted or scattered by any heterogeneous medium –, there exists in theory a set of waves $p(\mathbf{r}, -t)$ that precisely retrace all of these complex paths and converge in synchrony, at the original source, as if time was going backwards. This idea gives the basis of time reversal acoustics.

Taking advantage of these two properties the concept of time reversal cavity (TRC) and time reversal mirror (TRM) has been developed and several devices have been built which illustrate the efficiency of this concept [1-3]. In such a device, an acoustic source, located inside a lossless medium, radiates a brief transient pulse that propagates and is distorted by the medium. If the acoustic field can be measured on every point of a closed surface surrounding the medium (acoustic retina), and retransmitted through the medium in a time-reversed chronology, then the wave will travel back to its source and recover its original shape. Note that it requires both time reversal invariance and spatial reciprocity [4] to reconstruct the exact time-reversed wave in the whole volume by means of a two-dimensional time-reversal operation. From an experimental point of view, a TRC consists of a two-dimensional piezoelectric transducer array that samples the wavefield over a closed surface. An array pitch of the order of $\lambda/2$, where λ is the smallest wavelength of the pressure field, is needed to insure the recording of all the information on the wavefield. Each transducer is connected to its own electronic circuitry that consists of a receiving amplifier, an A/D converter, a storage memory and a programmable transmitter able to synthesize a time-reversed version of the stored signal. In practice, TRCs are difficult to realize and the TR operation is usually performed on a limited angular area, thus limiting reversal and focusing quality. A TRM consists typically of some hundred elements, or time-reversal channels.

Two types of time-reversal experiments conducted with TRMs, will be discussed. It will be shown that the wave reversibility is improved if the wave traverses a multiply scattering medium before arriving on the transducer array. The multiple scattering processes allow to redirect one part of the initial wave towards the TRM, that normally miss the transducer array. After the time-reversal operation, the whole multiply scattering medium behaves as a coherent focusing source, with a large angular aperture for enhanced resolution. As a consequence, in multiply scattering media, one is able to reduce the size and the complexity of the TRM. The same kind of improvement may be obtained for waves propagating in highly reverberant medium such as closed cavities or waveguides. Multiple reflections at the medium boundaries significantly increase the apparent aperture of the TRM. A set of experiments conducted in chaotic cavities will be presented. It will be shown that, for a reflecting cavity with chaotic boundaries, a one channel time reversal mirror is sufficient to ensure reversibility and optimal focusing. Finally, connections between time-reversal experiments and coherent backscattering enhancement will be discussed for such geometry.

1. TIME REVERSAL CAVITIES AND MIRRORS

The basic theory employs a scalar wave formulation $\phi(\mathbf{r}, t)$ and, hence, is strictly applicable to acoustic or ultrasound propagation in fluid. However, the basic ingredients and conclusions apply equally well to elastic waves in solid and to electromagnetic fields.

In any propagation experiment, the acoustic sources and the boundary conditions determine a unique solution $\phi(\mathbf{r}, t)$ in the fluid. The goal, in time-reversal experiments, is to modify the initial conditions in order to generate the dual solution $\phi(\mathbf{r}, T-t)$ where T is a delay due to causality requirements. D. Dowling, D. Jackson and D. Cassereau [4, 5] have studied theoretically the conditions necessary to insure the generation of $\phi(\mathbf{r}, T-t)$ in the entire volume of interest.

1.1. THE TIME-REVERSAL CAVITY

Although reversible acoustic retinas usually consist of discrete elements, it is convenient to examine the behavior of idealized continuous retinas, defined by two-dimensional surfaces. In the case of a time-reversal cavity, we assume that the retina completely surrounds the source. The basic time-reversal experiment can be described in the following way :

In a first step, a point-like source located at $\mathbf{r}_0$ inside a volume V surrounded by the retina surface S, emits a pulse at $t = t_0 \geq 0$. The wave equation in a medium of density $\rho(\mathbf{r})$ and compressibility $\kappa(\mathbf{r})$ is given by

$$(L_r + L_t)\phi(\mathbf{r}, t) = -A\delta(\mathbf{r} - \mathbf{r}_0)\delta(t - t_0) \tag{1}$$

$$L_r = \nabla \cdot \left(\frac{1}{\rho(\mathbf{r})}\nabla\right) \text{ and } L_t = -\kappa(\mathbf{r})\partial_{tt},$$

where A is a dimensional constant that insures the compatibility of physical units between the two sides of the equation ; for simplicity reasons, this constant will be omitted in the following. The solution to Eq. (1) reduces to the Green's function $G(\mathbf{r}, t|\mathbf{r}_0, t_0)$. Classically, $G(\mathbf{r}, t|\mathbf{r}_0, t_0)$ is written as a diverging spherical wave (homogeneous and free space case) and additional terms that describe the interaction of the field itself with the inhomogeneities (multiple scattering) and the boundaries.

We assume that we are able to measure the pressure field and its normal derivative at any point on the surface S during the interval $[0, T]$. Since time-reversal experiments are based on a two-step process, the measurement step must be limited in time by a parameter T. In all the following, we suppose that the contribution of multiple scattering decreases with time, and that T is chosen such that the information loss can be considered as negligible inside the volume V.

During the second step of the time-reversal process, the initial source at r_0 is removed and we create on the surface of the cavity monopole and dipole sources that correspond to the time-reversal of those same components measured during the first step. The time-reversal operation is described by the transform $t \rightarrow T - t$ and the secondary sources are

$$\begin{cases} \phi_s(\mathbf{r}, t) = G(\mathbf{r}, T - t | \mathbf{r}_0, t_0) \\ \partial_n \phi_s(\mathbf{r}, t) = \partial_n G(\mathbf{r}, T - t | \mathbf{r}_0, t_0) . \end{cases} \tag{2}$$

In this equation, ∂_n is the normal derivative operator with respect to the normal direction $\mathbf{n}$ to S, oriented outward. Due to these secondary sources on S, a time-reversed pressure field $\phi_{tr}(\mathbf{r}_1, t_1)$ propagates inside the cavity. It can be calculated using a modified version of the Helmholtz-Kirchhoff integral

$$\phi_{tr}(\mathbf{r}_1, t_1) =$$

$$\int\limits_{-\infty}^{+\infty} dt \int\int\limits_{S} [G(\mathbf{r}_1, t_1 | \mathbf{r}, t) \partial_n \phi_s(\mathbf{r}, t) - \phi_s(\mathbf{r}, t) \partial_n G(\mathbf{r}_1, t_1 | \mathbf{r}, t)] \frac{d^2\mathbf{r}}{\rho(\mathbf{r})} . \tag{3}$$

Spatial reciprocity and time-reversal invariance of the wave equation (1) yield the following expression :

$$\phi_{tr}(\mathbf{r}_1, t_1) = G(\mathbf{r}_1, T - t_1 | \mathbf{r}_0, t_0) - G(\mathbf{r}_1, t_1 | \mathbf{r}_0, T - t_0) . \tag{4}$$

This equation can be interpreted as the superposition of incoming and outgoing spherical waves, centered on the initial source position. The incoming wave collapses at the origin and is always followed by a diverging wave. Thus the time-reversed field, observed as a function of time, from any location in the cavity, shows two wavefronts, where the second one is the exact replica of the first one, multiplied by -1.

If we assume that the retina does not perturb the propagation of the field (free-space assumption) and that the acoustic field propagates in a homogeneous fluid, the free-space Green's function G reduces to a diverging spherical impulse wave that propagates with a sound speed c. Introducing its expression in (4) yields the following formulation of the time-reversed field :

$$\phi_{tr}(\mathbf{r}_1, t_1) = K(\mathbf{r}_1 - \mathbf{r}_0, t_1 - T + t_0) , \tag{5}$$

where the kernel distribution $K(\mathbf{r}, t)$ is given by

$$K(\mathbf{r}, t) = \frac{1}{4\pi |\mathbf{r}|} \delta \left(t + \frac{|\mathbf{r}|}{c} \right) - \frac{1}{4\pi |\mathbf{r}|} \delta \left(t - \frac{|\mathbf{r}|}{c} \right) . \tag{6}$$

The kernel distribution $K(\mathbf{r}, t)$ corresponds to the difference between two impulse spherical waves that respectively converge to and diverge from the origin of the spatial coordinate system, i.e. the location of the initial source. It results from this superposition that the pressure field remains finite for all time throughout the cavity, although the converging and diverging spherical waves show a singularity at the origin. Note that this singularity occurs at time $t_1 = T - t_0$.

The time-reversed pressure field, observed as a function of time, shows two wavefronts, where the second one is the exact replica of the first one, multiplied by -1. If we consider a wide-band excitation function instead of a Dirac distribution $\delta(t)$, the two wavefronts overlap near the focal point, therefore resulting in a temporal distortion of the acoustic signal. It can be shown that this distortion yields a temporal derivation of the initial excitation function at the focal point.

If we now calculate the Fourier transform of (6) over the time variable t, we obtain

$$\tilde{K}(\mathbf{r}, \omega) = \frac{1}{2j\pi} \frac{\sin(\omega|\mathbf{r}|/c)}{|\mathbf{r}|} = \frac{1}{j\lambda} \frac{\sin(k|\mathbf{r}|)}{k|\mathbf{r}|}, \tag{7}$$

where λ and k are the wavelength and wavenumber, respectively. As a consequence, the time-reversal process results in a pressure field that is effectively focused on the initial source position, but with a focal spot size limited to one half-wavelength. The size of the focal spot is a direct consequence of the superposition of the two wavefronts and can be interpreted in terms of the diffraction limitations (loss of the evanescent components of the acoustic fields).

A similar interpretation can be given in the case of an inhomogeneous fluid, but the Green's function G now takes into account the interaction of the pressure field with the inhomogeneities of the medium. If we were able to create a film of the propagation of the acoustic field during the first step of the process, the final result could be interpreted as a projection of this film in the reverse order, immediately followed by a re-projection in the initial order.

The apparent failure of the time-reversed operation that leads to diffraction limitation can be interpreted in the following way : The second step described above is not strictly the time-reversal of the first step : During the second step of an ideal time-reversed experiment , the initial active source (that injects some energy into the system) must be replaced by a *sink* (the time-reversal of a source). An acoustic sink is a device that absorbs all arriving energy without reflecting it. J. de Rosny has recently built such a sink in our laboratory, using the source as a diverging wavefront canceller, and has observed a focal spot size quite below diffraction limits [6].

1.2.   THE TIME-REVERSAL MIRROR

This theoretical model of the closed time-reversal cavity is interesting since it affords an understanding of the basic limitations of the time-reversed self-focusing process; but it has some limitations, particularly compared to an experimental setup :

- It can be proven that it is not necessary to measure and time-reverse both the scalar field (acoustic pressure) and its normal derivative on the cavity surface : measuring the pressure field and re-emitting the time-reversed field in the backward direction yields the same results, on the condition that the evanescent parts of the acoustic fields have vanished (propagation along several wavelengths) [7]. This comes from the fact that each transducer element of the cavity records the incoming field from the forward direction, and retransmits it (after the time-reversal operation) in the backward direction (and not in the forward direction). The change between the forward and the *backward* directions replaces the measurement and the time reversal of the field normal derivative.

- From an experimental point of view, it is not possible to measure and re-emit the pressure field at any point of a 2D surface : experiments are carried out with transducer arrays that spatially sample the receiving and emitting surfaces. The spatial sampling of the TRC by a set of transducers may introduce grating lobes. These lobes can be avoided by using an array pitch smaller than $\lambda_{min}/2$, where λ is the smallest wavelength of the transient pressure field. In this latest case, each transducer senses all the wave vectors of the incident field.

- The temporal sampling of the data recorded and transmitted by the TRC has to be at least of the order of $T_{min}/8$ (T_{min} minimum period) to avoid secondary lobes [8].

- It is generally difficult to use acoustic arrays that surround completely the area of interest, and the closed cavity is usually replaced by a TRM of finite angular aperture. This yields an increase of the point spread function dimension that is usually related to the mirror angular aperture observed from the source.

2.   TIME REVERSAL EXPERIMENTS

2.1.   TIME-REVERSAL THROUGH RANDOM MEDIA

A. Derode *et al.* [9] carried out the first experimental demonstration of the reversibility of an acoustic wave propagating through a random collection of scatterers with strong multiple scattering contributions. In an experiment such

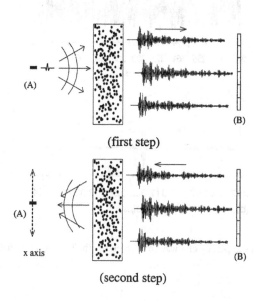

(first step)

(second step)

Figure 1.   Sketch of experiment.

as the one depicted on Fig. 1, a multiple scattering sample is placed between the source and an array made of 128 elements. The whole set-up is in a water tank. The scattering medium consists in a set of 2000 parallel steel rods (diameter 0.8 mm) randomly distributed. The sample thickness is $L = 40$ mm, and the average distance between rods is 2.3 mm. The source is 30 cm away from the TRM and transmits a short (1 μs) ultrasonic pulse (3 cycles of a 3.5 MHz). Figure 2a shows one part of the waveform received on the TRM by one of the element. It spreads over more than 300 ms, i.e $\sim$ 300 times the initial pulse duration. After the arrival of a first wavefront corresponding to the ballistic wave, a long incoherent wave is observed, which results from the multiply scattered contribution. In the second step of the experiment, the 128 signals are time-reversed and transmitted and an hydrophone measures the time reversed wave around the source location. Two different aspects of this problem have been studied : the property of the signal recreated at the source location (time compression) and the spatial property of the time-reversed wave around the source location (spatial focusing). The time-reversed wave traverses the rods back to the source, and the signal received on the source is represented on

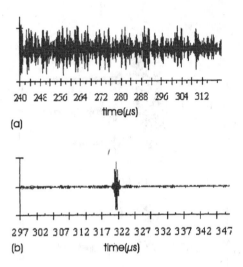

Figure 2.   (a) Typical portion of a waveform received. (b) Time-reversed field at the source transducer.

Fig. 2b: an impressive compression is observed, since the received signal lasts about 1 μs, against over 300 μs for the scattered signals. The pressure field is also measured around the source, in order to get the directivity pattern of the beam emerging from the rods after time-reversal and the results are plotted on Fig. 3. Surprisingly, multiple scattering has *not* degraded the resolution of the system : indeed, the resolution is found to be six times finer (solid line) than the classical diffraction limit (dotted line) ! This effect does not contradict the laws of diffraction, though. The intersection of the incoming wavefront with the sample has a typical size D. After time reversal, the waves travel on the same scattering paths and focus back on the source as if they were passing through a converging lens with size D. The angular aperture of this pseudo-lens is much wider than that of the array alone, hence an improvement in resolution. In other words, because of the scattering sample, the array is able to detect higher spatial frequencies than in a purely homogeneous medium. High spatial frequencies that would have been lost otherwise are redirected towards the array, due to the presence of the scatterers in a large area. This experiment shows also that the acoustic time-reversal experiments are surprisingly stable. The recorded signals have been sampled with 8-bit analog-to-digital converters that introduce quantization errors and the focusing process still works. This has to be compared to time-reversal experiments involving particles moving like balls on an elastic billiard of the same geometry. Computation of the direct

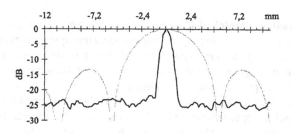

Figure 3.   Directivity patterns.

and reversed particle trajectory moving in a plane among a fixed array of some thousand concave obstacles (a Lorentz gas) shows that the complete trajectory is irreversible. Indeed, such a Lorentz gas is a well known example of chaotic system that is highly sensitive to initial conditions. The finite precision that occurs in the computer leads to an error in the trajectory of the time-reversed particle that grows exponentially with the number of scattering encounters.

Recently, R. Snieder and J. Scales [10] have performed numerical simulations to point out the fundamental difference between waves and particles in the presence of multiple scattering by random scatterers. In fact, they used time reversal as a diagnostic of wave and particle chaos : in a time reversal experiment, a complete focusing on the source will only take place if the velocity and positions are known exactly. The degree δ to which errors in theses quantities destroy the quality of focusing is a diagnostic of the stability of the wave or particle propagation. Intuitively, the consequences of a slight deviation δ in the trajectory of a billiard ball will become more and more obvious as time goes on, and as the ball undergoes more and more collisions. Waves are much less sensitive than particles to initial conditions. Precisely, in a multiple scattering situation, the critical length scale δ that causes a significant deviation at a time t in the future decreases exponentially with time in the case of particles, whereas it only decreases as the square-root of time for waves in the same situation. Waves and particles react in fundamentally different ways to perturbations of the initial conditions. The physical reason for this is that each particle follows a well-defined trajectory whereas waves travel along all possible trajectories, visiting all the scatterers in all possible combination. While a small error on the initial velocity or position makes the particle miss one obstacle and completely change its future trajectory, the wave amplitude is much more stable because it results from the interference of all the possible trajectories and small errors on the transducer operations will sum up in a linear way resulting in small perturbation.

Time-reversal as a matched-filter or time correlator.   As any linear and time-invariant process, wave propagation through a multiple scattering medium may be described as a linear system with different impulse responses. If a source, located at r_0 sends a Dirac pulse $\delta(t)$, the j-th transducer of the TRM will record the impulse response $h_j(t)$ that corresponds, for a point transducer, to the Green function $G(\mathbf{r}, t | \mathbf{r}_0, 0)$. Moreover, due to reciprocity, $h_j(t)$ is also the impulse response describing the propagation of a pulse from the j-th transducer to the source. Thus, neglecting the causal time delay T, the time-reversed signal at the source is equal to the convolution product $h_j(t) * h_j(-t)$.

This convolution product, in terms of signal analysis, is typical of a *matched filter*. Given a signal as input, a matched filter is a linear filter whose output is optimal in some sense. Whatever the impulse response $h_j(t)$, the convolution $h_j(t) * h_j(-t)$ is maximum at time $t = 0$. This maximum is always positive and equals $\int h_j^2(t)dt$, i.e the energy conveyed by $h_j(t)$. This has an important consequence. Indeed, with an N-elements array, the time-reversed signal recreated on the source writes as a sum

$$\phi_{tr}(\mathbf{r}_0, t) = \sum_{j=1}^{j=N} h_j(t) * h_j(-t) \,. \tag{8}$$

Even if the $h_j(t)$ are completely random and apparently uncorrelated signals, each term in this sum reaches its maximum at time $t = 0$. So all contributions add constructively around $t = 0$, whereas at earlier or later times uncorrelated contributions tend to destroy one another. Thus the re-creation of a sharp peak after time-reversal on a N-elements array can be viewed as an interference process between the N outputs of N matched filters. The robustness of the TRM can also be accounted for through the matched filter approach. If for some reason, the TRM does not exactly retransmits $h_j(-t)$ but rather $h_j(-t) + n_j(t)$, where $n_j(t)$ is an additional noise on channel j, then the re-created signal is written :

$$\sum_{j=1}^{j=N} h_j(t) * h_j(-t) + \sum_{j=1}^{j=N} h_j(t) * n(t) \,.$$

The time reversed signals $h_j(-t)$ are tailored to exactly match the medium impulse response, which results in a sharp peak. Whereas an additional small noise is not matched to the medium and, given the extremely long duration involved, it generates a low-level long-lasting background noise instead of a sharp peak.

Time reversal as a spatial correlator.   Another way to consider the focusing properties of the time-reversed wave is to follow the impulse response approach and treat the time-reversal process as a spatial correlator. If we note $h'_j(t)$ the propagation impulse response from the j-th element of the array to an observation point $\mathbf{r}_1$ different from the source location $\mathbf{r}_0$, the signal recreated in $\mathbf{r}_1$ at time $t_1 = 0$ writes :

$$\phi_{tr}(\mathbf{r}_1, 0) = \int h_j(t) h'_j(t) dt. \qquad (9)$$

Notice that this expression can be used as a way to define the directivity pattern of the time-reversed waves around the source. Now, due to reciprocity, the source S and the receiver can be exchanged, i.e. $h'_j(t)$ is also the signal that would be received in $\mathbf{r}_1$ if the source was the j-th element of the array. Therefore, we can imagine this array element is the source, and the transmitted field is observed at two points $\mathbf{r}_1$ and $\mathbf{r}_0$. The spatial correlation function of this wavefield would be $\left\langle h_j(t) h'_j(t) \right\rangle$ where the impulse response product is averaged on different realizations of the disorder. Therefore, Eq. (9) can be viewed as an estimator of this spatial correlation function. Note that in one time-reversal experiment we have only access to a single realization of the disorder. However, the ensemble average can be replaced by a time average, a frequency average or by a spatial average on a set of transducers.

In that sense, the spatial resolution of the time-reversal mirror (i.e. the -6 dB width of the directivity pattern) is simply an estimate of the correlation length of the scattered wavefield [11].

This has an important consequence. Indeed, if the resolution of the system essentially depends on correlation properties of the scattered wavefield, it should become independent from the array's aperture. This is confirmed by the experimental results. Fig. 4 presents the directivity patterns obtained through a 40 mm-thick multiple scattering sample, using either 1 array element or the whole array (122 elements) as a time-reversal mirror. In both cases, the spatial resolution at -6dB is the same : ~ 0.85 mm. In total contradiction with what happens in a homogeneous medium, enlarging the aperture of the array does not change the -6dB spatial resolution. However, even though the number N of active array elements does not influence the typical width of the focal spot, it has a strong impact on the background level of the directivity pattern ($\sim$ -12 dB for $N = 1$, $\sim$ -28 dB for $N = 122$), as can be seen on Fig. 4.

Finally, the fundamental properties of time-reversal in a random medium rely on the fact that it is both a space and time correlator, and the time-reversed waves can be viewed as an estimate of the space and time auto-correlation functions of the waves scattered by a random medium. The estimate becomes better with a large number of transducers in the mirror.

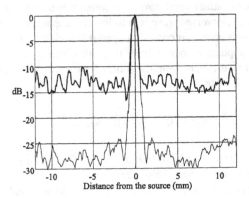

Figure 4.   Directivity patterns with $N = 122$ transducers (thin line) and $N = 1$ transducer (thick line).

Moreover, the system is not sensitive to a small perturbation since adding a small noise to the scattered signals (e.g. by digitizing them on a reduced number of bits) may alter the noise level but does not drastically change the correlation time or the correlation length of the scattered waves. Even in the extreme case where the scattered signals are digitized *on a single bit*, A. Derode has recently shown [12] that the time and space resolution of the TRM were practically unchanged, which is a striking evidence for the robustness of wave time-reversal in a random medium.

2.2.   TIME-REVERSAL IN BOUNDED MEDIA

In the time-reversal cavity approach, the transducer array samples a closed surface surrounding the acoustic source. In the last paragraph, we have seen how the multiple scattering processes in a large sample widen the effective TRM aperture. The same kind of improvement may be obtained for waves propagating in a waveguide or in a cavity. Multiple reflections at the medium boundaries significantly increase the apparent aperture of the TRM. The basic idea is to replace one part of the TRC transducers by reflecting boundaries that redirect one part of the incident wave towards the TRM aperture. Thus spatial information is converted into the time domain and the reversal quality crucially depends on the duration of the time-reversal window, i.e. the length of the recording to be reversed.

Time-reversal in waveguides.   Experiments conducted by P. Roux in rect-angular ultrasonic waveguides have shown the effectiveness of the TR processing to compensate for multi-path effects [13]. Impressive time recompression has been observed, that compensate for reverberation and dispersion. Besides, as in the multiple scattering experiment, the TR beam is focused on a spot which is much thinner than the one observed in free water. This can be interpreted by the theory of images in a medium bounded by two mirrors. For an observer, located at the source point, the TRM seems to be escorted by an infinite set of virtual images related to multi-path propagation, and effective aperture 10 times larger than the real aperture have been observed.

Acoustic waveguides are also currently found in underwater acoustic, especially in shallow water, and TRMs can compensate for the multi-path propagation in oceans that limits the capacity of underwater communication systems. The problem arises because acoustic transmission in shallow water bounces off the ocean surface and floor, so that a transmitted pulse gives rise to multiple copies of it that arrive at the receiver. Recently, underwater acoustic experiments have been conducted by W. Kuperman and his group from San Diego University in a sea water channel of 120 m depth, with a 24 elements TRM working at 500 Hz and 3.5 kHz. They observed focusing and multi-path compensation at distances up to 30 km [14].

Time-reversal in chaotic cavities.   In this paragraph, we are interested in another aspect of multiply reflected waves : waves confined in closed reflecting cavities such as elastic waves propagating in a silicon wafer. With such boundary conditions, no information can escape from the system and a reverberant acoustic field is created. Moreover, if the cavity shows ergodic and mixing properties and negligible absorption, one may hope to collect all information at only one point. C. Draeger *et al.* [15-17] have shown experimentally and theoretically that, in this particular case, a time-reversal can be obtained *using only one TR channel.* The field is measured at one point over a long period of time and the time-reversed signal is re-emitted at the same position.

The experiment is 2D and has been carried out by using elastic surface waves propagating along a D-shaped mono-crystalline silicon wafer. This geometry is chosen to avoid quasi-periodic orbits. Silicon was selected for its weak absorption. The elastic waves which propagate in such a plate are Lamb waves. An aluminum cone coupled to a longitudinal transducer generates these waves at one point of the cavity. A second transducer is used as a receiver. The central frequency of the transducers is 1 MHz and its bandwidth is 100 %. At this frequency, only three Lamb modes are possible (one flexural, two extensional). The source is isotropic and considered as point-like because the cone tip is much smaller than the central wavelength. A heterodyne laser interferometer measures the displacement field as a function of time at different points on the cavity. Assuming that there are nearly no mode conversion between the flexural

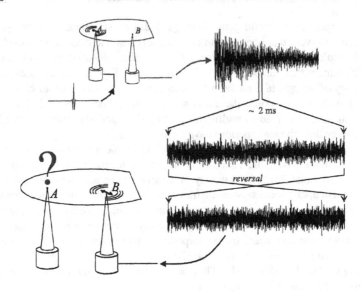

Figure 5.   Experimental procedure: see details in the text.

mode and other modes at the boundaries, we only have to deal with one mode, the flexural-scalar field.

The experiment is a two step-procedure as described on Fig. 5: In the first step, one of the transducers, located at point A, transmits a short omnidirectional signal of duration 0.5 μs into the wafer. Another transducer, located at B, records a very long (50 ms) chaotic signal, which results from multiple reflections (some hundred) of the incident pulse at the boundary of the cavity. A 2-ms-long portion of the signal is selected, time-reversed and re-emitted at point B. As the time reversed wave is a flexural wave that induces vertical displacements of the silicon surface, it can be detected using an optical interferometer that scans the surface around point A (Fig. 5).

One observes both an impressive time recompression at point A and a refocusing of the time-reversed wave around the origin (Figs. 6a and 6b), with a focal spot whose radial dimension is equal to half the wavelength of the flexural wave. Due to the reflections at the boundary, the time-reversed wave field converges towards the origin from all directions and gives a circular spot, like the one that could be obtained with a closed time reversal cavity covered with transducers. The 2 ms time-reversed waveform is the temporal sequence needed to focus exactly on point A.

The success of this time-reversal experiment is particularly interesting in two respects. Firstly, it proves again the feasibility of time-reversal in wave

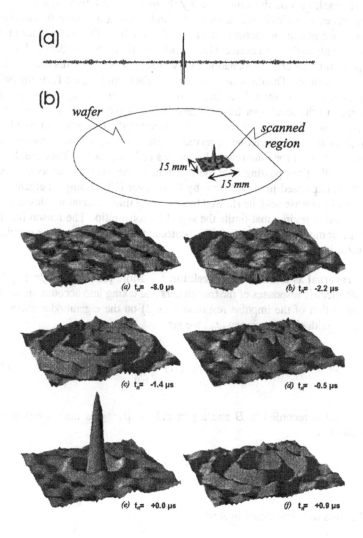

Figure 6.   (a) Time-reversed signal observed at point A. The observed signal is 210 μs long. (b) Time-reversed wavefield observed at different times around point A on a square of 15 mm × 15 mm.

systems with chaotic ray dynamics. Paradoxically, in the case of one-channel time-reversal, chaotic dynamics is not only harmless but also even useful, as it guarantees ergodicity and mixing. Secondly, using a source of vanishing aperture, we obtain an almost perfect focusing quality. The performance here is comparable to that of a closed TRC, which has an aperture of $360°$. Hence, a one-point time-reversal in a chaotic cavity produces better results than a TRM in an open system. Thanks to the reflections at the boundary, the focusing quality is not aperture limited. In addition, the time-reversed collapsing wavefront converges to the focal spot from all directions. Although one obtains excellent focusing, a one-channel time-reversal is not perfect, as a weak noise level throughout the system can be observed. Residual temporal and spatial sidelobes persist even for time-reversal windows of infinite size. They are due to multiple reflections passing over the locations of the TR transducers and they have been expressed in closed form by C. Draeger [17]. Using an eigenmode analysis of the wavefield, he showed that, for long time reversal windows, there is a saturation regime that limits the signal-to-noise ratio. The reason for this limit is that modes with nodes exactly located at the detector or emitter positions are missed.

Time-reversal as a temporal correlator.   More precisely, neglecting the acousto-electric responses of the transducers and taking into account the modal decomposition of the impulse response $h_{AB}(t)$ on the eigenmodes $\psi_n(\mathbf{x})$ of the cavity with eigenfrequency ω_n , we get

$$h_{AB}(t) = \sum_n \psi_n(A)\psi_n(B)\frac{\sin(\omega_n t)}{\omega_n} \quad (t > 0). \qquad (10)$$

This signal is recorded in B and a part $\Delta T = [t_1; t_2]$ is time-reversed and re-emitted as

$$h_{AB}^{\Delta T}(-t) = \begin{cases} h(-t) & t \in [-t_2, -t_1] \\ 0 & \text{elsewhere} \end{cases} \qquad (11)$$

So the time-reversed signal in A reads:

$$\phi_{tr}^{\Delta T}(t) = \int_{t_1}^{t_2} d\tau\, h_{AB}(t_R + \tau)h_{AB}(\tau)$$

$$= \sum_n 1/\omega_n \psi_n(A)\psi_n(B) \sum_m 1/\omega_m \psi_m(A)\psi_m(B).I_{mn} , \qquad (12)$$

with I_{mn} equal to :

$$I_{mn} = \int_{t_1}^{t_2} d\tau \sin(\omega_m \tau) \sin(\omega_m(\tau + t))$$

$$= \tfrac{1}{2} \sin(\omega_n t) \int_{t_1}^{t_2} d\tau \left[\sin\left((\omega_m - \omega_n)\tau\right) + \sin\left((\omega_m + \omega_n)\tau\right)\right] \qquad (13)$$

$$+ \tfrac{1}{2} \cos(\omega_n t) \int_{t_1}^{t_2} d\tau \left[\cos\left((\omega_m - \omega_n)\tau\right) - \cos\left((\omega_m + \omega_n)\tau\right)\right].$$

The second term of each integral gives a contribution of order $1/(\omega_m + \omega_n) \ll \Delta T$ which can be neglected. Thus we obtain

$$I_{mn} \cong$$

$$\begin{cases} \tfrac{1}{2} \frac{\sin(\omega_n t)}{\omega_m - \omega_n} \left[-\cos\left((\omega_m - \omega_n)t_2\right) + \cos\left((\omega_m - \omega_n)t_1\right)\right] & \\ \qquad\qquad\qquad & \text{if } \omega_m \neq \omega_n \\ + \tfrac{1}{2} \frac{\cos(\omega_n t)}{\omega_m - \omega_n} \left[\sin\left((\omega_m - \omega_n)t_2\right) - \sin\left((\omega_m - \omega_n)t_1\right)\right] & \\ \qquad\qquad \tfrac{1}{2}\Delta T \cos(\omega_n t) & \text{if } \omega_m = \omega_n \end{cases}$$

$$(14)$$

Under the assumption that the eigenmodes are not degenerated, then $\omega_n = \omega_m \Leftrightarrow n = m$, and the second term represents the diagonal elements of the sum over n and m. The first term is only important if the difference $\omega_n - \omega_m$ is small, i.e. for neighboring eigenfrequencies. In the case of a chaotic cavity, nearest neighbors tend to repulse each other and if the characteristic distance $\Delta \omega$ is sufficiently large so that $\Delta T \gg 1/\Delta \omega$, the non-diagonal terms of I_{mn} are negligible compared to the diagonal contributions and one obtains

$$I_{mn} = \tfrac{1}{2}\delta_{mn}\Delta T \cos(\omega_n t) + O\left(1/\Delta \omega\right) . \qquad (15)$$

In the limit $\Delta T \to \infty$, the time-reversed signal observed in A by a reversal in B is given by

$$\phi_{tr}^{\Delta T}(t) = \tfrac{1}{2}\Delta T \sum_n \frac{1}{\omega_n^2} \psi_n^2(A)\psi_n^2(B) \cos(\omega_n t) . \qquad (16)$$

This expression gives a simple interpretation of the residual temporal lobes which are observed experimentally in Fig. 6. The time-reversed signal observed at the origin cannot be simply reduced to a Dirac distribution $\delta(t)$, but is equal, even for $\Delta T \gg 1/\Delta\omega$, to another cross-correlation product:

$$C_{\Delta T}(t) = h_{AA}(-t) \underset{(\Delta T)}{*} h_{BB}(t)$$

$$= \int_{t_1}^{t_2} d\tau\, h_{BB}(t+\tau) h_{AA}(\tau) \tag{17}$$

$$= \sum_n 1/\omega_n \psi_n^2(B) \sum_m 1/\omega_m \psi_m^2(A) I_{mn},$$

where the impulse responses h_{AA} and h_{BB} describe the backscattering properties of A and B due to the boundaries of the cavity. Each impulse response is composed of a first peak at $t = 0$ followed by multiple reflections that pass over the source point even after the excitation has ended. Hence, the signal observed in A, after a Dirac excitation, can be described by $h_{AA}(t)$. Therefore, a perfect time-reversal operation (i.e. we simply count time backwards) would give in A the signal $h_{AA}(-t)$, i.e. some multiple reflections with a final peak at $t = 0$. For the same reason, the reversed point B cannot exactly transmit any waveform in the cavity. Due to the boundaries, a Dirac excitation at B will also give rise to a transmitted signal $h_{BB}(t)$. So, in the limit of a very long time-reversal window we get for a one channel time-reversal experiment the cavity formula deduced by C. Draeger

$$h_{AB}(-t) * h_{BA}(t) = h_{AA}(-t) * h_{BB}(t). \tag{18}$$

Time reversal as a spatial correlator.   As for the multiple scattering medium, focusing properties of the time-reversed wave can be calculated using the spatial correlator approach. If we note $h_{A'B}(t)$ the propagation impulse response from point B to an observation point A' (with coordinates $\mathbf{r}_1$) different from the source location A the time-reversed signal recreated at A' at time $t_1 = 0$ writes:

$$\phi_{tr}(\mathbf{r}_1, 0) = \int h_{AB}(t)\, h_{A'B}(t)\, dt. \tag{19}$$

Thus the directivity pattern of the time-reversed wavefield is given by the cross correlation of the Green functions that can be developed on the eigenmodes of the cavity

$$\phi_{tr}(\mathbf{r}_1, 0) = \sum_n \frac{1}{\omega_n^2} \psi_n(A)\psi_n(\mathbf{r}_1)\psi_n^2(B). \tag{20}$$

Note that in a real experiment one has to take into account the limited bandwidth of the transducers, so a spectral function $F(\omega)$ centered on frequency ω_c, with bandwidth $\Delta\omega$, must be introduced and we can write Eq. (20) in the form

$$\phi_{tr}(\mathbf{r}_1, 0) = \sum_n \frac{1}{\omega_n^2} \psi_n(A) \psi_n(\mathbf{r}_1) \psi_n^2(B) F(\omega_n). \tag{21}$$

Thus the summation is limited to a finite number of modes, that is typically of the order of a few hundreds in our experiment. As we do not know the exact eigenmode distribution for each chaotic cavity, we cannot evaluate this expression directly. However, one may use a statistical approach and consider the average over different realizations, which consists in summing over different cavity realizations. So we replace in Eq. (20) the eigenmodes product by their expectation values $\langle ... \rangle$. We use also a qualitative argument proposed by Berry [18-20] to characterize irregular modes in chaotic system. If chaotic rays support an irregular mode, it can be considered as a superposition of a large number of plane waves with random direction and phase. This implies that the amplitude of an eigenmode has a Gaussian distribution with $\langle \Psi_n^2 \rangle = \sigma^2$ and a short-range isotropic correlation function given by a Bessel function that writes:

$$\langle \psi_n(A) \psi_n(\mathbf{r}_1) \rangle = J_0(2\pi |\mathbf{r}_1 - \mathbf{r}_0| / \lambda_n), \tag{22}$$

where λ_n is the wavelength corresponding to ω_n. If A and A' are sufficiently far apart from B not to be correlated, then

$$\langle \psi_n(A) \psi_n(\mathbf{r}_1) \psi_n^2(B) \rangle = \langle \psi_n(A) \psi_n(\mathbf{r}_1) \rangle \langle \psi_n^2(B) \rangle. \tag{23}$$

One obtains finally:

$$\langle \phi_{tr}(\mathbf{r}_1, 0) \rangle = \sum_n \frac{1}{\omega_n^2} J_0(2\pi |\mathbf{r}_1 - \mathbf{r}_0| / \lambda_n) \sigma^2 F(\omega_n). \tag{24}$$

We have shown that the experimental results agree quite well with this prediction and that, in a chaotic cavity, the spatial resolution is independent of the time reversal mirror aperture. Indeed, with a one-channel time-reversal mirror, the directivity pattern at $t = 0$ is close to the Bessel function $J_0(2\pi |\mathbf{r}_1 - \mathbf{r}_0| / \lambda_c)$ where λ_c is the wavelength corresponding to the central frequency of the transducers.

One can also see on Fig. 6b that a very good estimate of the eigenmode correlation function is experimentally obtained with only one cavity realization. A one-channel omnidirectional transducer is able to perfectly refocus a wave in a chaotic cavity, and we have not averaged the data on different cavities or on different positions of the transducer B.

Time reversal versus Phase conjugation.   This interesting result emphasizes the great interest of time-reversal experiments, compared to phase conjugated experiments. In *phase conjugation*, one works only with monochromatic waves and not with broadband pulses. For example, if one works only at frequency ω_n, so that there is only one term in Eq. (21), one cannot refocus a wave at point A. An omnidirectional transducer, located at any position B, working in monochromatic mode, will send a diverging wave in the cavity which has no reason to refocus on point A. The refocusing process works only with broadband pulses, with a large number of eigenmodes in the transducer bandwidth. Here, the averaging process that gives a good estimate of the spatial correlation function is not obtained by summing over different realizations of the cavity, like in Eq.(24), but by a sum over "pseudo-realizations" which corresponds to the different modes in the same cavity. This come from the fact that in a chaotic cavity, one may assume a statistical independence between the different eigenmodes. As the number of eigenmodes available in the transducer bandwidth increases, the refocusing quality becomes better and the focal spot pattern becomes closer to the ideal Bessel function. Hence, the signal to noise level should increase as the square-root of the number of modes in the transducer bandwidth.

A similar result has also been observed in the time-reversal experiment conducted in multiply scattering medium. A clear refocusing was obtained with only one single array element. (Fig. 4). The focusing process works with broadband pulses (the transducer center frequency is 3.5 MHz with a 50 % bandwidth at -6 dB), while for each individual frequency there is no focusing. The key of time-reversal focusing is the statistical independence of wavefields at different frequencies. Thus for a large bandwidth the time-reversed field is self-averaging and focusing with one transducer is obtained for one realization of the disorder, which is not possible in a phase conjugating experiment.

2.3.   TIME REVERSAL AND BACKSCATTERING ENHANCEMENT

Time-reversal experiments are not the only ones to bring to the fore coherent interference effects in random media. Different experiments have shown that coherent interference effects survive in random media despite disorder. In particular, they are manifested in the coherent backscattering effect, which was first observed in optics in 1985 [21], and later in acoustics in 1997 [22]. It is well explained by constructive interference between a multiple scattering path and its reciprocal counterpart in the exact backscattering direction, which yields a factor of two in the averaged intensity compared to other directions. To observe the backscattering enhancement with monochromatic signals, averaging over realizations is performed. Similar results have also been obtained for elastic waves propagating in a 2D chaotic cavity [23, 24]. However the main difference with optics is that, due to broadband pulsed excitation, coherent backscattering effect can be observed with a single realization. By assuming that a Dirac pulse

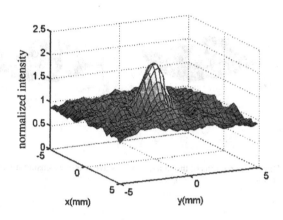

Figure 7.   Spatial distribution of the normalized intensity.

is transmitted in a 2D chaotic cavity by a point transducer located at point A (Fig 5), the field intensity $I(A, B)$ can be measured at different points B of the cavity with an optical probe scanning the surface. For one realization, the spatial distribution of the time integrated squared impulse responses $h_{AB}(t)$ is computed as:

$$I(A, B) = \int h_{AB}^2 (t)dt \qquad (25)$$

and plotted on Fig 7. The intensity is stronger at the source point ($B = A$) than the background intensity ($A \neq B$) by a factor about 2 as expected from reciprocity arguments.

A comparison between Eq. (25) and Eq. (19) shows that the field intensity $I(A, B)$ is nothing else than the amplitude of the wave observed at the source position A at time $t = 0$ after a TR process from point B. Hence the coherent backscattering effect answers the following question : where is the best point to time reverse a wave in a chaotic cavity with only one point transducer? The answer is that the peak of the time reversed wave is twice as high if the initial source point A and the time reversal point B are at the same location. This is clearly shown on Fig. 8. The amplitude of the time-reversed wave is nearly

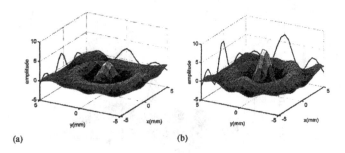

(a)   (b)

Figure 8.   Directivity patterns of the time-reversed wave at the refocusing time for $A \neq B$ (a) and for $A = B$ (b)

2 times greater when $A = B$. C. Draeger has also shown that another reason for choosing point A as the time-reversed point is that the residual background noise around the focal spot is also weaker than in the case $A \neq B$ [16].

Conclusion

In this paper, we have presented a study of how multiple scattering in random media and multiple reflection in chaotic cavities enhance resolution in time-reversal acoustics. We have also shown that working in the time domain allows self averaging of the spatial correlation. Multiple scattering makes the effective size of the TRM much larger than its physical size. This study has proven the feasibility of acoustic time-reversal in media with chaotic ray dynamics. Paradoxically, chaotic dynamics is useful as it reduces the number of transducers needed to insure an accurate time-reversal experiment. The connection between time-reversal experiments and coherent backscattering enhancement has also been discussed.

Acknowledgments

The authors wish to recall that most of the work on time reversal in chaotic cavities was done by Carsten Draeger during his doctoral studies in Paris. He was an outstanding researcher, full of imagination and curiosity. He made the first experimental demonstration of reversibility in a chaotic cavity and then gave a very clear theoretical interpretation of these effects. Carsten Draeger died in April 1999 during his post doc at Rochester (USA) and we dedicate this review to his memory.

References

[1] M. Fink, C. Prada, F. Wu and D. Cassereau, IEEE Ultrasonics Symposium Proceedings, Montreal , Vol 1, pp 681-686, (1989).

[2] M. Fink, IEEE Trans. Ultrason. Ferroelec. Freq. Contr, **39** (5) 555-566 (1992).

[3] M. Fink, Physics Today, **50**, 3, 34-40 (1997).

[4] D. R. Jackson and D. R. Dowling, J. Acoust. Soc. Am **89** (1), 171(1991).

[5] D. Cassereau and M. Fink, IEEE Trans. Ultrason. Ferroelec. Freq. Contr, **39** (5), 579-592 (1992).

[6] J. de Rosny and M. Fink to be published.

[7] D. Cassereau and M. Fink, J Acoust. Soc Am, **96** (5), 3145- 3154 (1994).

[8] G. S. Kino, *Acoustics Waves*, Prentice Hall, Signal Processing Series, 1987.

[9] A. Derode, P. Roux and M. Fink, Phys. Rev. Lett. **75**, 23, 4206-4209 (1995)

[10] R.Snieder, J. Scales, Phys. Rev. E **58** (5), 5668-5675 (1998).

[11] A. Derode, A. Tourin and M. Fink, J. Acoust. Soc. Am. **107**, 6, 2987-2998 (2000).

[12] A. Derode, A. Tourin and M. Fink, J. Appl. Phys., **85** (9), p.6343-6352 (1999).

[13] P. Roux, B. Roman and M. Fink, Appl. Phys. Lett., **70**, p.1811-1813 (1997).

[14] W. A. Kupperman, W. S. Hodgkiss, H. C. Song, T. Akal, T. Ferla and D. Jackson, J .Acoust. Soc. Am, **103**, 25-40 (1998).

[15] C. Draeger and M. Fink, Phys. Rev. Lett., **79** (3), 407-410, (1997).

[16] C. Draeger and M. Fink, J. Acoust Soc Am, **105** (2), p.618-625, (1999).

[17] C. Draeger and M. Fink, J. Acoust Soc Am, **105** (2), p.611-617, (1999).

[18] M. V. Berry in Les Houches 1981 - *chaotic behaviour of deterministic systems* (North Holland, Amsterdam, 1983), pp. 171-271.

[19] S. W. McDonald and A. N. Kaufman, Phys. Rev. A **37** (8), 3067-3086 (1988).

[20] R. L. Weaver and J. Burkhardt, J. Acoust. Soc. Am. **96**, 3186 (1994).

[21] M.P. van Albada and A. Lagendijk, Phys. Rev. Lett. **55**, 2692 (1985); P. E. Wolf and G. Maret, Phys. Rev. Lett. **55**, 2696 (1985).

[22] A. Tourin, A. Derode, P. Roux, B.A. Van Tiggelen and M. Fink, Phys. Rev. Lett. **79**, 3637 (1997).

[23] J. de Rosny, A. Tourin and M. Fink, Phys. Rev. Lett. **84**, 1693 (2000).

[24] R.L. Weaver and O.I. Lobkis, Phys. Rev. Lett. **84**, 4942 (2000).

IV

IMAGING IN HETEROGENEOUS MEDIA: BALLISTIC PHOTONS

Chapter 1

IMAGING BIOLOGICAL TISSUE USING PHOTOREFRACTIVE HOLOGRAPHY AND FLUORESCENCE LIFETIME

N. P. Barry, M. J. Cole, M. J. Dayel, K. Dowling, P. M. W. French, S. C. W. Hyde, R. Jones, D. Parsons-Karavassilis and M. Tziraki
Femtosecond Optics Group, Physics Dept.
Imperial College
London SW7 2BZ, U.K.
Tel: +44-171-594 7784, Fax: +44-171-594 7782
email: paul.french@ic.ac.uk

M. J. Lever
Department of Biological and Medical Systems
Imperial College
London SW7 2BY, U.K.

K. M. Kwolek, D. D. Nolte and M. R. Melloch
Purdue University
West Lafayette, IN 47907-1396, USA

M. A. A. Neil, R. Juškaitis and T. Wilson
Department of Engineering Science
University of Oxford
Parks Road, Oxford OX1 3PJ, U.K.

A. K. L. Dymoke-Bradshaw and J. D. Hares
Kentech Instruments Ltd., Unit 9, Hall Farm Workshops
South Moreton, Didcot, Oxon., OX11 9AG, U.K.

213

P. Sebbah (ed.), Waves and Imaging through Complex Media, 213–234.
© 2001 *Kluwer Academic Publishers. Printed in the Netherlands.*

Abstract   This article reviews two approaches to biomedical imaging, namely photorefractive holography as a means of realising depth-resolved imaging through turbid media and fluorescence lifetime imaging as a spectroscopic imaging modality.

Keywords: holography, imaging, photorefractive, fluorescence lifetime, microscopy, 3-D imaging, ultrafast lasers

Introduction

This chapter reviews our work at Imperial College towards the development of biomedical optical technologies for realizing functional imaging in tissue. Photorefractive holography is intended to provide a means of achieving 3-D imaging in scattering media using ballistic (unscattered) light in the near infra-red. It aims to contrast different tissue components by their absorption and scattering properties. Fluorescence lifetime imaging is an approach that provides an additional means of contrasting tissues that complements spectral and intensity information. This technique may be used with ballistic or scattered light and has been used in 2-D and 3-D imaging systems.

1.   PHOTOREFRACTIVE HOLOGRAPHY

Much research has focused on methods to form images through thin tissue depths using ballistic (unscattered) light, which is normally obscured by the scattered diffuse background. The work reported here has been motivated by a desire to develop imaging systems for "in vivo" applications such as screening for skin cancer, for which rapid image acquisition time is a priority. To this end we have concentrated on developing "whole-field" imaging techniques, which acquire all the pixel information in parallel, while accepting compromises in achievable resolution and sensitivity compared to confocal scanning techniques. The imaging technique discussed here uses holography in photorefractive media [2] as a means of discriminating in favour of the unscattered light that retains coherence with a reference beam. This is related to coherence gating techniques previously reported by various groups including optical heterodyne detection [3] and optical coherence tomography [4], as well as to conventional, electronic [5] and light-in-flight holography [6, 7]. It fundamentally differs from other techniques, however, in that this coherent detection system is, to first order, insensitive to the incoherent diffuse light background. Images may be acquired using back-scattered light in time-of-flight reflection geometry, exploiting low coherence interferometry to reconstruct 3-D images. The use of reusable photorefractive holographic media, such as bulk Rhodium-doped barium titanate ($Rh:BaTiO_3$) crystals or photorefractive multiple quantum well (MQW) devices allows for fast holographic image recording and the potential for real-time read-out at frame rates significantly exceeding video rate. Time-gated holography was applied to biomedical imaging by Spears *et al.* [8], following earlier experiments in light-in flight holography by Abramson [6]. Depth-resolved photorefractive holography through turbid media was first demonstrated at Imperial College by Hyde *et al.* [9, 10]. Initial experiments performed using a crystal of Rhodium-doped Barium Titanate (Rh: $BaTiO_3$), which entails the hologram being read out by a Bragg-matched beam and recorded on a CCD camera. The high dynamic range of $Rh:BaTiO_3$ permits imaging through relatively high (16MFP) scattering media [11]. However, the long response time of the $Rh:BaTiO_3$ means that image acquisition times as high as 100's of seconds must be used when imaging through strong scattering media and this is clearly unacceptable for most biomedical applications. There is therefore a requirement to use photorefractive media with a much faster response. One possible candidate is a photorefractive GaAs/AlGaAs multiple quantum well (MQW) device. The latter have been the subject of our recent research, which is described in the next section.

1.1. PHOTOREFRACTIVE HOLOGRAPHY WITH Rh:BaTiO₃

Rhodium-doped barium titanate is a semi-insulating crystal that is also an electro-optic medium. The recording of photorefractive holograms in $Rh:BaTiO_3$ is illustrated in Fig. 1. The incident object and reference beams interfere to

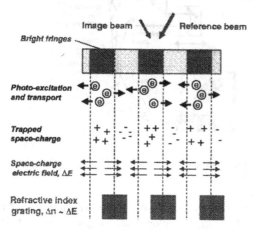

Figure 1. Holographic recording in an electro-optic medium such as Rhodium-doped Barium titanate

form a fringe pattern on the photorefractive crystal. In the light regions, photo-carriers (electrons) are generated that drift until they are trapped by defects in the crystal. This results in an accumulation of positive charge in the bright regions and negative charge in the dark regions. This space charge produces a distribution of local electric fields that modify the local refractive index through the electro-optic effect. Thus the hologram is recorded as a refractive index grating in the medium. This process requires the spatial modulation of the incident light field, i.e. the photorefractive effect increases as the spatial derivative of the incident light intensity distribution, rather than with the intensity as is the case for most integrating detectors such as photodiodes, CCD arrays or photographic film. A uniform intensity distribution would produce no space charge and hence no modulation of the refractive index in a photorefractive recording medium. In this manner it is possible to record a weak hologram in the presence of a strong (uniform) background of diffuse scattered light that would saturate a conventional (integrating) detector.

We first demonstrated depth-resolved photorefractive holography using the apparatus shown in Fig. 2, with a mode-locked Ti:Sapphire laser producing 3 ps, or later 100 fs [10] pulses used as the source. The holograms were recorded in

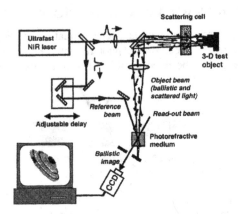

Figure 2.   Experimental configuration for holographic imaging through a scattering medium using a bulk Rh:BaTiO$_3$ crystal

Rh:BaTiO$_3$ and read out by a Bragg-matched read-out beam and recorded onto a CCD camera. As with any holographic imaging system, holograms could only be written when the interferometer arm lengths were matched to within the coherence length of the source ($\sim 30\ \mu$m). Using a short coherence length source therefore provided depth resolution, which we demonstrated using a 3-D test object consisting of a series of concentric aluminium cylinders, ranging from 1 to 5 mm in diameter and separated in depth by 100 μm. This object was unpolished and provided no significant specular reflection, but back-scattered $\sim 8\%$ of the incident light. By varying the delay of the reference arm, a set of depth-resolved images sufficient to reconstruct a 3-D image of the object could be obtained. The depth resolution of the images was 100 μm, while the transverse spatial resolution was approximately 30 μm as predicted from the pulse duration and beam geometry [11]. Figure 3 shows reconstructed images of the 3-D test object recorded through an aqueous suspension of polystyrene spheres of (a) 0, (b) 12 and (c) 14 MFP scattering depth. It can be seen that while the signal to noise ratio of the image decreases with scattering depth, the spatial resolution does not. Using a USAF test chart as the object, we obtained an image of 100 μm bars through 16 MFP of scattering solution. The minimum intensity required in the object beam to record a usable hologram was 10 nW/cm^2 (for 300 seconds integration).

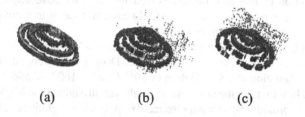

(a)   (b)   (c)

Figure 3. Computer reconstructions of 3-D objects using depth resolved holographic images recorded in Rh:BaTiO$_3$. Images recorded through scattering medium of (a) 0 (b) 12 and, (c) 14 MFP scattering depths. [From [12], Fig. 5].

The discrimination against a diffuse background noise signal was investigated quantitatively by simulating an incoherent scattered light background using a third beam for which 100 fs pulses were delayed by more than their coherence length, relative to the signal and reference beam pulses. It was possible to record a useful hologram in the presence of an incoherent background that was 10^6 times more powerful than the signal beam [13]. This dynamic range exceeds that of many detectors and was ultimately limited by scattering off inhomogeneities in the crystal and by beam fanning. It illustrates the potential of photorefractive holography to provide a unique means of rejecting unwanted scattered light when imaging through turbid media. Unfortunately, the long response time of Rh:BaTiO$_3$ is unacceptable for most biomedical applications and there is therefore a requirement to use photorefractive media with a much faster response. Two such media are bulk semiconductors, such as Cadmium Telluride (CdTe), and photorefractive GaAs/AlGaAs multiple quantum well (MQW) devices. The latter have been the subject of our recent research, which is described in the next section.

1.2.   PHOTOREFRACTIVE HOLOGRAPHY WITH GaAs/AlGaAs MQW DEVICES

The strong, narrow excitonic absorption features of MQWs lead to a large optical non-linearity [14] when an electric field is applied, parallel or perpendicular to the wells, due to the Franz-Keldysh or the quantum confined Stark effects respectively. The photorefractive MQW devices used in our experiments were developed with the Franz-Keldysh geometry at Purdue University. Semiconductors usually have higher sensitivities compared to oxide materials and the high carrier mobility's in MQW devices result in very fast response times. We have measured the response time of these devices while imaging through scattering media to be faster than 0.4 ms [15]. This permits images to be captured much faster than standard video rate (30 frames/sec). As a simple demonstration of real-time depth-resolved imaging, we have acquired holographic images reading an analogue CCD camera directly to a conventional videocassette recorder with no frame-grabber.

The photorefractive MQW devices were grown in a Varian GEN-II molecular beam epitaxial chamber on a GaAs substrate. Deep defects were introduced by proton implantation at a double dose of 10^{12} / cm^2 at 160 keV and 5×10^{11} / cm^2 at 80 keV to make the device semi-insulating throughout the MQW region. The defects provide traps for photorefractive space-charge gratings. The MQW surface was protected with wax and lift-off of the film was performed using a highly selective (>10^8) 12% HF etch to dissolve the AlGaAs layer. The sample was then van der Waals bonded to a glass substrate. Two gold contacts were evaporated on the top of the sample to apply a transverse electric field parallel to the quantum-well layers. None of these processing steps are considered exotic and these devices can potentially be manufactured at low cost.

The photorefractive holographic recording process for these MQW devices, used in the transverse Franz-Keldysh geometry, with a large applied sinusoidal field of the order of 10's of kV/cm applied in the plane of the quantum wells, is analogous to that of Rh:BaTiO$_3$. In this case, however, there is an external applied electric field and the space charge produces the refractive index grating via the Franz-Keldysh effect. During photorefractive holographic recording, carriers are produced at the bright regions of the intensity gratings and drift to shield the applied field in these regions. The Franz-Keldysh effect is the broadening and lowering of the exciton peak via field-ionisation, which is due to the large applied field. In the regions where the carriers have shielded the external field, the net applied field is reduced and the absorption is affected accordingly. In the darker regions, where the net field is stronger, the absorption decreases more significantly at the exciton peak. Thus the optical interference pattern is transferred to a local spatial modulation of absorption and, via the Kramers-Kronig relation, of refractive index.

The experimental configuration used to record and reconstruct 3-D holograms in a MQW device was similar to that shown in Fig. 2. The beam from a mode-locked Ti:Sapphire laser (100fs at 830nm) passed through a beam splitter, dividing into the reference and the image beams. The image beam passed through the scattering material, backscattered off the test object and passed once more through the scattering cell. A quarter wave plate and another beam splitter were used to separate the reflected, image-bearing beam from the incident beam. Another half wave plate was used before the polarising beam splitter to ensure that the image and the reference beam had the same polarisation. We used a 3:1 demagnifying telescope and a 1:1 image lens to relay the image beam onto the 3 mm wide aperture of the MQW device where it interfered with the reference beam. These holographic writing beams were separated at an angle of 1.4° that corresponded to grating spacing of 31 μm.

1.3.   REAL-TIME DEPTH-RESOLVED IMAGING

To evaluate the transverse spatial resolution of the holographic system, holograms of an USAF test chart were recorded in the absence of any scattering media. The reconstructed image was measured to have a transverse resolution of 19 μm. Figure 4a shows the reconstructed image of an USAF test chart, as it was read-out from the MQW device and digitally stored to a frame-grabber. This image has not been corrected for static background "noise" (which arises from light scattered by fixed inhomogeneities in the MQW device surface due to contamination of the epoxy used). The fast response times of the MQW devices allowed us to observe the test chart moving in real-time and to record, high resolution holograms directly to a conventional video cassette recorder with no signal processing [16]. Figure 4b shows an image grabbed from a video signal also recorded without any background subtraction. It will be seen that no significant difference can be observed comparing the Fig. 4a, recorded with the frame-grabber, and the corresponding video image of Fig. 4b. The real-time depth-resolved imaging is possible because no computational signal processing is required, as in the case of electronic holography [5]. Video clips of real time holograms, including through scattering media are presented at Optics Express [17].

1.4.   IMAGING THROUGH LIQUID AND SOLID SCATTERING MEDIA

We investigated the capability of the system to image through turbid media by imaging the USAF test chart through a double pass of 13 MFP scattering solution. As is shown in Fig. 5a, no image was observed when we imaged directly through the scattering cell due to the high amount of diffuse background. Using our holographic system described above, the image shown in Fig. 5b

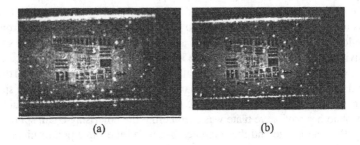

(a)   (b)

Figure 4.   Holographic reconstruction of an USAF test chart showing 50 μm bars recorded (a) via a frame-grabber to a CCD camera,(b) directly to a video cassette recorder. [From [16] Fig. 3a].

was obtained, demonstrating a transverse resolution of 50 μm. Figure 5c is the same hologram recorded through 16 MFP of scattering solution, for which the transverse resolution was 70 μm. This scattering depth compares reasonably

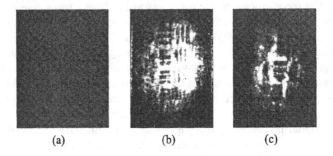

(a)   (b)   (c)

Figure 5.   USAF test chart imaged (a) directly, (b) holographically through 13 MFP and (c) holographically through 16 MFP of liquid scattering medium.

well with those achieved by other whole-field coherent imaging techniques but unfortunately it is not directly applicable to tissue, which is a largely static scattering medium (on the timescale of the image acquisition time. The images presented in Fig. 5 have benefited from the Brownian motion of the scattering centres, which results in *time-varying* noise intensity distributions that average to a uniform background level over the image acquisition time. As has been discussed above, the photorefractive media respond to a spatial derivative of intensity, rather than its magnitude, so this uniform background does not affect the reconstructed image. For a static scattering medium such as tissue, the 'instantaneous' noise speckle patterns would not average to a uniform background and therefore could degrade the reconstructed image. This can be considered as "inter-pixel cross talk" and is a problem for *any* whole-field imaging technique. It is not a problem when imaging in transmission mode since the coherent ballistic light can usually be temporally separated from the scattered light. An approach to counter this cross-talk is the use of low spatial coherence sources [2], which limit the extent scattered photons can move transversely away from their original position before they become incoherent with the reference beam. In effect this limits the speckle size and improves the image quality.

A first approximation to a spatially incoherent source can be realised by introducing a rotating diffuser into a laser beam. A 10% transmission rotating diffuser was introduced into the Ti:Sapphire laser beam to provide an effective extended light source (area of intersection of the laser beam with the diffuser) of 5 mm in diameter. A 100 mm focal length collimating lens then resulted in a source divergence of 5°. Initially the USAF test-chart was imaged without the rotating diffuser in place through a solid phantom of e^{-7} ballistic light attenuation in the double pass that consists of a 400 mm thick layer of sandstone. When viewing the test chart directly through the sandstone, no clear image of the test chart could be observed, as shown in Fig. 6a. Using photorefractive

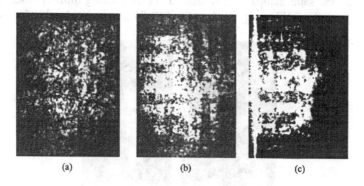

(a)   (b)   (c)

Figure 6.   (a) Direct image and holographic reconstruction of the USAF test chart recorded through the sandstone using (b) the spatially coherent laser source and (c) the spatially incoherent source with the rotating diffuser. [From [20], Fig. 2(a,b) and Fig. 3].

holography, however, the image shown in Fig. 6b was obtained. While the 140 μm bars of the test chart are resolved, the image quality is highly degraded by speckle noise. When the rotating diffuser was introduced into the laser beam, the hologram shown in Fig. 6c was recorded with the low spatial coherence laser beam passing through the sandstone. The last one is superior to Fig. 6b in terms of noise suppression although some residual speckle patterns are still observable. We note that the highest speed of the diffuser rotation was 12 rps, and the corresponding decorrelation time is estimated to be ~1 ms. The response time of the MQW is ∼100 μs and so each recorded hologram effectively contains a 'frozen' speckle pattern. There was not, therefore, any speckle averaging in the photorefractive recording process. The integration time of the CCD camera however, was > 10 ms and so the speckle-corrupted holograms were averaged during a single camera frame acquisition, yielding an improved image, albeit with a background level that compromised the dynamic range. Had the speckle been averaged within the photorefractive holography integration time, however, the resulting background signal would not have been recorded in the MQW device and would not have compromised the CCD camera.

In order to demonstrate photorefractive holography with a source that should be incoherent on the timescale of the photorefractive process, we investigated imaging through solid scattering media using a low cost, moderate power (∼ 10 mW) Hitachi LED operating c.w. at 760 nm with a bandwidth of 45 nm. We note that LED's provide extremely broadband radiation and therefore high depth resolution (∼ 10 μm) owing to their short coherence length [21]. The set-up used for imaging with the LED was a Michelson interferometer incorporating a 4-F imaging system (details are in reference [20]). The test chart was imaged through the same sample of sandstone as in the rotating diffuser case. The resulting holographic image is shown in Fig. 7a. A restricted field of view of

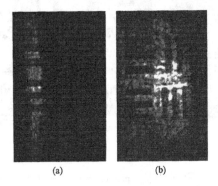

(a) (b)

Figure 7. Holographic images of the test chart recorded through the sandstone sample using (a) LED source and (b) LED and a 10 nm band pass interference filter. [From [20], Fig. 4(a,b)].

the test chart is observed in this hologram. This is due to the very short temporal coherence length of the LED leading to 'walk-off'. To overcome this problem we either have to make the interfering beams closer to collinear or we must increase the temporal coherence using an interference filter. The first approach is of limited use since the resulting large fringe period will reduce the angle between the first diffracted order and the zero order and make it difficult to stop the zero order beam from reaching the CCD camera. The second solution results in a reduction of the spectral width, thereby increasing the field of view at the expense of depth resolution. Figure 7b shows a holographic image of the object, recorded as in Fig. 4a but using an interference filter with a spectral width of 10 nm. This appears broader than Fig. 7a but dimmer, owing to the low transmittance of the filter (10%). The writing beam intensities at the MQW device were 40 µW/cm^2 and 4 µW/cm^2 respectively. It is interesting to compare Fig. 6b and Fig. 7b and note that the spectrally filtered LED provides images through solid scattering media that are superior to those obtained with the laser, although both sources have approximately the same spectral width. As expected, the reduced spatial coherence source produced a holographic image with significantly less speckle noise.

Holographic images of the test chart were also recorded through biological tissue using the LED source. Owing to its relatively low power, we were restricted to using samples of low scattering depth. The images depicted in Fig. 8 were recorded though a 600 µm thick sample of chicken breast, for

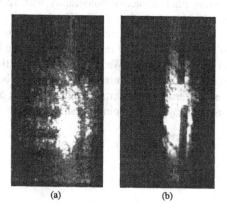

(a) (b)

Figure 8.   Holographic images of the test chart recorded through the chicken tissue sample using (a) Ti:Sapphire source and (b) LED source. [From [20], Fig. 5(a,b)].

which the ballistic light attenuation was measured to be e^{-5} in double pass. For comparison, holograms were recorded with the same intensity (1.8 mW/cm^2 incidence on the tissue) using the Ti:Sapphire laser or the LED. The speckle noise is clearly reduced when imaging through the tissue using the low spatial coherence LED source. These initial results suggest that low cost LED sources can be used for imaging through biological tissue to reduce speckle noise.

1.5.   OUTLOOK FOR PHOTOREFRACTIVE HOLOGRAPHIC IMAGING

We have demonstrated a real-time holographic imaging system that provides better than 50 μm depth and transverse spatial resolution. This technique uses time-gated holography with photorefractive MQW devices to provide whole 2-D image field acquisition without the need for pixel-by-pixel transverse scanning. Direct depth-resolved image acquisition to a video recorder has been demonstrated. We anticipate many applications for rapid high-resolution 3-D imaging including 3-D microscopy e.g. of living subjects and motion analysis. To these ends, further improvements in the field of view and optical quality are anticipated as the technology matures. We have applied this coherence gating system to image through up to 16 MFP of scattering medium. We note that 16 scattering MFP "penetration depth" corresponds approximately to a depth in tissue of $\leq$ 1 mm. This could be increased in practice by using a higher incident intensity at the tissue or by optimising the wavelength of the radiation. We have also demonstrated imaging through weakly scattering solid media, including biological tissue, using photorefractive holography with spatially incoherent light from a diffuse laser beam and an LED. We observed a reduction in the speckle noise that arises from inter-pixel cross talk. The use of high power (> 500 mW) LED's would permit 3-D real-time holographic imaging through thicker samples of biological tissue. We note that the reduction in the field of view due to walk-off needs to be minimized, either by using band pass interference filters or, more preferably, by alternative methods that do not reduce the source power. This is a subject of ongoing research.

2. FLUORESCENCE LIFETIME IMAGING FOR BIOMEDICAL RESEARCH

One of the most widespread biomedical research tools is the optical microscope, which allows researchers to study biological processes down to the cellular level. Using incoherent or laser illumination, spectroscopic imaging of the absorption or fluorescence of a sample is routinely achieved using appropriate filters, spectrographs, or simply using colour cameras. In recent years, optical microscopy has greatly advanced in its ability to extract information from biological samples, in terms of both increased functional specificity and improved resolution. It is of great interest to apply functional microscopy in vivo.

Fluorescence imaging

Aim: to image different types of tissue using fluorescence to achieve *contrast*

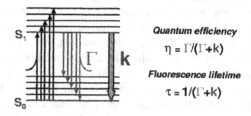

Quantum efficiency

$$\eta = \Gamma/(\Gamma + k)$$

Fluorescence lifetime

$$\tau = 1/(\Gamma + k)$$

Γ = *radiative decay rate*
   - a property of the fluorophore
k = *non-radiative decay rate*
   - a function of fluorophore environment

Figure 9.   Schematic of fluorescence

2.1. FUNCTIONAL IMAGING USING FLUORESCENCE

Fluorescence microscopy, in which the sample absorbs incident photons and emits light (fluorescence) at a longer wavelength, is a particularly powerful technique for spectroscopic imaging. The wavelength of the fluorescence may be used to distinguish between different molecules in the sample and the efficiency of the fluorescence process can provide information about the local environment of the fluorescent molecules (fluorophores). The latter is usually parameterized by the quantum efficiency, η, which varies according to how the fluorophore environment affects the non-radiative decay rate, **k**.

For some fluorophores **k** is sensitive to the local pH or to calcium ion concentration or to physical factors such as the viscosity. Distributions of changes in **k** can be monitored by appropriate imaging of fluorescence intensity distributions to obtain maps of how η varies from pixel to pixel, although care must be taken to factor out artefacts due to non-uniform fluorophore concentration or excitation flux. In practice it is often necessary to employ special exogenous fluorophores, usually dyes known as "probes", whose fluorescence spectra change in a predictable way according to the strength of the local environmental perturbation (e.g. $[Ca^{2+}]$). By incorporating these dyes into the sample and recording fluorescence intensity images at two or more wavelengths, it is possible to produce a map of, e.g. $[Ca^{2+}]$ distributions to study the firing of nerve synapses. This functional imaging technique, known as "wavelength-ratiometric imaging", is increasingly deployed in many areas of biology but is limited to those instances for which suitable ratiometric probes are available.

As well as exploiting the wavelength dependence of fluorescence, it is also possible to use the fluorescence lifetime, τ, to identify or distinguish between different fluorophore molecules, or to image perturbations in the local fluorophore environment by determining τ for each pixel in the field of view. This may be done using only relative measurements of the temporal decay of the fluorescence signal after excitation. Fluorescence lifetime imaging (FLIM) can be used instead of wavelength ratio-metric imaging, or the two approaches may be combined. FLIM is useful when there is no suitable wavelength–ratiometric probe available – as is often the case - and is often able to make use of naturally occurring fluorophores in biological samples. The fluorescence lifetimes of biological fluorophores can be as short as a few picoseconds and many exogenous fluorescence probes exhibit lifetimes in the nanosecond range. These ultrafast timescales have not been accessible to most biomedical researchers but recent advances in ultrafast laser technology and in high speed imaging detectors, such as gated CCD cameras and gated microchannel plate optical intensifiers, are prompting many groups to adopt FLIM as a means of functional imaging. The determination of fluorescence lifetime requires only relative measurements of intensity and so is especially useful for biomedical samples in which the heterogeneous nature of tissue and autofluorescence cause significant problems. Since fluorescence lifetime is dependent upon both radiative and non-radiative decay rates, it may be used to distinguish between different fluorophore molecules (with different radiative decay rates) and to monitor local environmental perturbations that affect the non-radiative decay rate. Fluorescence lifetime probes have been demonstrated for many biologically significant analytes including $[O_2]$, $[Ca^{2+}]$ and pH [19]. Fluorescence lifetime imaging (FLIM) can be applied to almost any optical imaging modality, including microscopy and potentially to non-invasive optical biopsy [22]. Fluorescence lifetime data may be acquired in the frequency or time domain [23, 24]. Usually the frequency domain instruments are considered to be simpler and cheaper but the time-domain instruments often provide more information and higher temporal resolution. In the time domain the fluorescence decay is observed directly while in the frequency domain fluorescence lifetime is typically obtained by measuring the phase shift between a sinusoidally modulated laser excitation signal and the resulting fluorescence light, which will be modulated at the same frequency (10). If the fluorophores exhibit a single exponential decay constant, τ, then the phase shift will simply be given by $\tan \phi = \omega \tau$. For a single point measurement, this technique is straightforward to implement for modulation frequencies up to ~ 1 GHz and this is suitable for nanosecond fluorescence lifetimes. Frequency domain fluorescence lifetime imaging is achieved using a modulated 2-D detector such as a CCD camera or multi-channel optical intensifier.

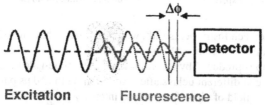

Excitation   **Fluorescence**

Figure 10.   Frequency domain measurement of fluorescence lifetime

2.2.   TIME-DOMAIN 2D FLUORESCENCE LIFETIME IMAGING

The recent development of user-friendly and relatively portable ultrafast laser technology and the availability of ultrafast gated microchannel plate image intensifiers enable the development of relatively inexpensive time domain FLIM instruments that may be deployed outside specialist laser laboratories. We have reported a time domain FLIM system [25, 26] that exploits an all-solid-state diode-pumped Cr:LiSGAF oscillator-amplifier system combined with a gated optical image intensifier (GOI). This laser system requires only four (commercially available) 500 mW pump diodes and will run off a domestic power outlet with minimal water-cooling requirements. The apparatus used for time-domain FLIM is shown in Fig. 11. Our initial FLIM system was based around a commercial argon ion laser pumping a Ti:Sapphire laser (Spectra Physics Tsunami) operating at 830nm. The 100fs pulses from this system were used to seed a home-built argon ion laser-pumped Cr:LiSAF regenerative amplifier. The 10 µJ, 10 ps output of the amplifier (at 10 kHz repetition rate) was frequency doubled in BBO to produce a 1 µJ excitation source at 415 nm. The fluorescence from the sample was imaged onto a gated optical image intensifier

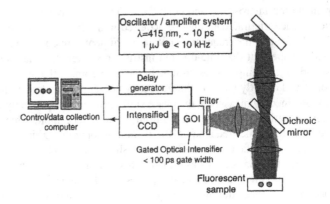

Figure 11. Experimental set up for time domain FLIM. [From [26], Fig. 1]

(GOI), from Kentech Instruments Ltd., which acquires whole-field 2-D images with a gate width as short as ∼ 90 ps.

A FLIM map is produced by recording a series of time-gated 2-D images of the fluorescence at different delays after excitation by the 10 ps pulses and, for each pixel in the field of view, fitting these intensity profiles to an exponential decay (see Fig. 12). When imaging simple fluorophore distributions, such as shown in Fig. 13, we have demonstrated near real-time FLIM, with an update time of less than 3 seconds [27].

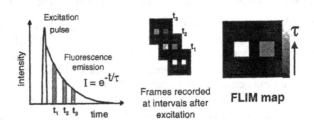

Figure 12. Data acquisition schematic for fluorescence lifetime imaging. [From [26], Fig. 2].

2.3. FLIM OF LASER DYE SAMPLES

Figure 13 shows an example of a typical FLIM map of three pipettes (internal diameter ∼ 1 mm) of Coumarin-314 laser dye (80 μM in ethanol) interlaced with pipettes of the saturable absorber dye DASPI (80 μM dissolved in a 50:50 mixture of ethanol:glycerol). Both fluorophores exhibit mono-exponential decays and their (spatially averaged) lifetimes were determined to be 3.46±0.02 ns and 143±5 ps, demonstrating the excellent lifetime dynamic range of this system. These lifetimes were confirmed using a photon-counting system and a streak camera. It is also possible to resolve fluorescence from multiple samples both temporally and spectrally in a single acquisition by use of a monochromator before the GOI.

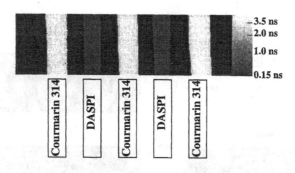

Figure 13.   Schematic of pipettes of Coumarin 314 interlaced with pipettes of DASPI (in 50:50 ethanol glycerol solvent) and FLIM map with gray scale from 0 to 4 ns. [From [26], Fig. 3].

In a different experiment, the variation of fluorescence lifetime with viscosity has been clearly demonstrated in samples of DASPI in solvents of varying viscosity (ethanol:glycerol ratio). The non-radiative decay rate (and hence the fluorescence lifetime) of the dye DASPI is a sensitive function of the viscosity of its solvent [28] and so can be adjusted by preparing solutions in different mixtures of ethanol and glycerol. We tested the FLIM system's lifetime discrimination in this manner and demonstrated a sub-10 ps lifetime discrimination [29], which is comparable with that of non-imaging detectors, such as streak cameras and single-photon-counting instruments.

2.4. FLIM OF BIOLOGICAL TISSUE EXPLOITING AUTOFLUORESCENCE

Although exogenous fluorophores may be used to contrast different tissue types, in the clinical environment, there is also potential to exploit the autofluorescence of the tissue itself to determine tissue properties. It may thus be possible to distinguish between healthy and diseased tissue e.g. [30]. We have demonstrated that FLIM may be used to distinguish between aorta, elastin and collagen extracted from rat tissue. Figure 14 shows that, while elastin and collagen are not contrasted by their time-averaged fluorescence spectra (Fig. 14a) or by a fluorescence intensity image (Fig. 14b) for excitation at 415 nm, they may be contrasted by fluorescence lifetime. For this excitation wavelength we found that the autofluoresence decays approximated well to a double exponential fit. Figure 14c is the FLIM map showing the distribution of the short-lived τ_1 fluorescence components whilst Fig. 14d corresponds to the longer lived τ_2 components. There is clearly contrast in fluorescence lifetime between the elastin, collagen and aorta, which is particularly pronounced in the τ_2 component.

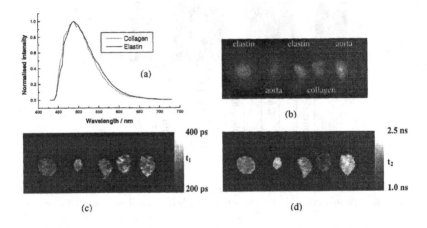

Figure 14.   FLIM of tissue constituents (a) time integrated spectral fluorescence profile, (b) time-gated fluorescence image recorded immediately following excitation, FLIM maps of (c) the short-lived (τ_1) and (c) long-lived (τ_2) components of fluorescence decay. Samples (L-R) elastin, aorta, elastin, collagen, and aorta. [From [26], Fig. 6].

Using our system, 10 μm thick sections of rat ear containing various tissue components including elastic cartilage and blood vessels were imaged within a single field of view. The FLIM maps have the capability to show considerable contrast between the veins and elastic cartilage and the surrounding tissue. In rat ear sections, FLIM images of the artery show a clear fluorescence lifetime contrast between the vessel wall and the clotted blood in the interior [26].

2.5.   WHOLE-FIELD 3D IMAGING USING STRUCTURED LIGHT

3-D optical sectioning is elegantly realised in the confocal scanning optical microscope but the need to acquire images by scanning pixels sequentially could make the acquisition of sectioned FLIM images prohibitively slow. A whole field, light-efficient imaging technique with sectioning capability is clearly desirable for FLIM and we have begun to investigate the approach proposed by

Neil *et al.* [31]. The principle, illustrated in Fig. 15, exploits the fact that in
a conventional microscope it is only the *zero* spatial frequency that does not
attenuate with defocus. By using a transmission grating to project a structured excitation field onto the sample to produce a strongly spatially modulated
fluorescence field, it is possible to extract a whole-field depth-resolved fluorescence image by recording images at three different grating positions and using
an appropriate computer algorithm to remove the spatial modulation. In our

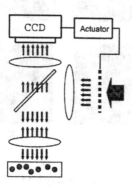

Figure 15.   Schematic for whole-field depth resolved imaging using structured light. [From
[26], Fig. 9].

experiments we imaged a grating of 2 line pairs per mm onto the sample using
a x10 microscope objective at three different transverse positions (i.e. spatial
phase changes of 0, $2\pi/3$ and $4\pi/3$) acquiring a set of time-gated fluorescence
images at each of these positions. Our 3-D fluorescence test object comprised
cotton wool fibres stained with solutions of Coumarin-314 and DASPI. By combining the FLIM data acquired at each spatial phase position we were able to
demonstrate whole field sectioned FLIM images. Figures 16a, 16e show conventional fluorescence intensity images, separated by 315 μm in the z direction,
which are significant blurred by out of focus light, and Fig. 16b, 16f are the
corresponding FLIM maps in which the out of focus blur is clearly a problem,
although the longer lifetime Coumarin-314 is contrasted from the shorter lifetime DASPI. Figures 16c, 16g show the sectioned intensity images for which
there is some residual spatial modulation of the image, attributed to uncertainty
in the grating mark:space ratio, but the out of focus light has been eliminated.
In the sectioned FLIM images of Fig. 16d, 16h, the out of focus blur has been
eliminated and the fluorescence lifetime contrast has been preserved. There
is still some residual spatial modulation but this problem should be resolved
with improved apparatus. The resolution of this technique is comparable to that

of the confocal microscope, achieving sub-micron sectioning with the use of appropriate objectives.

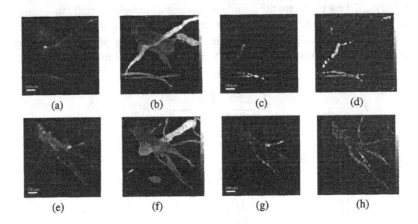

Figure 16. Microscope images (x, y plane) of cotton wool stained with Coumarin-314 and DASPI for two different z positions, (a)-(d) and (e)-(h), differing by 315 mm. Figures (a), (e) are conventional microscope intensity images;(b), (f) are corresponding conventional FLIM maps; (c), (g) are sectioned intensity images; (d), (h) are sectioned FLIM maps. Lifetime gray scales span from 700ps (black) to 2.6ns (white) in each case. [From [26], Fig. 10].

2.6. OUTLOOK FOR FLUORESCENCE LIFETIME IMAGING

We have demonstrated a time-domain fluorescence lifetime imaging system with high temporal dynamic range that is capable of imaging lifetime differences as small as 10ps. Fluorescence lifetime imaging of local fluorophore environment (solvent viscosity) has also been demonstrated. This FLIM system has been used to investigate the autofluorescence properties of tissue constituents and tissue sections. We have developed a diode-pumped FLIM instrument, which has the potential to be inexpensive and portable, and have applied it to in vitro imaging of tissue sections. Optically sectioned FLIM images have been obtained using structured illumination. Future work will include further characterisation of the 3D imaging capabilities of this FLIM system as well as application to endoscopic imaging, two photon excitation and imaging through scattering media. We anticipate many biomedical applications of fluorescence lifetime imaging, including non-invasive optical biopsy. Ultimately this will entail combining FLIM with techniques to image through strongly scattering media. This presents an exciting challenge.

Acknowledgments

This research was funded by the UK Engineering and Physical Sciences Research Council (EPSRC), the Biotechnology and Biological Sciences Research Council (BBSRC) and the Paul Instrument Fund of the Royal Society. M.J. Cole and K. Dowling acknowledge EPSRC Co-operative Award in Science and Engineering (CASE) studentships supported by the Royal Marsden National Health Service Trust. M. Tziraki acknowledges a Marie-Curie Fellowship supported by the EU Training and Mobility of Researchers (TMR) program. The authors gratefully acknowledge the assistance of Patrick Sebbah with the preparation of this manuscript.

References

[1] B. Devaraj, M. Usa, K. P. Chan, T. Akatsuka and H. Inaba, IEEE J STQE-2, 1008 (1996).

[2] A. V. Mamaev, L. I. Ivleva, N. M. Polozkov, V. V. Shkunov, Conf. Lasers and Electro-Optics 1993, paper CFK6, pp 632-634.

[3] A. Knuttel, J. M. Schmitt and J. R. Knutson, Opt. Lett., 19, pp 302-304 (1994).

[4] D. Huang, E. A. Swanson, C. P. Lin, J. S. Schuman, W. G. Stinson, M. R. Hee, T. Flotte, K. Gregory, C. A. Puliafito and J. G. Fujimoto, Science 254, pp 1178-1181 (1991).

[5] E. Leith, C.Chen, H. Chen, Y. Chen, D.Dilworth, J. Lopez, J. Rudd, P-C. Sun, J.Valdmanis, and G.Vossler: J. Opt.Soc.Am. A, 9, 1148 (1992).

[6] N. Abramson, Opt. Lett. 3, pp 121-123 (1978).

[7] A. Rebane and J. Feinberg, Nature 351, pp 378 (1991).

[8] K. G. Spears, J. Serafin, N. Abramson, X. Zhu and H. Bjelkhagen, IEEE Trans. Biomed. Eng. 36, pp 1210-1231 (1989).

[9] S. C. W. Hyde, N. P. Barry, R. Jones, J. C. Dainty, P. M. W. French, M. B. Klein and B. A. Wechsler, Opt. Lett. 20, pp 1331-1333 (1995).

[10] S. C. W. Hyde, R. Jones, N. P. Barry, J. C. Dainty and P. M. W. French, Opt. Lett. 20, 2330 (1995).

[11] S. C. W. Hyde, N. P. Barry, R. Jones, J. C. Dainty, P. M. W. French, K. M. Kwolek, D. D. Nolte and M. R. Melloch, IEEE JSTQE Special Issue on Lasers in Medicine and Biology 2, 965 (1996).

[12] R. Jones, N. P. Barry, S. C. W. Hyde, m. Tziraki, J. C. Dainty, P. M. W. French, D. D. Nolte, K. M. Kwolek and M. R. Melloch, IEEE JSTQE Special Issue on Ultrafast Optics, 4, 360-369 (1998).

[13] N. P. Barry, S. C. W. Hyde, R. Jones, J. C. Dainty and P. M. W. French, Electron. Lett. 33, pp1732-3 (1997).

[14] D. D. Nolte and M. R. Melloch, *Photorefractive effects and materials*, chapter 7, D. D. Nolte ed. (Kluwer Academic Publishers, Dordrecht, 1995).

[15] R. Jones, S. C. W. Hyde, M. J. Lynn, N. P. Barry, J. C. Dainty, P. M. W. French, K. M. Kwolek, D. D. Nolte and M. R. Melloch, Appl. Phys. Lett. **69**, 1837 (1996).

[16] R. Jones, N. P. Barry, S. C. W. Hyde, P. M. W. French, K. M. Kwolek, D. D. Nolte and M. R. Melloch, Opt. Lett. **23**, 103-105 (1998).

[17] R. Jones, M. Tziraki, D. Parsons-Karavassilis and P. M. W. French, K. M. Kwolek and D. D. Nolte, M. R. Melloch, Opt. Exp. **2** (1998).

[18] M. Tziraki, R. Jones, P. M. W. French, D. D. Nolte, and M. R. Melloch: App. Phys. Lett. **75**,1363 (1999).

[19] J. R. Lakowicz, *Principles of Fluorescence Spectroscopy*, (Plenum Press, 1983).

[20] M. Tziraki, R. Jones, P. M. W. French, M. R. Melloch and D. D. Nolte, Appl. Phys. **B 70**, 151-154 (2000).

[21] E. Beaurepaire, A.C. Boccara, M. Lebec, L. Blanchot, and H. Saint-Jalmes: Opt. Lett. **23**, 224 (1998).

[22] B. B. Das, Feng Liu and R. R. Alfano, Rep. Prog. Phys. **60**, 227 (1997).

[23] H. Szmancinski, J. R. Lackowicz, and M. L. Johnson, Methods Enzymol. **240**, 723 (1994).

[24] R. Cubeddu, P. Taroni, and G. Valentini, Opt. Eng. **32**, 320 (1993).

[25] R. Jones K. Dowling, M. J. Cole, D. Parsons-Karavassilis, M.J. Lever, P.M.W. French, J.D. Hares and A.K.L. Dymoke-Bradshaw, Electron Lett. **35**, 256 (1999).

[26] M. J. Cole, K. Dowling, P. M. W. French, R. Jones, D. Parsons-Karavassilis, M. J. Lever, A. K. L. Dymoke-Bradshaw, J. D. Hares, M. A. A. Neil, R. Juskaitis and T. Wilson, Proceedings of the In Vivo Optical Imaging Workshop, National Institute of Health, Washington 1999, OSA TOPS.

[27] K. Dowling, M.J. Dayel, S.C.W. Hyde, P.M.W. French, M.J. Lever, J.D. Hares, A.K.L. Dymoke-Bradshaw, J. of Mod. Opt. **46**, 199 (1999).

[28] W. Sibbett and J.R. Taylor, J. Lumin. **28**, 367 (1983).

[29] K. Dowling, M. J. Dayel, M.J. Lever, P.M.W. French, J.D. Hares and A.K.L. Dymoke-Bradshaw, Opt. Lett. **23**, 810 (1998).

[30] B.B. Das, Liu Feng and R.R. Alfano, Rep. Prog. Phys. **60**, 227 (1997).

[31] M.A.A. Neil, R. Juškaitis, and T. Wilson, Opt. Lett. **22** 1905-1907 (1997).

Chapter 2

SIMULTANEOUS OPTICAL COHERENCE AND TWO-PHOTON FLUORESCENCE MICROSCOPY

E. Beaurepaire, L. Moreaux and J. Mertz

Laboratoire de neurophysiologie et nouvelles microscopies INSERM EPI 00-02,
Ecole supérieure de physique et chimie industrielles
10 rue Vauquelin 75005 Paris
emmanuel.beaurepaire@espci.fr

Abstract   We demonstrate simultaneous non-fluorescence and fluorescence imaging with sub-micron resolution in tissues by combining optical coherence tomography (OCT) and two-photon-excited fluorescence (TPEF) microscopy.

Keywords: optical coherence tomography, non-linear microscopy, biomedical imaging.

Introduction

Optical techniques for imaging in scattering media typically rely on the detection of either ballistic or diffuse light. Microscopy based on the detection of ballistic light is limited to near-surface imaging (a few scattering mean free paths) and can provide information with a spatial resolution comparable to an optical wavelength. Diffuse light imaging techniques rely on a model of photon propagation and can probe scattering media more deeply, at the cost of a reduced spatial resolution. Since the development of the confocal scanning microscope [1], three-dimensional microscopy has become an ubiquitous tool in biology. In confocal microscopy, a spatial filter is used to discriminate light originating from outside the focal plane. In scattering media such as biological tissues, the penetration depth of this technique is limited either by background originating from superficial regions of the sample, or by the detector sensitivity (since the signal is damped exponentially with depth, for a given incident power). In recent years, alternative microscopies have been developed with superior penetration depths. We report here the combination of two such techniques, namely two-photon-excited fluorescence (TPEF) microscopy and optical coherence to-

P. Sebbah (ed.), Waves and Imaging through Complex Media, 235–241.
© *2001 Kluwer Academic Publishers. Printed in the Netherlands.*

mography (OCT).

Two-photon-excited fluorescence (TPEF) microscopy is an alternative to fluorescence confocal microscopy where fluorescent molecules are excited nonlinearly by simultaneous absorption of two photons [2]. TPEF microscopy performs better than standard confocal fluorescence microscopy in scattering media because

- (i) longer excitation wavelengths (near IR) provide deeper penetration;

- (ii) little background is generated from out-of-focus planes since fluorescence excitation is proportional to the square of the intensity distribution. In order to create a background photon, two diffuse excitation photons have to be incident on the same molecule at the same time;

- (iii) fluorescence need not to be imaged, meaning that the technique is insensitive to the diffusion of signal light.

Optical coherence tomography [3] is an alternative to reflection confocal microscopy where a coherence gate rather than a spatial filter is used to reject background light (Fig. 1a). OCT can probe more deeply than confocal reflection imaging because

- (i) coherence gating outperforms confocal filtering in the rejection of background photons originating from superficial regions of the sample (i.e. far from the focal plane); the axial response of an interferometer using a source with a Gaussian spectrum falls as $exp(-z^2)$ where z is the distance from the focal plane, whereas the axial response of a confocal microscope falls as $1/z^4$;

- (ii) coherent detection yields shot-noise-limited measurements even at low light levels.

The axial resolution of an OCT scanner is usually prescribed by the coherence length of the source, and is typically 7-15 μm for commonly available sources. The most straightforward way to increase this resolution is to reduce the coherence length of the source. For example, OCT imaging with 1 μm axial resolution has been reported recently using a laser delivering <5 fs pulses [4]. An alternative way to improve the resolution of an OCT scanner is to take advantage of a confocal geometry and use high aperture optics in an en-face (XY) geometry rather than the usual XZ geometry of OCT systems (see Fig. 1). This approach is usually called optical coherence microscopy (OCM) [5]. In this configuration, the resolution is principally defined by the confocal nature of the detection, and coherence gating supplements the spatial filter in the rejection of background photons.

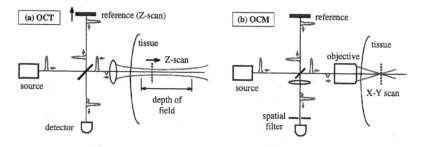

Figure 1. Possible geometries for low-coherence imaging systems. (a) Optical coherence tomography (OCT) : XZ-geometry. (b) Optical coherence microscopy (OCM) : transverse (XY) geometry, high aperture optics.

1. COMBINED SYSTEM

The contrast mechanisms for the two imaging modalities are intrinsically different in that OCT is based on the detection of backscattered illumination light whereas TPE imaging is based on the detection of fluorescence light generated by endogenous or exogenous markers. In practice, OCT usually provides information on sample morphology whereas fluorescence microscopy can provide information on sample functionality. We devised a microscope that can provide simultaneously both pieces of information [6].

In our layout (Fig. 2), a mode-locked Ti:Sapphire laser (100 fs, 80 Mhz) is

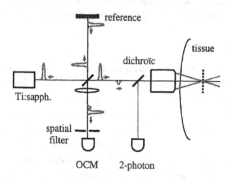

Figure 2. Schematic of the combined OCM-TPEF microscope.

focussed and scanned in three dimensions through a fixed sample, generating both backscattered and fluorescence light which are independently detected.

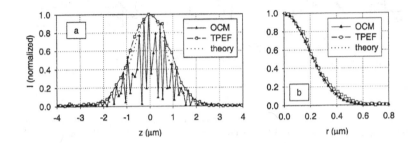

Figure 3.   (a)Axial and (b) transverse resolution of the combined microscope measured by acquiring 3D images of a 0.3 μm diameter fluorescent polystyrene bead immobilized in a gel. Axial and radial resolutions are 2 μm and 0.4 μm respectively (FWHM) using a 0.9 NA water immersion objective.

We use an en-face (XY) imaging geometry and a high NA (0.9) water immersion objective lens, which allows us to maintain high intensities at the focal plane even when imaging deep inside the sample. XY raster scanning is performed using fast galvanometers and our pixel acquisition rate is up to 1 MHz. A pinhole filter providing true confocal operation is used in combination with the coherent detection of the backscattered light, resulting in a similar micron 3D resolution for both OCT and TPE imaging modes (Fig. 3).

2.   RESULTS

To illustrate the enhanced background rejection of OCM compared to standard confocal microscopy, we imaged a metallic object through a scattering suspension of polystyrene spheres. Direct confocal detection is performed by blocking the reference arm in our system. Fig. 4 shows images obtained with standard confocal and coherent (OCM) detection through 7 scattering mean free paths (single pass) of a suspension of polystyrene spheres. Detected background completely erodes the contrast when confocal detection alone is used, whereas coherence gating enables one to record a contrasted image of the sample.

We demonstrate simultaneous micron-resolution imaging with the two modalities in a drosophila embryo labeled with the fluorescent marker Texas Red. Fig. 5 shows that two images can be obtained based on independent contrast

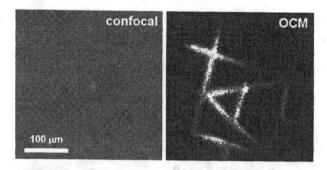

Figure 4.   Metallic "A" imaged through 7 scattering mean free paths (single pass) of a scattering suspension using standard confocal and coherent (OCM) detection.

mechanisms.

In order to illustrate the complementarity of the two modes of contrast, we

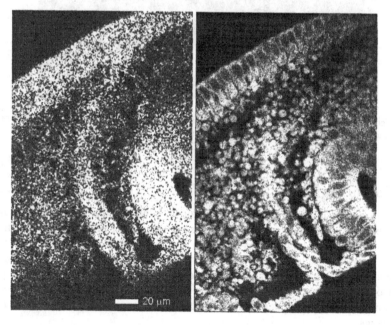

Figure 5.   Simultaneous OCM (left) and TPEF (right) images of a drosophila embryo labeled with Texas Red. Pixel rate: 150 kHz.

injected fluorescein in the blood circulation of a rat and recorded images in the neocortex of the anesthetized animal after cranotomia. Fig. 6 is a composite image in which one can identify two veins (TPEF image, smooth) as well as the unlabeled tissue environment (OCM image, speckled). Cell bodies appear as dark non-scattering structures in the OCM image.

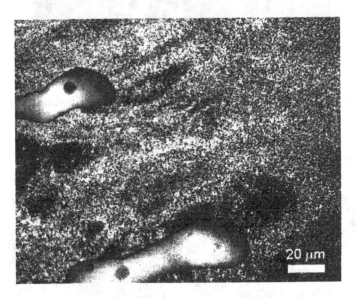

Figure 6.   Composite OCT (speckled) and TPEF image (smooth) recorded in the neocortex of a rat after injection of fluorescein in the blood circulation.

Conclusion

In conclusion, our results demonstrate that the two techniques of optical coherence microscopy and TPEF microscopy may be combined in a single instrument. The combination is natural in that the two modalities can be run off the same laser source and are compatible in terms of resolution and acquisition speed. Because they are based on independent contrast mechanisms, they provide complementary information. In particular, they allow the possibility of simultaneous morphological and functional microscopic imaging in biological tissues.

Acknowledgments

The authors gratefully acknowledge financial support by the Institut Curie and by the Centre National de la Recherche Scientifique (CNRS). We thank E.

Farge and S. Charpak for scientific assistance, and C. Boccara for fruitful discussions. Part of this study was conducted while the authors were affiliated with the UMR 7637 of the CNRS. E.B. acknowledges the Ministère de l'Education Nationale, de la Recherche et de la Technologie for a postdoctoral fellowship (Télémédecine).

References

[1] M. Minsky, US Patent 3013467 (1957).

[2] W. Denk, J.H. Strickler and W.W. Webb, Science **248**, 73 (1990).

[3] D. Huang, E.A. Swanson, C.P. Lin, J.S. Schuman, W.G. Stinson, W. Chang, M.R. Hee, T. Flotte, K. Gregory, C.A. Puliafito and J.G. Fujimoto, Science **254**, 1178 (1991).

[4] W. Drexler, U. Morgner, F.X. Kärtner, C. Pitris, S.A. Boppart, X.D. Li, E.P. Ippen and J.G. Fujimoto, Opt. Lett. **24**, 1221 (1999).

[5] J.A. Izatt, M.D. Kulkarni, H.-W. Wang, K. Kobayashi and M.V. Sivak Jr, IEEE J. Sel. Topics Quantum Electron. **2**, 1017 (1996).

[6] E. Beaurepaire, L. Moreaux, F. Amblard and J. Mertz, Opt. Lett. **24**, 696 (1999).

Chapter 3

LOW COHERENCE
INTERFEROMETRIC TECHNIQUE

G. Brun, I. Verrier, K. Ben Houcine, C. Veillas and J. P. Goure

Laboratoire TSI UMR CNRS 5516

Université Jean Monnet

23 rue du Docteur Paul Michelon, 42023 Saint-Etienne cedex 2, France

brun@univ-st-etienne.fr

Abstract   We propose an original technique based on low coherence interferometry to detect objects in strongly scattering media. In order to use broadband light, a special interferometer called the SISAM correlator is used to allow coherent superposition of the reference and tested beams. Direct measurement of the time intercorrelation function is obtained. Image reconstruction of a 3D phase object embedded in a latex beads suspension is demonstrated. Experiments are successfully compared to the theoretical analysis of the method.

Keywords: Scattering medium, Coherent detection, SISAM Interferometer.

Introduction

Low coherence interferometry offers an interesting route for depth-resolved imaging in turbid media by enabling the discrimination between the ballistic photons and the diffuse light which scatters in the forward direction. In this contribution, we propose a new method connected to various techniques proposed in the literature (see [1-8]) which allows to record interferences with broadband light and is promising for object detection in scattering media. In the first part, principle of the method based on the SISAM (Interferential Spectrometer by Selection of Amplitude Modulation) correlator is described [9] and the expression of the intensity profile at the output of the interferometer is derived. In the second part, experimental setup is detailed and the reconstruction of a simple phase object imbedded in a latex beads suspension is demonstrated. Good

P. Sebbah (ed.), Waves and Imaging through Complex Media, 243–250.
© 2001 *Kluwer Academic Publishers. Printed in the Netherlands.*

agreement is found between the theory and the experiment. A careful analysis of setup and experimental limits is carried out.

1. PRINCIPLE OF THE METHOD

1.1. THE SISAM CORRELATOR

In the method presented in this work, interferences with broadband light are obtained by means of a Mach-Zehnder interferometer in which the scattering medium can be set. Before adding the fields issued from each arm of the interferometer, a particular setup is used (Fig. 1). On each arm of the interferometer,

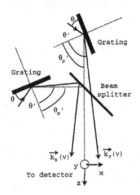

Figure 1. The SISAM correlator.

gratings are set in order to supply appropriate delays for reference and tested beams, which are combined on the detector only afterwards by a beam-splitter. This device constitutes the SISAM (Interferential Spectrometer by Selection of Amplitude Modulation) interferometer. The angle between the normal to the grating and the first order diffracted beam is given by $\theta' = \theta'(\nu)$. A value ν_p of the frequency, corresponding to the angle $\theta'_p = \theta'(\nu_p)$, exists for which the wave vectors $\mathbf{k}_R$ and $\mathbf{k}_T$ of the diffracted reference and tested beams are parallel. At a frequency $\nu \neq \nu_p$, the expression of the wave vectors in the referential (x, y, z) bound to the detector is given by:

$$
\begin{aligned}
\mathbf{k}_T &= \frac{2\pi\nu}{c} \left\{ \sin\left(\theta' - \theta'_p\right), 0, \cos\left(\theta' - \theta'_p\right) \right\}, \\
\mathbf{k}_R &= \frac{2\pi\nu}{c} \left\{ -\sin\left(\theta' - \theta'_p\right), 0, \cos\left(\theta' - \theta'_p\right) \right\}.
\end{aligned}
\tag{1}
$$

The phase difference introduced by the correlator is a function of the optical frequency ν and the diffraction angles θ'_p and $\theta' = \theta'(\nu)$.

1.2. INTERFERENCES WITH BROADBAND LIGHT

For one given wavelength of the light source, the combination of the reference and tested beams of an interferometer gives rise to fringes of interference. The spatial frequency of the fringes depends on the wave vectors modulus and directions, k_R and k_T. When all wavelengths are taken into account, the intensity at the output of any standard interferometer for the whole spectral bandwidth $\Delta\nu$ writes:

$$E = K + \int_{\nu=\nu_0-\frac{\Delta\nu}{2}}^{\nu=\nu_0+\frac{\Delta\nu}{2}} \{S_R(\mathbf{r},\nu)S_T(\mathbf{r},\nu)exp(-i(\mathbf{k}_T-\mathbf{k}_R)\mathbf{r})\}\, d\nu, \quad (2)$$

where K is a constant due to light background, $\Delta\Phi(\mathbf{r},\nu) = (\mathbf{k}_T-\mathbf{k}_R)\mathbf{r}$ is the phase difference introduced by the wave vectors. The resulting signal is the temporal intercorrelation function between the reference field $S_R(\mathbf{r},\nu)$ and the tested beam $S_T(\mathbf{r},\nu)$ and is maximum when the optical difference between them equals zero.

1.3. THE INTENSITY PROFILE

Using Eq. (2) and considering now the phase difference introduced by the SISAM interferometer give the intensity profile in the transverse output plane (x,y) of the correlator:

$$E(x,y) = I_0 + \Delta I \mathrm{sinc}\left\{\pi\Delta\nu Q(x+\frac{\delta(x,y)}{cQ})\right\}.$$
$$\cos\left\{2\pi\nu_0(\frac{\delta(x,y)}{c}+\frac{2}{c}\sin(\theta_0'-\theta_p'))\right\}, \quad (3)$$

where I_0 is a constant background, ΔI is the signal envelop, $\delta(x,y)$ is the optical path difference between the reference and the tested beam, $\nu_0 = c/\lambda_0$ and $\Delta\nu$ are the central frequency and the frequency bandwidth of the laser, $\theta_0' = \theta'(\nu_0)$ and Q is a parameter introduced by the SISAM correlator given by:

$$Q = \frac{2}{c}\sin(\theta_p'-\theta_0') - \frac{2N_0}{\nu_0}\frac{\cos(\theta_p'-\theta_0')}{\cos\theta_0'}. \quad (4)$$

In Eq. (3), the "sinc" function corresponds to the correlation signal centered at position $\frac{\delta(x,y)}{cQ}$ on the detector. The cosine term modulates the intercorrelation signal and can be adjusted via the optical (ν_p) and geometrical (θ_p') parameters. If $\nu_p = \nu_0$ and $\theta_p' = \theta_0'$, this modulation is cancelled. This can be done experimentally by properly setting the end beam-splitter.

Equation (3) has been used to compute the intensity profile when an object constituted of two phase plates with different widths and indices (see Fig. 4) is

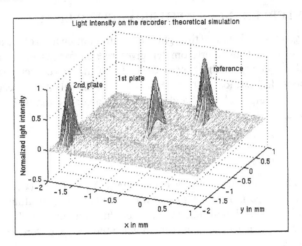

Figure 2. Simulation for three optical paths

scanned across the tested arm. This profile is presented on Fig. 2. The separation between correlation peaks is a mark of the different optical pathlengths experienced by the beam across the object and the surrounding medium.

2. PHASE OBJECT RECONSTRUCTION

2.1. EXPERIMENTAL SETUP

Experimental setup is shown on Fig 3. The light source is a dye laser, tunable over several nanometers around central wavelength $\lambda_0 = 660$ nm. One arm of the Mach-Zehnder interferometer contains a delay line to adjust the path length of the reference signal. The cell (object + scattering medium) mounted on a translation stage is set in the other arm. The SISAM correlator described before combines coherently the two beams. The lens behind the beam-splitter permits to restore the intercorrelation signal onto a CCD camera plane. Specific softwares are used for image acquisition and analysis.

The scattering medium is a solution of 0.05 μm-radius latex beads. Their concentration can be changed to modify the scattering characteristics of the medium. The mean free path versus concentration can be computed from Mie theory. For example, a concentration of 2.10^{19} part/m^3 corresponds to a mean free path of roughly $\ell = 2$ mm. In a 1 cm-thick cell, the number of mean free paths will be about 5. In the present experiment, $\ell = 1.5$ mm (average refractive index $n_0 = 1.33$)

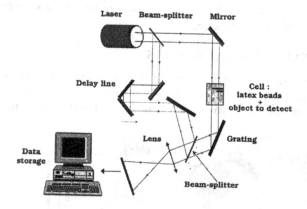

Figure 3. Experimental setup.

The object immersed in the scattering medium is schematically represented on Fig. 4. It is composed of two glass plates ($2 \times 10 \times 30$ mm and $1 \times 5 \times 27$ mm, refractive index $n_1=1.55$ and $n_2=1.75$) stuck together.

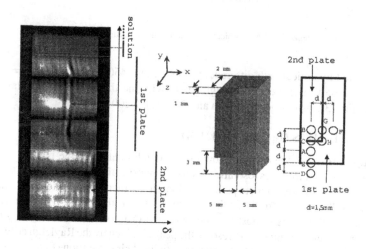

Figure 4. Phase object reconstruction : principle

The acquisition is performed by scanning the object every 1.5 mm (see Fig. 4). Each image of correlation recorded for each scanned area corresponds to an optical path $\delta(x, y)$. So object reconstruction is realized by plotting the optical path versus the step of the scan.

2.2.   EXPERIMENTAL RESULTS

The result of the object reconstruction is presented on Fig. 5. The lower part corresponds to the latex beads solution, the next part to the large glass plate and the upper part to the small glass plate. The different parts are respectively

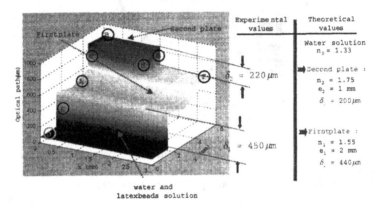

Figure 5.   Phase object reconstruction : experimental results

separated by optical path differences of $\delta_1(x, y) = 450$ μm and $\delta_2(x, y) = 220$ μm. These values are in good agreement with the theoretical ones obtained from Eq. (3): $\delta_{1th}(x, y) = 440$ μm and $\delta_{2th}(x, y) = 200$ μm.

2.3.   LIMITS OF THE METHOD

setup limits.   We consider now the various limitation of the method. The setup by itself introduces resolution limits. Supposing that the tested beam scans an area with different optical paths to be measured, then several intercorrelation peaks will appear. To discriminate between them, the peaks (which have a spatial width of $Q\Delta\nu$) must be well separated (Fig. 6a) to be detected by the camera. The first limitation to resolve the peaks is given by the Rayleigh criteria: the width of the peak must be smaller than twice their separation (Fig. 6b). This condition implies that the path difference is large enough. In our case, the path difference between the two signals must be greater than 5 μm. The second limitation factor is the size of the camera's pixels (8 μm in our case) (Fig. 6c). The correlated peaks have to be separated by more than three pixels in order to be measured by the detector. It supposes a minimum optical path difference of

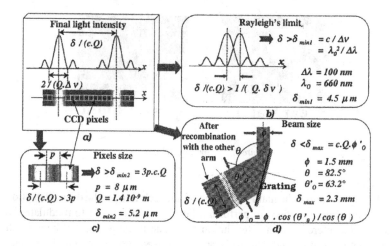

Figure 6. Theoretical setup limits : a. Separation between two peaks ; b. Rayleigh criteria ; c. pixels size ; d. beam size

5 μm. The last point is related to the dynamic of the system and gives the upper limit of the optical path difference (Fig. 6d). The optical difference should not exceed the maximum value corresponding to the camera field. In our setup, the optical path difference $\delta(x, y)$ must not exceed 2.3 mm.

Experimental limits.   The ability of this method to detect object embedded in turbid media depends also on the limited amount of coherent light that can cross the scattering environment. To evaluate the depth resolution of this technique, we have performed measurements of the intercorrelation between the reference beam and the tested one propagating through the scattering medium (alone), for different concentrations of the latex beads. We showed that, although the correlation peak stays always at the same place, the contrast decreases with increasing concentration. The detection limit is reached for $\ell = 1.12$ mm (about 7 mean free paths for a 8 mm-thick cell). Figure 7 compares the images acquired for a concentration corresponding to this value of the mean free path to the case of clear water. Image filtering and treatment are under development to improve this resolution.

Conclusion

We have proposed a new low coherence interferometric method based on the SISAM correlator and have demonstrated its ability to image phase object in

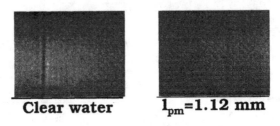

Clear water $l_{pm}=1.12$ **mm**

Figure 7.   Experimental limits

scattering medium. The main advantage of this technique is the direct detection of the signal with no need to perform a Fourier Transform or to modulate the path on the reference arm. Resolution limit due to finite spectral bandwidth of the optical source and detector's pixel size needs however to be improved. Future work will concentrate on more complex object reconstruction in different scattering media.

References

[1] J. Schmitt, S. Lee and K. Yunk, Opt. Com. 142, 203–207 (1997).

[2] C. Thomson, K. Webb and A. Weiner, J. Opt. Soc. Am. A **14**, 2269–2277 (1997).

[3] H. Brunner, R. Lazar, R. Seschek, T. Meier, and D. Benardon, in *Proc. SPIE : Handbook of Theoretical Computer Science, Vol. B*, volume 3194, 205–211 (Elsevier Science Publishers, North-Holland, 1998).

[4] M. Toida, M. Kondo, T. Ichimura and H. Inaba, Appl. Phys. **B52**, 391–394 (1991).

[5] H. Chiang and J. Wang, Opt. Com. **130**, 317–326 (1996).

[6] J. Schmitt, J. of Biomedical Optics 3, 66–75, (1998).

[7] D. Hoelscher, C. Kemmer, F. Rupp and V. Blazek, in *Proc. SPIE Photonics West (San Jose)*, volume 3915/10, (2000).

[8] S. Xiang, Y. Zhao, Z. Chen and J. Nelson, in *Proc. SPIE Photonics West (San Jose)*, volume 3915/12, (2000).

[9] G. Brun, I. Verrier, D. Troadec and C. Veillas, Opt. Com. **168**, 261–275 (1999).

Chapter 4

LASER OPTICAL FEEDBACK TOMOGRAPHY

E. Lacot, R. Day and F. Stoeckel

Laboratoire de Spectrométrie Physique, Université Joseph Fourier de Grenoble,
B.P. 87, 38402 Saint -Martin-d'Hères Cedex, France

Eric.Lacot@ujf-grenoble.fr

Abstract   We present a new technique for imaging in turbid media called Laser Optical Feedback Tomography (LOFT). This technique is based on the detection of retrodiffused light reinjected into the emitting laser cavity. The advantages of the system are the spatial and coherent filtering, the easy optical alignment and the resonant amplification of the weak reinjected signal.

Keywords: Feedback, Laser, Dynamics

Introduction

The non-invasive imaging of objects in diffusing media is a widely studied subject with many different applications, such as medical imaging of living tissue [1], undersea visibility and visibility in cloud. A number of methods have been developed for imaging in turbid media such as homodyne and heterodyne scanning confocal microscopy [2], photon density-wave propagation [1], optical coherence tomography [3], and time-resolved experiments, including time-resolved holography. The confocal techniques have the advantage of avoiding the inverse problem which is a complex issue intrinsic to tomography. All techniques, however, using ballistic photon propagation are limited by the rapid decay of photon density with increasing penetration into the diffusing medium. The objective is therefore to have a system which has the maximum of sensitivity to retrodiffused photons while retaining an experimental set-up that is simple and cost-effective.

In this chapter we describe a new method [4] called Laser Optical Feedback Tomography (LOFT). This technique is based on the same principles used in

P. Sebbah (ed.), Waves and Imaging through Complex Media, 251–257.

heterodyne scanning confocal microscopy but where a small cavity laser acts as the source, the detector, a spatial and coherent filter and an optical amplifier.

1. OUTLINE

Figure 1 shows a comparison between the set-up for traditional confocal microscopy and the LOFT technique. In confocal microscopy a laser beam is focused in or onto the turbid medium of interest. There is a small quantity of ballistic photons that is retrodiffused from points located near the focus and which comes back along the same optical path. These photons are then directed via a beam splitter through a spatial filter and onto a photodetector.

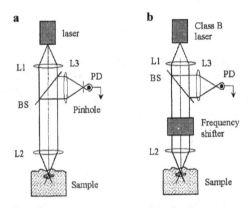

Figure 1.   Schematic diagram of (a) traditional confocal microscopy and (b) LOFT technique. BS, beam splitter; PD, photodetector; L1-L3, lenses.

In the LOFT technique we use a class B laser as the source, the properties of which will be explained later. The ballistic photons retrodiffused from points located near the focus are shifted in optical frequency by $F = \Omega/2\pi$ and reinjected by mode matching into the laser. In the laser cavity there is a beating effect between the intra-cavity field and the reinjected frequency shifted field. This beating at frequency F is detected on the laser output intensity using an external photodiode. The amplitude and phase of the beating signal is measured by means of a lock-in amplifier or a spectrum analyzer. Using a translation unit we move the sample to obtain a one-, two- or three-dimensional image. The signal is recorded and processed by a PC.

2. THEORY

A class B laser is defined as that having a cavity damping rate γ_c greater than the population-inversion damping rate γ_1. The output intensity of such a laser is very sensitive to small perturbations of the laser parameters and exhibits a resonant frequency.

In order to understand how the laser is able to amplify the weak feedback we will consider the conventional rate equations for a reinjected laser [5]. We will limit our calculation to the approximation of weak feedback with a shift in optical frequency. The frequency shifter is placed close to the laser to simplify the calculated expression.

We start with the conventional rate equations for a reinjected laser,

$$
\begin{aligned}
\frac{dN}{dt} &= \gamma_1(N_0 - N) - BN|E|^2 + F_N(t), \\
\frac{dE(t)}{dt} &= \left[i(\omega_c - \omega) + \frac{1}{2}(BN - \gamma_c) \right] E(t) \\
&\quad + \gamma_{ext} E_{inj}(t) + F_E(t),
\end{aligned}
\tag{1}
$$

where N is the population inversion, E is the electric field inside the laser cavity, γ_1 is the decay rate of the population inversion, γ_c is the photon decay rate inside the cavity, $\gamma_1 N_0$ is the pumping rate, ω is the optical running frequency, ω_c is the laser-cavity frequency, B is the Einstein coefficient, $F_N(t)$ and $F_E(t)$ describe, respectively, the population and the electric-field fluctuations that are due to quantum noise and $\gamma_{ext} = \gamma_c \sqrt{R_{eff}}$ where R_{eff} is the effective power reflectivity of the object.

The frequency shifted feedback that is reinjected back into the laser can be expressed as

$$
\begin{aligned}
E_{inj}(t) &= E_c(t - \tau) \exp[-i(\omega + \Omega)\tau] \\
&\quad \times \exp[i\Phi_c(t - \tau)] \exp(i\Omega t),
\end{aligned}
\tag{2}
$$

where τ is the round-trip time of flight outside the cavity.

After separation of real and imaginary parts, Eqs. (1) become

$$
\begin{aligned}
\frac{dN}{dt} &= \gamma_1(N_0 - N) - BN|E|^2 + F_N(t), \\
\frac{dE_c(t)}{dt} &= \frac{1}{2}(BN - \gamma_c)E_c(t) + \gamma_{ext} \cos[\Omega t - \Phi_c(t) \\
&\quad + \Phi_c(t - \tau) - (\omega + \Omega)\tau]E_c(t - \tau) + F_{E_c}(t), \\
\frac{d\Phi_c}{dt} &= (\omega_c - \omega) + \gamma_{ext} \sin[\Omega t - \Phi_c(t) + \Phi_c(t - \tau) \\
&\quad - (\omega + \Omega)\tau]\frac{E_c(t - \tau)}{E_c(t)} + F_{\phi_c}(t),
\end{aligned}
\tag{3}
$$

where $E_c(t)$ and $\Phi_c(t)$ are, respectively, the amplitude and phase of the electric field in the cavity.

The cosine and sine terms that appear in Eqs. (3) represent the coherent interaction between the intra-cavity and the reinjected field. This results in a modulation of the net laser gain at the frequency $\Omega/2\pi$.

In the absence of feedback ($\gamma_{ext} = 0$), the steady states are given by $N_S = \gamma_c/B$, $\Phi_S = 0$ (assuming $\omega = \omega_c$) and $E_S^2 = E_{sat}^2(\eta - 1)$, where the saturation intensity is $E_{sat}^2 = \gamma_1/B$ and the normalized pumping parameter is $\eta = BN_0/\gamma_c$.

In the case of weak feedback ($\gamma_{ext} \ll \gamma_c$), and assuming that the round-trip flight time is much smaller than the frequency shift period ($\Omega\tau \ll 1$), Eq. (3) can be written as

$$\frac{d}{dt}\Delta n = -(\gamma_1 + B|E|^2)\Delta n - 2BN_S E_S \Delta e,$$

$$\frac{d}{dt}\Delta e = \frac{1}{2}BE_s\Delta n + \gamma_{ext}E_S \cos[\Omega t - (\omega + \Omega)\tau],$$

$$\frac{d}{dt}\Delta\Phi = \gamma_{ext}\sin[\Omega t - (\omega + \Omega)\tau], \tag{4}$$

where Δn, Δe and $\Delta\Phi$ are the small deviations from the steady state values N_S, E_S, and Φ_S, respectively. If we solve Eqs. (4), we obtain an intensity varying at frequency $\Omega/2\pi$ with an amplitude given by

$$\left|\frac{\Delta I(\Omega)}{I_S}\right| = \left|\frac{2\Delta e(\Omega)}{E_S}\right|$$

$$= 2\gamma_{ext}\sqrt{\frac{\eta^2\gamma_1^2 + \Omega^2}{\eta^2\gamma_1^2\Omega^2 + (\omega_R^2 - \Omega^2)^2}}, \tag{5}$$

where the steady state intensity is $I_S = E_S^2$ and the laser resonant frequency is $\omega_R = \sqrt{(\eta - 1)\gamma_1\gamma_c}$. We see from Eq. (5) that the maximum amplitude is obtained when $\Omega = \omega_R$. By considering the condition $\gamma_c \gg \gamma_1$, which is the case for a class B laser, we can write the maximum of the relative amplitude of modulation of the output beam at frequency Ω as

$$\frac{[\Delta I(\Omega = \omega_R)]}{I_S} = \frac{2\sqrt{R_{eff}}}{\eta}\frac{\gamma_c}{\gamma_1}. \tag{6}$$

For a microchip laser $\gamma_c/\gamma_1 \approx 10^6$ so allowing a considerable enhancement of the weak reinjected signal. Given a pumping parameter $\eta = 2$, an effective power reflectivity $R_{eff} = 10^{-16}$ results in a relative intensity modulation of 1%, which is easily measurable.

3.   EXPERIMENTAL RESULTS

The results that we will present here were obtained using a Nd^{3+} : YAG microchip laser with an 800 μm thick cavity, which was pumped by a 810 nm, 100 mW diode laser.

Experimentally, we modulated the field intensity using an acousto-optic modulator. This gives a shift in optical frequency of $\pm F$. This frequency shift ranged typically from a few hundred Hz to 1 MHz.

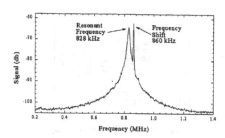

Figure 2.   Typical power spectrum of the output intensity of a Nd:YAG microchip laser with frequency shifted feedback.

Figure 2 shows an example of the spectrum of the signal detected by the photodiode in the presence of feedback. We see that the quantum noise follows the laser resonance which is here about 828 kHz. The peak at 860 kHz corresponds to the amplified frequency shifted feedback from the diffusing target.

Figure 3 shows, on a logarithmic scale, the amplitude of modulation at frequency F while scanning the focal point of the laser through a suspension of silica powder in water, in which a mirror is submerged. The peak at $\Delta z \approx 0$ mm is due to specular reflection at the surface of the water. The peak at $\Delta z = -1.25$ mm is the reflection of the immersed mirror. The number of photon mean free paths between the liquid surface and the mirror is estimated from the results to be about 5.2, if we neglect the absorption. The longitudinal resolution was calculated experimentally to be $\sim$6 μm (HWHM). The lens L2 used to focus into the medium had a focal length of 30 mm and a numerical aperture of 0.1. The longitudinal and transverse resolution is, as in traditional confocal microscopy, determined by the numerical aperture of the focusing lens.

In figure 4 we present a LOFT experiment applied to 2D imaging in diffusing media. A French 5 centimes piece was immersed in 10 mm of milk having a mean free path of 1.5 mm. The z position was fixed to obtain the maximum in amplitude for the flat part of the coin on the right hand side. We then scanned in

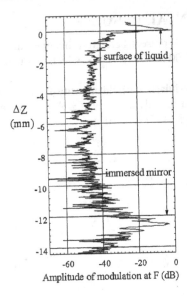

Figure 3. Detected amplitude using LOFT for a scan of a mirror in a diffusing liquid.

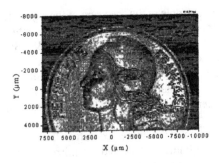

Figure 4. Two dimensional image of a French five centimes piece immersed in 1 cm of milk using LOFT.

the X and Y direction with a sample step of 120 μm. The contrast in amplitude is due to both defocalisation and the inclination of the probed surface.

Summary

We have demonstrated a method of using frequency shifted light reinjected into a type class B laser for imaging in diffusing media. The principal advantage of LOFT is the resonant amplification of the coherent feedback. This results in the quantum noise becoming more significant than the technical noise, yielding a shot-noise limited system without the need of a high performance detector. As the laser acts as both the source and detector, the system is self-aligned and behaves as a spatial and coherent filter.

Applications, taking account of phase variation, may be envisaged in many different fields where the imaging of or in diffusing media is required such as vibrometry, velocimetry and distance measurement. These applications are currently under investigation.

Acknowledgments

This study was supported by the Department of Research-Industry of the Université J. Fourier of Grenoble. The authors thank E. Engin of the Laboratoire d'Electronique de Technologie et d'Instrumentation/Commissariat a l'Energie Atomique for providing the microchip lasers.

References

[1] A. Yodh and B. Chance, Phys. Today **48**(3), 34 (1994).

[2] J.M. Schmitt, A. Knüttel, and M. Yadlowsky, J. Opt. Soc. Am. A **11,** 226 (1994).

[3] D. Huang, J. Wang, C. P. Lin, J. S. Shuman, W. G. Stinson, W. Chang, M. R. Hee, T. Flotte, K. Gregory, C. A. Puliafito, and J. G. Fujimoto, Science **254,** 1178 (1991).

[4] E. Lacot, and F. Stoeckel, "Detecteur optique actif", patent 97/12391 (September 30, 1997), patent PCT/FR98/02092.

[5] A. E. Siegman, *Lasers* University Science Books, Mill Valley Ca, (1986).

Chapter 5

IMAGERY OF DIFFUSING MEDIA BY HETERODYNE HOLOGRAPHY

M. Gross, F. Le Clerc

Laboratoire Kastler Brossel de L'Ecole Normale Supérieure
CNRS- UMR 8552 -Univertité-CNRS Paris 6,
24 rue Lhomond 75231 Paris Cedex 05, France
gross@lkb.ens.fr, leclerc@lkb.ens.fr

L. Collot

Thomson CSF Optronique,
Rue Guynemer BP 55, 78 283 Guyancourt, France
lcollot@club-internet.fr

Abstract   Heterodyne holography is an original method for digital holography that relies on two-dimensional heterodyne detection to record the amplitude and the phase of a field. Knowing the field on the detector allows determining the field in all points of space. Among the many possible applications, heterodyne holography open the way to a new imagery technique of diffusing media. The idea is to acquire a maximum of information on the diffused field, and to extract then the pertinent one. First result in selecting ballistic photons is presented.

Keywords: heterodyne holography, digital holography, turbid media

1.   HETERODYNE HOLOGRAPHY

Demonstrated by Gabor [1] in the early 50's, the purpose of holography is to record on a 2D detector the phase and the amplitude of the light coming from an object under coherent illumination. Thin film holography (see Fig. 1) does not provide a direct access to the recorded data. Numerical holography [2] replaces the holographic film by a 2D electronic detector allowing quantitative numerical analysis. This simple idea needed the recent development of computer and

P. Sebbah (ed.), *Waves and Imaging through Complex Media*, 259–266.
© 2001 *Kluwer Academic Publishers. Printed in the Netherlands.*

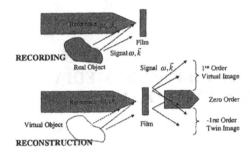

Figure 1.　Thin film off-axis Holography: $\omega = \omega_0$, $k \neq k_0$

video technology to be experimentally demonstrated [3]. In both thin film and numerical holography the system records only one interference phase state, that allows to calculate only one quadrature of the complex field. This incomplete measurement yields to a virtual image ($order = 1$) of the object that is superposed [4] with a ghost twin image (order=-1) and with the remaining part of the reference field (order=0). A solution to this problem (see Fig. 1) is to tilt the reference beam in respect to the signal beam [5] in order to separate physically the three images and to select the wanted image. This off-axis method reduces the useful angular field of view and restricts measurement to the far field region where the 3 images are spatially separated.

In order to avoid these problems, it is necessary to get more information by recording more than one phase state of the interference holographic pattern. A possible method is to record several holograms while shifting the phase of the reference beam with a PZT mirror[6]. We have developed an alternate technique that we call *heterodyne holography* [7] where we record on a CCD camera the interference of the signal field with a reference field, which is frequency shifted by δf. Each pixel of the CCD camera performs thus heterodyne detection of the signal field. To make the demodulation procedure easier the heterodyne frequency $\delta f = 2\pi(\omega_0 - \omega) = 6.25$ Hz is chosen equal to $1/4$ of the 25 Hz video frame rate. If the reference is a plane wave, the complex signal field E_s is proportional to $(I0 - I2) + i.(I3 - I4)$ where $I0, I1, I2, I3$ are four successive video images.

Our heterodyne holography experimental setup is shown in Fig. 2. Although we consider here a transmission setup, most of the following remain valid in a reflection configuration. The coherent source is an Helium Neon laser L. The two acousto-optic modulators AOM1 and AOM2, which are working at $80MHz$ and $80MHz + 6.25Hz$ respectively, provide the frequency shift of the reference beam. The reference and signal beam are expanded by 2 beam expanders BE1 and BE2. The 2 beams are combined by the beam splitter

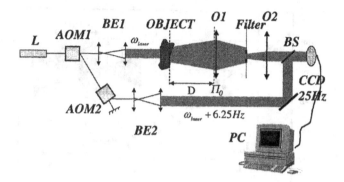

Figure 2.   Heterodyne Holography Setup: L HeNe laser, AOM1 and AOM2 acousto-optic modulator, BE1 and BE2 beam expanders, O1 and O2 confocal objectives, BS beam splitter and CCD detector.

BS and the interference pattern is recorded by the CCD camera. A frame grabber transfers the information to the PC. Since the CCD pixel spacing is finite (typically $10\mu m$), the CCD camera performs a spatial sampling of the field. The sampling theorem limits thus the field of view angle θ for valid measurements to:

$$|\theta| \leq \theta_{\max} = \lambda/(2d_{pixel}),\qquad(1)$$

where d_{pixel} is the CCD pixel spacing. One can notice that this sampling condition is common to both on and off-axis digital holography. In order to fulfill Eq. (1), we have selected, in our experimental setup, the near axis photons by a spatial filter system ($O1, O2$, Filter on Fig. 2).

Detailed tests and discussions on our system for holography are given in [7]. We show that heterodyne holography performs within the spatial filter selected region, a complete measurement of the signal field without information loss. Since the technique is sensitive to the field amplitude (heterodyne detection), one photon detection is possible. The dynamic range is limited by the number $n \approx 3.10^5$ of electrons that can be stored on each CCD pixel, and corresponds for each pixel to about $n/\sqrt{n} \approx 5.10^2$, i.e to 9 bit data, or to 54 dB. Another important point is information. Our system grabs in one second 12.10^6 words with 16 bit, and extracts 6.25 complex field images with 5.10^5 pixels.

2.   APPLICATION TO DIFFUSING MEDIA

In tissues, diffusion of light is mainly related to the small ($\approx$ 5 to 10%) change of the refractive index within each cell. As these changes occur over distances larger than the wavelength, scattering is highly directive [8] in the

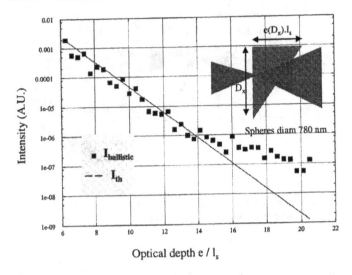

Figure 3.   Ballistic photons in gel.

forward direction and the 2 lengths l_s and l'_s that govern scattering are very different. $l_s \simeq 50$ μm is the scattering length, i.e. the length beyond which the light phase is lost, and $l'_s \simeq 1$ mm is the light transport mean free path, i.e. the length for loosing the incident direction of propagation.

In this context, heterodyne detection has been used to select the photons transmitted through a diffusing media, which remain coherent. Using this technique Inaba [9] gets quite nice images when scanning a mono-pixel detector (photodiode). Here we go further and perform heterodyne detection on 2Ddetectors by using our heterodyne holography technique. Contrarily to Inaba, we record holograms where the pixel to pixel relative phase remains meaningful. Our idea is to acquire a maximum amount of information on a diffusing object, in order to extract later useful pertinent results.

To illustrate this, we have performed an heterodyne holography experiment on latex sphere solutions in gel and in liquid , and we have selected in the transmitted signal field, the ballistic component that corresponds to the photons, which have passed through the solution without interacting. In gel, the concentration of the 780 nm sphere solution is kept constant while the light is focused on the surface of the prism shaped solution cell. Translating the cell in the transverse direction changes the medium effective depth (see Fig. 3). In liquid the cell that contains the 480 nm sphere solution is rectangular. The cell

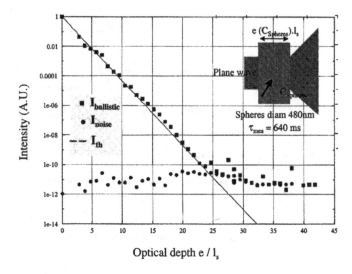

Figure 4.   Ballistic photons in liquid.

is illuminated by plane light beam and the effective depth of the medium is altered by changing the concentration of the solution (see Fig. 4).

The field transmitted through the diffusing medium is first measured. The weight of the ballistic photons components is then determined by calculating the correlation of the measured field, which is the sum of a ballistic and a diffused component, with the pure ballistic reference field, that corresponds to the field measured without scatters. In Fig. 3 and Fig. 4 the relative (with respect to the incident field) weight of the ballistic component (black squares) is plotted in log scale (y-axis), as a function of the depth of the medium in scattering length unit l_s (x-axis). As expected, for both liquid and gel, the ballistic component decreases exponentially with a length parameter l_s (dotted line in Fig. 3 and Fig. 4). The noise floor is about 70 dB (10^{-7}) and 110 dB (10^{-11}) below the incident field for gel and liquid, respectively. This correspond to an effective depth of about 16 l_s and 25 l_s, respectively. These results can be compared favorably with [10], where equivalent experiments were performed using pinhole selection of the ballistic photons.

Our results can be understood quite easily. As the diffusing depth is large, most of the incident photons are back reflected, and the small transmitted component is spread over a large solid angle. A small part p of the incident photon ($p \approx 10^{-2}$ to 10^{-3}) passes thus through the spatial filter (that corresponds to Eq. (1) sampling condition), reaches the CCD, and is detected. By Fourier

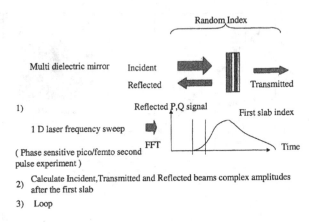

Figure 5.   Solving the 1D diffusion inverse problem

transform, we have calculated the decomposition of the detected signal over the k-modes. The diffused component is randomly spread with equal weight over all the k-modes, while the ballistic component remains in the k-mode that corresponds to the incident field. As the number of k-modes N is simply equal to the number of pixels (5.10^5), the noise floor, which corresponds to the weight of the diffused component within the ballistic k-mode (or any k-mode), is expected to be simply $p/N \approx 2.10^{-7}$ to 2.10^{-8}. This result is in good agreement with the gel experiment.

In liquids, the experiment overcomes this limit, because the Brownian motion of the scatters shifts the frequency of the scattered light. As our heterodyne system has an extremely narrow detection bandwidth (that is equal to the inverse of the measurement time: 640 ms), most of the diffused photons are not detected. We have plotted (black circle on Fig. 4), for liquid, the average value of the noise per mode. As expected, the ballistic noise floor corresponds to this one mode noise.

Heterodyne holography may also be used in a more ambitious way. The idea is to perform many measurements on a diffusing object in order to solve the inverse diffusion problem. Let us discuss this point on the simpler 1D "gedanken" experiment . Let us perform heterodyne holography on a 1D diffusing object as, for example, a random index flat multi-layer mirror (see Fig. 5). The 2D detector becomes here a 0-D mono-pixel detector. To calculate the refractive index for the M slabs of the mirror, one needs M independent measurements (or more) and a mono detector measurement is not sufficient. It is thus required to add one dimension (1D) in the measurement process. Let us sweep, for example, the source wavelength while measuring -as a network analyzer does-the response to light (attenuation and phase change) of the mirror. Making a frequency to time Fourier transform allows to determine the time response of the mirror to ultra short pulses. This sweeping technique is expected to be more sensitive (one photon sensitivity), more precise (phase sensitive) and must work with a higher dynamic range (since it detects an amplitude and not an intensity) than the short laser pulse technique [11]. Solving the 1D diffusion problem is then possible. Let us consider the reflection configuration. The first reflected signal allows to determine the refractive index of the first slab. One can thus calculate the second slab incident field at any time by accounting of the first slab reflection. Calculation of the successive slabs index can thus be done by iteration of this stripping process.

This 1D "gedanken" experiment illustrate the advantage of heterodyne holography in performing a complete measurement of the field that allows a further powerful data analysis. It also shows that sweeping the coherent source wavelength is an essential ingredient to go further. We must notice that in the gel experiment (Fig. 3), the observed noise floor is related to speckle. Speckle comes from the multi scattered photons, which remain coherent in time (because our Helium Neon source have a long coherent length), and which are detected efficiently by our heterodyne system. OCT and short laser pulse experiments [11] use low time coherent source to filter off in time the multi-scattered photons. We intend to include time filtering in heterodyne holography to suppress most of the diffused photons. This can be done either by sweeping the source wavelength, as mentioned above, or by using a low coherent source. With this technique, heterodyne holography, which is already an interesting tool, is expected to be very powerful in studying diffusing media.

Acknowledgments

This work was supported by Thomson-CSF Optronique and funded by DGA under contract n°98 10 11A.000.

References

[1] D. Gabor, Proc. R. Soc. A **197**, 454 (1949).

[2] A. Macovsky, Optica Acta **22**, 1268 (1971).

[3] U. Schnars, J. Opt. Soc. Am. A **11**, 977 (1994).

[4] T. Kreis, W. Juptner, and J. Geldmacher, SPIE **3478**, 45 (1988).

[5] E. Leith and J. Upatnieks, JOSA **52**, 1123 (1962).

[6] I. Yamaguchi and T. Zhang, Optics Letters **18**, 31 (1997).

[7] F. LeClerc, L. Collot, and M. Gross, Opt. Lett. **25**, 716 (2000).

[8] H. van de Hulst, *Light Scattering by Small Particles* (John Wiley and Sons, Inc., N.Y., 1957).

[9] B. Devaraj, M. Takeda, M. Kobayashi, K. Chan, Y. Watanabe, T. Yuasa, T. Akatsuka, M. Yamada, and H. Inaba, Appl. Phys. Lett. **69**, 3671 (1996).

[10] M. Kempe, A. Genack, W. Rudolph, and P. Dorn, J. Opt. Soc. Am. A **14**, 216 (1997).

[11] L. Wang, P. Ho, C. Liu, G. Zhang, and R. Alfano, Science **253**, 769 (1991).

Chapter 6

IMAGING THROUGH DIFFUSING MEDIA BY IMAGE PARAMETRIC AMPLIFICATION

E. Lantz, G. Le Tolguenec and F. Devaux

Laboratoire d'Optique P.M. Duffieux
U.M.R. 6603 CNRS Université de Franche-Comté
25030 Besançon cedex, France
tel: (33) 3 81 66 64 27
Fax:(33) 3 81 66 64 23
elantz@univ-fcomte.fr

Abstract   Imaging through thick diffusing media requires selection of the least scattered light, because ballistic light is only transmitted up to a few millimeters. Hence, temporal gating, i.e. isolation of the front part of a femto or a picosecond pulse appears as a promising tool to form images of objects embedded in thick media, like tumors in human breast. We have used parametric image amplification to obtain images with a time gate duration of about 20 ps. At first, we used latex beads solutions, whose diffusing properties are well known, to characterize the method and its dynamic range for selection of the ballistic light. In a second step, objects embedded in biological tissues, like chicken meat, were imaged. The best results have been obtained by forming the image without lenses, by exploiting the fact that the idler is phase-conjugated with respect to the signal in the transverse directions, while it propagates forward. A 1 cm³ piece of liver embedded in a 4 cm thick chicken breast tissue has been detected.

Keywords: imaging, parametric amplification, mammography

Introduction

The ability of near infrared light, that experiences very small absorption in the therapeutic window (600-1300 nm), to form images of tumors through human breast was recognized more than 70 years ago. Because of its poor resolution due to the strong scattering of light in biological media, this technique, called diaphanography [1, 2], was progressively replaced by X-ray mammography, despite its ionizing character. However, in the last decade a renewal of interest

P. Sebbah (ed.), Waves and Imaging through Complex Media, 267–273.
© 2001 *Kluwer Academic Publishers. Printed in the Netherlands.*

has occurred for optical methods that involve a selection of the non or least scattered photons to improve the resolution. In relatively thin media, a part of the incident light, called ballistic, is not scattered and images are now routinely obtained by using optical coherence tomography, that detects this ballistic light by interferometry [3]. In thicker media, like human breast, imaging tumors remains challenging, because the thickness of the diffusing medium is so important that ballistic light does not exist any more and images must be formed with the scattered photons. Since diffusion in a biological medium is strongly anisotropic, the photons are preferentially scattered in the forward direction and the front part of a short pulse, largely stretched by the medium, is formed with the least scattered photons that have traveled on the shortest paths. Hence, the front part still carries some amount of spatial information, with a resolution that depends on the thickness of the diffusing medium. We present here our results about parametric image amplification as a method to separate this front part from the multi-diffused light. The paper is organized as follows. We recall in section 1 the bases of parametric image amplification. Our characterization experiments with latex beads are described in section 2. Last, experiments in thick biological media are presented in section 3.

1. PRINCIPLE OF PARAMETRIC IMAGE AMPLIFICATION

When a strong pump pulse interacts with a weak signal pulse in a non linear χ^2 crystal, the signal becomes amplified and a third wave called idler is generated, provided that two conditions are fulfilled. First, the relation between the involved frequencies means that the pump photons split in two and create pairs of signal and idler photons. The second condition, called phase matching, expresses the equality between the pump wavevector and the sum of the signal and the idler wavevectors. Phase matching is necessary to avoid a back conversion of the signal and idler photons into pump photons. The most classical method to obtain phase matching is using a birefringent crystal. To amplify an image that is formed on the input face of the crystal, phase matching must be conserved for the different directions of the signal wavevectors corresponding to the spatial frequencies of the image. We showed [4] that phase matching can be conserved for a cone of wavevectors, giving an image of 80x80 resolved points amplified with a 40 dB gain in a 7x7x20 mm KTP crystal. Moreover, the idler image, that is cross-polarized with the signal in a type 2 interaction and can be separated from the signal by a simple polarizer, is time-gated because it is generated only during the interaction with the pump, and is phase conjugated with the signal wave in the transverse direction [5]. These two properties will be used to the detection through scattering media.

2. IMAGING THROUGH LATEX BEADS : STUDY OF THE RESOLUTION AND THE SIGNAL-TO-NOISE RATIO

2.1. EXPERIMENTAL SETUP

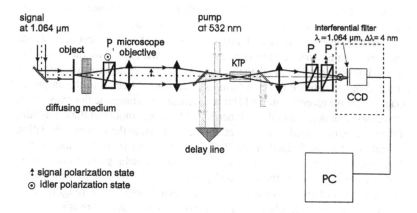

Figure 1. Experimental set-up. The signal is horizontally polarized by P_1 while P_2 and P_3 select the vertically polarized idler for time-gating

To characterize our method, we used latex beads, whose diffusing properties are well known. Figure 1 shows the experimental setup [6]. An infrared pulse at 1.064 μm is delivered by a mode-locked Nd:YAG laser, with a duration of approximately 55 ps (FWHM) at a repetition rate of 10 Hz. The radiation is partially frequency-doubled in a KDP crystal. The remaining of the infrared pulse is separated from the green and is horizontally polarized. It illuminates a resolution chart (the object) placed before the latex beads solution. The object is imaged in a 20 mm long, 7×7 mm wide KTP crystal with a telescopic system, including a microscope objective, whose magnification is 6.7. A Glan polarizer set after the scattering medium selects the horizontal polarization of the input signal photons. The green radiation used as the pump beam is superimposed to the infrared pulse in the KTP crystal with a delay line. The amplified image is then formed onto a CCD camera. Two Glan-polarizers placed between the crystal and the imaging lens select the vertical polarization corresponding to the idler. The measured rejection rate of the two polarizers was found to be 35 dB. The scattering media that we used were solutions of latex microspheres of different optical thicknesses. The scattering coefficients, and the anisotropy parameter were calculated using the results of Mie theory. The diameter of the particles was 0.302 μm with an initial particle density of $7.1 \times 10^{12}\,\text{cm}^{-3}$. At

1.064 μm, absorption is very small compared to scattering and can be neglected. The thickness of the cell was 10 mm. Solutions of 12, 15, 20 and 22 mean free paths were tested. The anisotropy parameter was 0.25 at 1.064 μm.

2.2. COHERENCE CONSIDERATIONS

The detector was a cooled CCD camera that permits detection of a very low flux. Hence, the noise levels due to the camera were much lower than the lowest flux detected, even in the case of a time integration of several seconds. On the other hand, parametric fluorescence generated in the crystal by the pump can have an intensity level higher than the signal level. Hence, an image of this noise recorded in the absence of the infrared signal was subtracted to the amplified image. We have recently shown [7] that the mean of parametric fluorescence in a resolution cell is equal to the number NT of temporal modes in the bandwidth of the parametric amplifier multiplied by the gain, while the standard deviation is equal to this mean divided by $\sqrt{NT}$. For one shot, NT is given by the product of the amplifier bandwidth (limited either by phase-matching or by a chromatic filter) by the duration of the temporal gate. To conclude this paragraph, the minimum level of signal that can lead to successful detection is roughly equal to $\sqrt{NT}$ photons per resolution cell. It is worth noting that there is much less than one photon per spatio-temporal mode in an incoherent (thermal) signal. Hence, the signal issued from the scattering medium appears still partially coherent, though the diffusion process.

2.3. RESULTS

We present in Fig. 2 images obtained with the attenuation of 22 mfp corresponding to an attenuation of the ballistic signal of 95 dB. Since the aperture time of the camera was 4 s, these images were obtained by summation of 40 laser shots. Figure 2a is the amplified image of the resolution chart without the diffusing medium. The maximum gain is nearly equal to 33 dB. Figure 2b is the non-amplified image of the resolution chart through the diffusing medium. No line is resolved. Figures 2c to 2i present the amplified images of the resolution chart through the diffusing medium for different temporal positions of the pump, where the time origin corresponds to the pump and signal pulses synchronized in the crystal in the absence of biological tissue.

The transition between the ballistic regime and the diffusing regime can be clearly observed on the images. In the image 2c, no line is resolved. Only a very weak signal is detected, formed by the diffused signal that has not been rejected by the Glan polarizers. For that position of the pump, the amplification is not high enough to compensate for the level of the non-rejected signal. For pump delays t ranging from -58 ps to -31 ps (images 2d to 2h), lines are resolved and the contrast is maximum for a pump delay of -44 ps (image 2f). For this image,

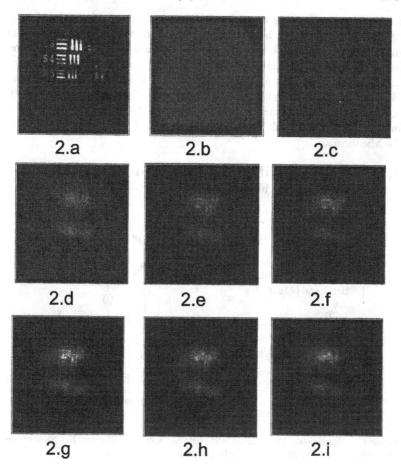

Figure 2. Parametric amplification through a latex bead solution of 22 mfp. (a) : image of the resolution chart with amplification, without the scattering medium, (b) : image of the resolution chart with the scattering medium of 22 mfp, without amplification ; no line is resolved, (c) to (i) : images of the resolution chart with amplification and with the diffusing medium. Between two images, the pump delay is shifted by 6.6 ps. (f) corresponds to the best signal to noise ratio of 1.85.

the signal to noise ratio was measured to be 1.85 and the resolution is similar to the resolution without diffusing medium, that means that the ballistic part of the signal has not completely disappeared and forms the usable part of the image. For *t* shorter than -24 ps (image 2i), the diffusion is too high to enable an image to be extracted. Note that for all images, the green pulse is always in advance of the ballistic part of the infrared pulse (negative delays). This means that the maximum amplification is not attained. For the case where the pump and the infrared signals are perfectly superimposed to give a maximum amplification, the diffusion is too high to allow the signal to be extracted from the noise. We fitted the experimental signal-to-noise ratio for different attenuations with a model where the parameters were the width of the diffused pulse, its peak intensity and its shift with the ballistic pulse. Results were in agreement with the Mie theory.

3.   IMAGING THROUGH THICK BIOLOGICAL TISSUES BY FORWARD PHASE-CONJUGATION

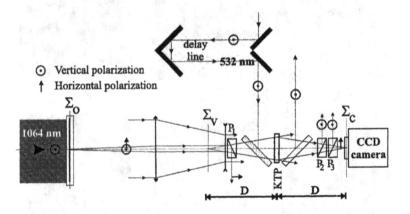

Figure 3.   Experimental set-up. Σ_o: object plane, Σ_v: virtual object plane, Σ_c: phase conjugated image plane. P_1, P_2, P_3 Glan-Taylor polarizers.

At the exit of thick biological media, ballistic light no more exists and images must be formed with the least scattered photons, with a resolution that depends of the medium thickness. The most efficient scheme, shown in Fig. 3, combines time-gating and forward phase conjugation for a better rejection of the diffused light [8]. In order to increase the field of view, a virtual image is formed with a 0.1 magnification in the plane symmetrical of the C.C.D. camera with respect to the crystal. After the crystal, the signal is rejected by the polarizers and the image is formed by the forward phase-conjugated idler onto the camera plane without any lenses. Figure 4 shows experimental results. Figure 4a presents an amplified image of the object, a cross formed by two 9 mm wide metal strips, without biological tissue. Figures 4b and 4c show images of this cross embedded in 4 cm thick chicken breast tissues. The delays are respectively 0

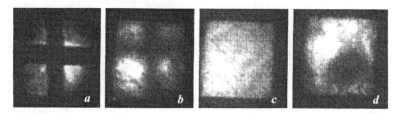

Figure 4.   a, amplified image of the strips without biological tissue. b-c, amplified images through 4 cm thick chicken breast tissue. The delays are respectively 0 and 158 ps. In b, the strips are resolved and the best SNR $\simeq$ 2 is obtained. d, resolved image of a 1 cm^3 piece of liver embedded in the same chicken breast sample.

and 158ps. In Fig. 4b the 9 mm wide metal strips are resolved and the best signal-to-noise ratio is obtained. When the delay is greater than 33ps, the strips are no longer resolved and only diffused light is amplified (Fig 4c). Figure 4d shows a resolved image of a 1 cm^3 piece of liver embedded in the same biological sample [9]. These results are very promising because the thickness of the biological tissues corresponds to that of human breast in mammography. Though the resolution is in the range of several millimeters, the non-ionizing and non-invasive character of the method could be very useful for routine preventive exams. Using parametric image amplification, images in a reflection configuration have been obtained recently at the IOTA with femtosecond pulses [10]. This pulse duration allows precise tomography, but for thinner samples.

References

[1] M. Cutler, Surg. Gynecol. Obstet **48,** 721(1929).

[2] E. B. De Haller, J. Biom. Opt. **1,** 7(1996).

[3] B. Bouma, G. J. Tearney, S. A. Boppart, M. R. Hee, M. E. Brezinski, and J. G. Fujimoto,Opt. Lett. **20,** 1486(1995).

[4] F. Devaux, E. Lantz, A. Lacourt, D.Gindre, H. Maillotte, P.A. Doreau, T. Laurent, Nonlinear Optics **11,** 25(1995).

[5] F. Devaux, E. Guiot, and E. Lantz, Opt. Lett. **23,** 1597 (1998).

[6] G. Le Tolguenec, E. Lantz, and F. Devaux, Appl. Opt. **36,** 8292 (1997).

[7] F. Devaux, E. Lantz, European Physical Journal D **8,** 117(2000).

[8] G. Le Tolguenec, F. Devaux, E. Lantz, Opt. Lett. **24,** 1047(1999).

[9] F. Devaux, G. Le Tolguenec, and E. Lantz, Light for Life Meeting Cancun, Mexico (july 1999).

[10] C. Doulé, T. Lépine, P. Georges and A. Brun, Opt. Lett. **25,** 353 (2000).

V

IMAGING IN HETEROGENEOUS MEDIA: DIFFUSE LIGHT

Chapter 1

DETECTION OF MULTIPLY SCATTERED LIGHT IN OPTICAL COHERENCE MICROSCOPY

K. K. Bizheva, A. M. Siegel
Dept. of Physics and Astronomy
Tufts University, Massachusetts, USA

A. K. Dunn and D. A. Boas
Nuclear Magnetic Resonance Center
Massachusetts General Hospital, Harvard Medical School
13th St., Bldg. 149, Charlestown, Massachusetts 02129, USA
dboas@nmr.mgh.harvard.edu

Abstract   In this study a number of experimental and computational methods were employed to examine the effect of multiply scattered light detection on image quality and penetration depth of optical coherence microscopy (OCM) in turbid media. Measurements of the OCM resolution at different optical depths in the scattering media reveal that detection of multiply scattered light causes significant degradation of the lateral and axial resolution and that the deterioration effect is dependent on the turbid media scattering anisotropy and the numerical aperture (NA) of the imaging objective.

Keywords: scattering, optical coherence microscopy, imaging, diffusion

Introduction

A major problem in biomedical optical imaging is the fact that image quality degrades with depth penetration: the deeper the imaged object is buried in turbid media like tissue, the less precise is the determination of it's location, size, shape and optical properties. Various microscopy techniques like confocal (CM) and optical coherence microscopy / tomography (OCM/OCT) [4-6] can produce high- resolution images (1-10 μm) in tissue by selectively detecting mainly ballistic and single scattered light. The image penetration depth though, is

P. Sebbah (ed.), Waves and Imaging through Complex Media, 277–298.

typically limited to a range of a few hundred microns up to a millimeter. Other methods like diffuse optical tomography (DOT) [7, 8] utilize the statistical properties of light thus permitting imaging of objects buried at a depth of a few centimeters at the expense of significantly reduced image quality (5-10 mm resolution).

Currently, the intermediate regime spanning between optical microscopy and DOT, where light is multiply scattered (a few photon scattering events) but not yet diffuse, is still relatively unexplored. Multiply scattered light reflected off an object embedded in a turbid medium carries a significant amount of information about the object location and properties. Therefore, development of new imaging modalities utilizing multiply scattered light will depend on how accurately this information can be interpreted. Proper interpretation though, will require thorough knowledge of the process of light interaction with the imaged object, and the turbid background for the imaging system employed. The detection of multiply scattered light and its deteriorating effect on image contrast, resolution and depth penetration has been partially investigated for both CM and OCM/OCT [9-13]. All of these studies employed measurements of back-scattered light in homogeneous or inhomogeneous samples as a function of the imaging lens focal depth or the total photon path length in the medium. A typical intensity profile acquired with an OCM system in a homogeneous scattering medium is shown in Fig. 1. For shallow depths of the focal point, the detected light is mainly single scattered resulting in an exponential decay (Beer-Lambert law) of the measured intensity (linear on a semi-log scale). As the objective focal point is translated deeper into the turbid medium, the contribution of multiply scattered light to the total intensity becomes more pronounced and after a certain depth the only light collected is diffuse . Since the transition from single scattered to diffuse light in the intensity profile (Fig. 1) is smooth, it is difficult to distinguish clearly the boundaries between regions where single scattered, multiply scattered and diffuse light are the dominant components of the measured intensity. The intensity profile in Fig. 1 poses various questions: when does detection of light that has scattered a few times become significant; when does the detected light become completely diffuse; how does partial detection of multiply scattered and diffuse light affect image contrast and resolution; how does the cumulative detection of multiply scattered light depend on the media optical properties, the instrument imaging geometry and the applied photon gating methods; is OCM imaging signal-to-noise or signal-to-background limited and can we define penetration depth limits as functions of the imaging geometry parameters and sample optical properties?

The main goal of this chapter is to provide answers to some of these questions. To do that a number of experimental and computational methods were utilized. To aid the investigation of the transition from single scattered to diffuse light detection in imaging systems, a new method, Dynamic Low Coherence

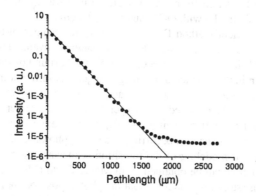

Figure 1.   OCM intensity as a function of photon pathlength in a homogeneous scattering suspension. The straight line through the data points represents the expected attenuation of single scattered light.

Interferometry (DLCI) was developed in our lab [14, 15]. The idea for DLCI was derived from the Quasi-Elastic Light Scattering (QELS) [16, 17] and Diffuse Wave Spectroscopy (DWS) [18, 19, 20, 21, 22] theories and utilizes the pathlength selectivity of an OCT/OCM system. DLCI measures cumulative changes in the photon total momentum transfer as a function of the photon total pathlength in homogeneous, dynamic random media therefore it is more sensitive to the number of photon scattering events than a regular intensity measurement. DLCI is limited to measurements in dynamic systems, therefore additional experimental and computational techniques were employed to study the effect of multiply scattered light detection on the quality of the OCM point-spread function (PSF).

The information presented in this book chapter is organized as follows: section 1 reviews the methods used to study the cumulative detection of multiply scattered light and the subsequent image degradation in OCM. Section 2 describes in detail the experimental and computational results, which are later discussed in section 3. Last section summarizes the accomplishments of this study and presents an outline of the future work.

1.   METHODS FOR INVESTIGATING OF THE TRANSITION FROM SINGLE SCATTERING TO LIGHT DIFFUSION

The following section reviews the experimental (DLCI and OCM resolution measurements) and computational (Monte Carlo models) methods used to investigate the transition from single scattered to diffuse light for the case of imaging in turbid media with OCM and OCT systems.

As a measurement method DLCI is sensitive to changes in the photon total momentum transfer, occuring after each scattering event. Therefore it can provide valuable insight on the effect of a turbid background on the OCM point-spread function in the absence of an imaged object. Additional information on the deteriorating effect of a scattering background on image quality (contrast and resolution) can be obtained from experimental measurements of the OCM lateral and axial resolution. Unfortunately, often in experimental studies detailed information on changes occurring at a single photon scale, necessary for understanding and modeling of the transition scattering regime, is lost. This type of information can be obtained from Monte Carlo (MC) models that can "label" photons, store information about the photon position, polarization state and momentum transfer after each scattering event and gate photons spatially (CM) and by their total path length (OCM).

1.1.   DYNAMIC LOW COHERENCE INTERFEROMETRY

The principle of operation of DLCI is the same as of any standard interferometer with the only exception that one of the mirrors is substituted with the examined sample (Fig. 2). In our case the single-mode fiber optic interferometer is illuminated with light from an 850 nm super-luminescent diode (25 nm spectral bandwidth, 1.2 mW output power). The optical properties of the sample generate a distribution of optical path lengths in the sample arm, while the path length in the reference arm is determined solely by the position of the retroreflector. Interference is observed only when the optical path length difference between the reference and the sample arms is within the coherence length of the source. Thus, a "coherence gate" is used to select specific path lengths within the sample. The amplitude of the interference signal is therefore proportional to the path length dependent reflection/scattering properties of the sample. The position of the reference mirror (retroreflector) is adjusted in such a way as to align the coherence gate with the beam waist, thus optimizing the rejection of out-of-focus background light. The instrument lateral and axial resolution in air is determined by the objective NA and the size and shape of the spatial filter applied to result in confocal detection. In our case spatial filtering was realized by using a single mode fiber system. The instrument axial resolution is further enhanced for small NA objectives by the presence of a coherence gate.

For a fixed position of the reference mirror the photocurrent at the detector can we written as:

$$i_d \propto I_r + I_s + I^{fluct}(z, t), \tag{1}$$

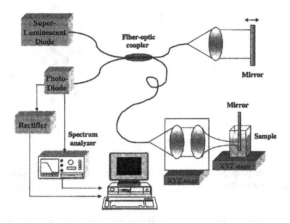

Figure 2.   Experimental setup used for the DLCI and OCM measurements.

where I_r (I_s) is the intensity from the reference (sample) arm and $I^{fluct}(z, t)$ is the cross term (or optical heterodyne term) resulting from the coherent mixing of the reference electric field E_r with the path length dependent electric field $E_s(z)$ of the scattered light. Temporal fluctuations in the cross term may result from changes in the amplitude and/or phase of E_r or E_s. When the reference mirror is held fixed, the amplitude and the phase of E_r do not change over time. Any dynamics in the turbid sample such as flow or Brownian motion, result in a distribution of Doppler frequency shifts in the light propagating through the medium, which causes phase variations in $E_s(z)$, resulting in fluctuations in the cross term intensity $I^{fluct}(z, t)$. The power spectrum (a Fourier transform of the photo detector current i_d) derived for a system consisting of light scattering particles undergoing Brownian motion is a Lorentzian [16, 17]:

$$P(\omega) = \frac{1}{\Omega} \frac{A}{1 + \left(\frac{\omega}{\Omega}\right)^2}, \tag{2}$$

where A is the amplitude of the power spectrum and Ω is the spectrum linewidth. This equation does not account for the $\Omega=0$ contribution of I_r and I_s. In an optically dilute suspension (Fig. 3a), light scatters only once before detection, therefore the scattering angle and polarization are well defined. For the case of single scattered light, assuming weak scattering and non-interacting particles, the power spectrum linewidth is proportional to the particle self-diffusion coefficient D_B:

$$\Omega = q^2 D_B = 4k_0^2 D_B \sin^2(\theta/2), \tag{3}$$

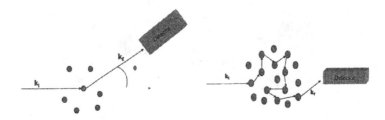

Figure 3.   (a) Light propagation through a dilute scattering medium. (b) Light propagation through a dense turbid medium.

where $D_B = k_B T/(3\pi \eta d)$, k_B is the Boltzman constant, T is the temperature of the sample, η is the viscosity of the suspending liquid (in our case H_2O $\eta = 1.0$ cps), and d is the hydrodynamic diameter of the scattering particle. Here q denotes the photon momentum transfer, $q = 2k_o \sin(\theta/2)$, where k is the wave number in the scattering medium and θ is the scattering angle. Note that eq. (2) is valid for optical heterodyne detection, while in the case of a homodyne measurement, the expression is multiplied by a factor of two.

In an optically dense medium (Fig. 3b), where light scatters multiple times before detection, the scattering angle, the polarization of the scattered wave and consequently the photon momentum transfer, q, changes with each scattering event. In the case of light diffusion through a turbid medium undergoing Brownian motion, the power spectrum of the detected diffuse light is also a Lorentzian, but with a linewidth dependent on the scattering properties of the suspension and the photon path length within it:

$$\Omega = 2k_0^2 D_B \mu_s (1-g)s. \tag{4}$$

Here s is the total path length traveled by the light in the medium, μ_s is the scattering coefficient and g is the scattering anisotropy of the turbid medium defined as an average of the cosine of the scattering angle. For polystyrene microspheres the scattering anisotropy factor and the scattering coefficient can be calculated using Mie theory and the Percus-Yevick structure factors for hard-spheres [21].

For our studies a spectrum analyzer (Stanford Research, SR760) was used to measure the power spectrum of light back scattered from concentrated turbid samples (see Fig. 2). The linewidth and the amplitude of the spectrum were determined by fitting the raw data with a Lorentzian function. The turbid samples used in all experiments were prepared by suspending polystyrene microspheres (Bangs Labs, Inc.) in distilled water. The particle volume fraction in each monodisperse suspension was chosen to result in a photon mean scattering length (l_s) of approximately 100 µm. Samples were kept at a constant room temperature, which was recorded prior to each experiment. For the measurements investigating the effect of the NA on the power spectrum broadening, the normalized spatial filter parameter ν_p ($\nu_p = 2\pi r_p NA/\lambda$), where r_p is the filter diameter was varied between 2 and 6.

1.2. EXPERIMENTAL MEASUREMENTS OF THE OPTICAL COHERENCE MICROSCOPE POINT SPREAD FUNCTION

A slightly modified version of the set-up employed in the Doppler experiments (Fig. 2) was used to measure the OCM axial and lateral resolution. A standard image resolution chart was fixed at a particular position in a scattering medium with known optical properties. The turbid sample was mounted on an XYZ translation stage, which made it possible to alter the thickness of the layer of scattering medium separating the chart from the front surface of the cuvette. The imaging optics in the sample arm of the interferometer was also mounted on an XYZ stage to allow for translation of the objective focal point in the medium. The OCM axial and lateral resolution was evaluated by scanning the focal point over the resolution chart in the X and Z directions and measuring the back-scattered light intensity. The turbid samples used for the OCM axial and lateral resolution measurements were prepared by suspending polystyrene microspheres ($d = 0.22$ µm, $g = 0.2$) in distilled water. The particle concentration was chosen to result in a photon mean scattering length (l_s) of approximately 200 µm. In this particular measurement, the combination of the objective NA and the confocal spatial filter resulted in $\nu_p = 6$.

1.3. MONTE CARLO SIMULATIONS

For our studies we have developed a Monte Carlo code to simulate light detection in OCM geometry for turbid samples with known optical properties. The general principles of our Monte Carlo code are similar to that described by Jacques and Wang [24]. A schematic of the imaging geometry used in all simulations is presented in Fig. 4. A non-uniform photon distribution at the sample surface is used to map out a focused Gaussian beam. The initial photon direction of propagation is determined by simulating a diffraction-limited volume centered at the chosen focal point position. Photons are propagated in the turbid medium according to the rules described by Jacques and Wang [23]. Since we were particularly interested in coherence detection, (OCM), the code was modified to simulate spatial filtering (CM) [24] and gate the detected photons by their path length (OCM).

To explore the changes in the OCM PSF due to a presence of a reflecting object within the medium positioned at the focus of the imaging lens, we have simulated measurements of the axial and lateral OCM resolution. This was done by moving a perfectly reflecting mirror along the optical axis and transverse to the axis at the beam waist. The Monte Carlo code designed to simulate OCM resolution measurements was used to investigate the dependence of the resolution loss on the objective NA and the sample scattering anisotropy g. Simulations were run for g=0.1, 0.9 and NA=0.1, 0.5. Diffraction theory was used to calculate the confocal lateral and axial resolution in air for the two simulated objectives. For a wavelength of 850 nm and NA=0.1 ($\Delta x = 5\,\mu$m, $\Delta z = 128\,\mu$m), and for NA=0.5 ($\Delta x = 1\,\mu$m, $\Delta z = 5\,\mu$m). The coherence gate parameter corresponding to the SLD coherence length was fixed to $L_c = 10\,\mu$m, which ultimately determined the measurement axial resolution for the small NA lens case. For all simulations the normalized pinhole parameter was fixed to $\nu_p = 7.4$ and the photon scattering length was kept constant ($l_s = 100\,\mu$m). The axial resolution was defined as the FWHM for the normalized intensity profile derived from the simulations, while the lateral resolution was determined as HWHM, where the half maximum (HW) in this case was defined as the middle point between 10% and 90% of the total intensity.

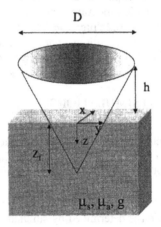

Figure 4.   Schematic representation of the Monte Carlo model. D is the diameter of the objective, h is the distance of the objective from the surface of the medium, z_f is the depth of the focal point in the medium.

2. EXPERIMENTAL AND COMPUTATIONAL RESULTS

The following section reviews all results obtained from the experimental measurements and Monte Carlo simulations. The data presented are organized in subsections depending on the method used.

2.1. DYNAMIC LOW COHERENCE INTERFEROMETRY

To demonstrate that DLCI can be used to image and quantify Brownian motion within highly scattering media in a way similar to DLS methods in very dilute media, we measured the power density spectrum for two separate monodisperse suspensions ($d = 0.22$ μm, $g = 0.2$ and $d = 1.02$ μm, $g = 0.9$) at a sample depth of 200 μm (see Fig. 5), i.e. at a depth of two scattering lengths within the highly scattering medium. This particular depth was chosen to avoid Fresnel reflections from the glass/suspension interface and to ensure detection of only single scattered light. The experimental data was fit with a Lorentzian (eq. (2)) as indicated by the solid lines through the data points. The linewidths obtained from these fits were used to determine the experimental Brownian diffusion coefficients and the corresponding particle sizes, assuming a backscattered angle of 180°. The calculated microsphere diameters (0.222 μm and 0.997 μm) were within 1-5% of the values cited by the manufacturer correspondingly. Multiple experiments with different size polystyrene microspheres showed the same accuracy.

To explore the effect of multiply scattered light detection on the spectrum linewidth we measured the power spectrum as a function of pathlength in a highly scattering, monodisperse suspension of 0.22 μm polystyrene microspheres in water ($g = 0.2$) with an objective with an NA = 0.2. Figure 6 shows the power spectra measured with the focal point positioned at 33 μm, 792 μm and 1089 μm below the surface of the sample, corresponding to photon total pathlengths of 0.066 mm and 1.584 mm and 2.178 mm respectively. The solid lines through the data points represent the Lorentzian fits. The spectra were normalized to facilitate their comparison. Note that for the measurements at longer pathlengths, where the signal was weaker, it was necessary to account for the 1/f electronic noise that was significant at lower frequencies. This was done by directly measuring the 1/f spectrum and including it in the Lorentzian fit.

The ability of DLCI to separate single scattered from multiply scattered light demonstrated with the results shown in Fig. 7a was further used to explore the transition from single scattered to diffuse light detection. To accomplish this goal the power spectrum was measured in steps of 33 μm inside the sample and all data was fit with a Lorentzian function. The results from the linewidth fitting are plotted as a function of the photon path length in the turbid medium.

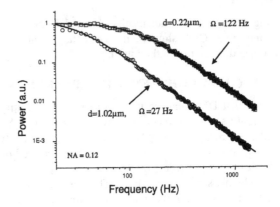

Figure 5.   Back-scattered light power spectra measured with DLCI at a depth of 200 μm for two different particle size suspensions. The solid lines through the data points represent the expected Lorentzian profiles. *d* is the diameter of polystyrene microsphere and Ω is the linewidth of the power spectrum.

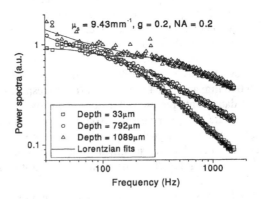

Figure 6.   Power spectra measured with DLCI at three different depths inside a highly scattering suspension of microspheres. The solid lines through the data points represent the Lorentzian fits.

The straight lines through the data points represent the linewidth behavior as predicted for the single scattering (eq. (2)) and diffusion (eq. (3)) regimes. The arrows on Fig. 7a mark the possible position of the boundaries between single scattering, multiply scattering, and diffusion regimes. The position of the arrows was defined the following way: an assumption was made that a change in the spectrum linewidth greater than 10% of the initial value (single scattered light detection) demonstrates a significant detection of multiply scattered light. The corresponding number of photon mean free paths in the medium, necessary to cause such a broadening of the linewidth was defined as the end of the single scattering regime. In a similar way a limit was established to define the beginning of the light diffusion regime.

Figure 7b illustrates the effect of the detection of multiply scattered light on the Lorentzian amplitude as determined from the fits. As expected, in our practically non-absorbing sample, in the single scattering regime the amplitude decayed exponentially with the path length with an extinction coefficient equal to the scattering coefficient μ_s of the medium ($\mu_s = 1/l_s$). The solid line through the data points represents the decay predicted from the single scattering theory. The arrows in Fig. 7 indicate the beginning and the end of the transition regime where light is scattered a few times as determined from the linewidth graph. The position of the arrows demonstrates clearly that a measurement of the total momentum transfer changes is a better method for separation of single scattered, multiply scattered and diffuse light than a regular intensity measurement.

To examine the dependence of the transition on the sample scattering aniso-tropy factor g, power spectra were measured in turbid suspensions with g=0.2, 0.39 and 0.75, while keeping the NA of the focusing optics fixed (NA=0.32). A summary of the experimental results is presented in Fig. 8. Note that Fig. 8 is a plot of the data as a function of the number of photon random walk steps ($l^* = l_s/(1-g)$) in the medium rather than scattering lengths (l_s) to utilize the s/l^* scaling predicted by DWS (eq. (3)). Furthermore, the data was normalized by the linewidth in the single scattering regime to facilitate the comparison.

To explore the transition dependence on the imaging lens NA, the power spectra were measured in a suspension with scattering anisotropy factor g=0.2 using three different microscope objectives (NA=0.12, 0.32 and 0.55). The ex-perimental results are shown in Fig. 9a, where the measured spectrum linewidth is plotted as a function of the photon pathlength scaled in terms of photon mean free paths in the medium. The results clearly demonstrate that an increase in the NA causes the transition to occur after a fewer number of photon mean free paths (MFP). Same measurements were repeated for samples with scattering anisotropy g=0.75 and a summary of the results is presented in Fig. 9b.

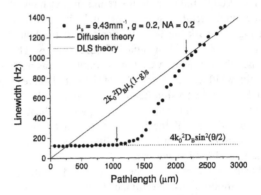

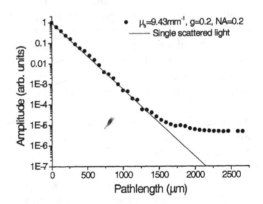

Figure 7.   (a) Lorentzian linewidth as a function of photon pathlength in a turbid medium. The dotted and the solid lines indicate the expected linewidth behavior in the single scattering and light diffusion regimes respectively. The arrows indicate the region of photon pathlengths corresponding to detection of multiply scattered but not yet diffuse light. (b) Lorentzian amplitude as a function of photon pathlength in the scattering medium. The solid line represents the expected extinction of single scattered light. The arrows indicate the region of photon pathlengths corresponding to detection of multiply scattered but not yet diffuse light.

2.2.   EXPERIMENTAL MEASUREMENTS OF THE OPTICAL COHERENCE MICROSCOPE AXIAL AND LATERAL RESOLUTION

Lateral scans obtained inside the turbid medium demonstrate a clear loss of resolution as the objective focal point is translated deeper into the sample. A

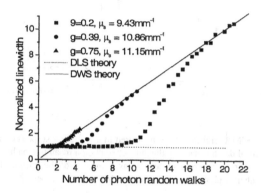

Figure 8. Normalized Lorentzian linewidth as a function of the number of photon random walks measured in media with three different scattering anisotropy factors (g). The dotted and the solid lines through the data points indicate the expected linewidth behavior in the single scattering and light diffusion regimes respectively.

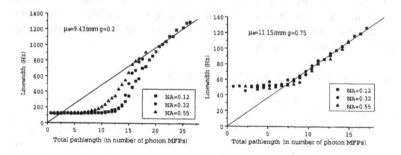

Figure 9. (a) Linewidth as a function of the number of photon mean free paths ($\mu_s s$) in a suspension with low scattering anisotropy (g=0.2) measured for three different NA's of the imaging lens. (b) Linewidth as a function of the number of photon mean free paths ($\mu_s s$) in a suspension with high scattering anisotropy (g=0.75) measured for three different NA's of the imaging lens.

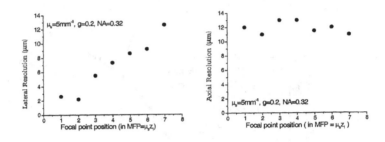

Figure 10.   (a) OCM lateral resolution measured in a scattering suspension as a function of the imaging lens focal depth (scaled by the number of photon mean free paths). (b) OCM axial resolution measured in a scattering suspension as a function of the imaging lens focal depth (scaled by the number of photon mean free paths). MFP = photon mean free path. z_f = depth of the focal point.

summary of the experimental results is presented in Fig. 10a where the measured lateral resolution is plotted as a function of the imaging lens focal depth in the scattering suspension in terms of photon MFPs.

A summary of the OCM axial resolution experimental results is presented in Fig. 10b where the measured axial resolution is plotted as a function of the focal point depth. Scans obtained at various depths within the turbid sample demonstrate no apparent change in the OCM axial resolution.

Due to measurement difficulties, no experimental data is currently available for turbid samples with different scattering properties therefore no definitive conclusions can be drawn for the effect of media scattering anisotropy on the OCM resolution degradation.

2.3.   MONTE CARLO SIMULATIONS

A summary of the lateral resolution simulation results is shown in Fig. 11a, where the OCM lateral resolution is plotted as a function of the focal depth (in photon MFPs). The OCM lateral resolution was determined as HWHM, where the half maximum (HW) in this case was defined as the middle point between 10% and 90% of the total intensity. The graph presents data for two cases of low and high imaging lens NA and two different suspensions with scattering anisotropy g=0.1 and 0.9.

Simulation results for the axial OCM resolution are summarized in Fig. 11b. The OCM axial resolution was defined as the FWHM for the normalized intensity profile derived from the simulations. The graph represents the axial resolution as a function of the focal depth for four possible combinations of objective lens NA and sample scattering anisotropy.

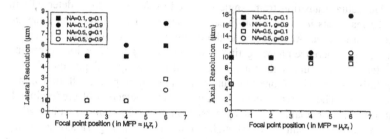

Figure 11.   (a) Simulated OCM lateral resolution as a function of the imaging lens focal depth (scaled by the number of photon mean free paths) in a scattering suspension. (b) Simulated OCM axial resolution as a function of the imaging lens focal depth (scaled by the number of photon mean free paths) in a scattering suspension.

3.   DISCUSSION

Initial measurements with DLCI at shallow depths (2 photon MFPs) in highly scattering suspensions reveal good agreement between experimental data and DLS theory (Fig. 5). This result demonstrates that at superficial depths DLCI is able to reject out of focus multiply scattered light and to assure precise determination of the size of the scattering particles. Hence DLCI can be applied as a non-invasive, accurate, flexible and inexpensive tool for real-time particle sizing in optically dense media.

For measurements deeper in the suspension though, there is a pronounced broadening of the power spectrum as demonstrated in Fig. 6. According to eq. (2), an increase in the spectrum linewidth can result from changes either in the particle diffusion coefficient or the total accumulated momentum transfer or both. Since the sample is a homogeneous, monodisperse suspension held at constant temperature, the observed broadening of the spectrum could only be attributed to changes in the accumulated photon momentum transfer resulting from detection of multiply scattered light. Comparison with DLS theory reveals that the spectrum acquired at a depth of 33 μm corresponds to detection of only single scattered light. The broadened spectrum measured at a depth of 1089 μm is in good agreement with that predicted by eq. (3) (DWS theory), thus indicating that at such depth the light collected with the DLCI is completely diffuse. The linewidth of the spectrum acquired at a depth of 792 μm does not agree with either the single scattering or light diffusion theories, which leads to the conclusion that in this case the detected light is multiply scattered but not completely diffuse. The data in Fig. 6 clearly demonstrates the ability of DLCI to distinguish between single scattered, multiply scattered and diffuse light.

The actual transition from single scattered to diffuse light as detected with DLCI manifests itself as a non-linear broadening of the Lorentzian linewidth shown in Fig. 7a. As indicated in the figure, the combination of confocal filtering and coherence photon gating results in a very good rejection of multiply scattered background light for measurement depths up to 5 photon MFPs in the turbid suspension (corresponding to a photon total pathlength of 1 mm in the figure). The linewidth data also demonstrates detection of predominantly diffuse light for imaging depths greater than 10 MFPs (photon pathlength of 2 mm). It is important to note though that the arrows in Fig. 7a indicate imaging depths at which multiply scattered and diffuse light become the dominant components of the detected light. The linewidth data exhibits between 2% and 10% deviation from the theoretically calculated value for imaging depths of 2 - 5 photon MFPs in the suspension. Considering that measurement error due to uncertainties in the particle size and concentration and small fluctuations in the sample temperature can amount up to 2%, the spectrum broadening observed for depths greater than 2 and smaller than 5 photon MFPs can be attributed to partial detection of multiply scattered light.

The transition from single scattered to diffuse light as detected with DLCI is also observed as a deviation in the Lorentzian amplitude data from the expected decay in the single scattering regime (Beer-Lambert law) (Fig. 7b). This deviation appears to occur for imaging depths greater than 7 photon MFPs in the scattering medium, though as observed from the linewidth data, at such depths a large portion of the total detected light is multiply scattered. To determine the significance of multiply scattered light detection at shallow imaging depths (2-7 photon MFPs) it is necessary to investigate its effect on the instrument PSF, which is related to image contrast and resolution. Comparison between the data presented in Fig. 7a and Fig. 7b demonstrates that an amplitude measurement lacks the sensitivity for early detection of multiply scattered light and the ability for discrimination between light that has scattered multiple times and light that is completely diffuse. The data in Fig. 7b also raises another question: what source of contrast should an object buried in homogeneous scattering media provide so that the light back-scattered from it raises above the diffuse light background? As determined from the graph, at an imaging depth of 10 photon MFPs (2 mm pathlength) the difference between single scattered and diffuse light intensity is more than two orders of magnitude and the signal light reflected from an object should at least match this difference in order to provide for a minimum image contrast.

Additional experiments performed with different NA's of the imaging objective in turbid samples with a variety of scattering anisotropy factors reveal that the transition from single scattered to diffuse light as detected with DLCI is dependent on the imaging geometry and sample scattering anisotropy. The experimental data shown in Fig. 8 demonstrates that the transition is dependent on the suspension scattering anisotropy, and for samples that scatter light almost isotropically (small g) it occurs after a greater number of photon random walks. In contrast, in turbid media with high scattering anisotropy, the transition from single scattered to diffuse light is very sharp and is centered at a depth of 1 photon random walk (total photon pathlength of 2 random walks as shown in Fig. 8. A quick referral to eq. (2) and eq. (3) shows that the linewidth in the diffuse regime (DWS) is equal to the single scattering linewidth when the depth (pathlength) in the sample is equal to 1 (2) random walk lengths. This is the shortest pathlength at which light can become diffuse. For the geometry of our measurement the photons may have to travel a greater pathlength before becoming diffuse due to the imposed confocal filter and coherence gate. Keeping this in mind, the linewidth behavior as shown in Fig. 8 can be explained as follows. In the case of highly anisotropic media, the lateral spread of the light beam is smaller than the NA of the imaging objective thus ensuring collection of all back-scattered light. In addition, the presence of a confocal spatial filter and a coherence gate causes rejection of only a small portion of the light considering the snake-like propagation of scattered photons close to the lens optical axis resulting in small changes of the photon total pathlength. Therefore, the transition to the diffusion regime in highly anisotropic media occurs at an imaging depth of 1 random walk length. In the case of an almost isotropically scattering media though, the presence of confocal filtering and coherence gating causes a significant rejection of multiply scattered, background light, and thus forces the transition to occur at greater imaging depths.

To test how sensitive the transition from single scattered to diffuse light is to changes in the objective NA for imaging in media with low scattering anisotropy, measurements were performed with NA=0.12, 032 and 0.55 (Fig. 9a). The data in Fig. 9a demonstrates that a five-fold increase in the imaging lens NA results in a shift of the transition boundary positions toward the sample surface by 5 photon MFPs (total pathlength). It is apparent that smaller NA lenses provide better rejection of multiply scattered background light for measurements in media with low scattering anisotropy. If this result is to be applied for imaging in scattering media though, it is necessary to consider the fact that the use of small NA objectives also results in poor spatial resolution and a poor signal-to-noise ratio. As expected, similar measurements performed in media with high scattering anisotropy show no apparent dependence of the transition on the lens NA (Fig. 9b). This result is of particular importance to microscopic imaging of biological tissue considering its high scattering anisotropy.

The results obtained from the DLCI linewidth measurements are summarized in Fig. 12. The solid lines in the graph represent the depth penetration limits for single scattered light detection in CM and OCM in turbid media as derived analytically by Izatt et al. [12]. The analytical model proposed in ref. [12] was used to make adjustments in the position of the solid lines in order to

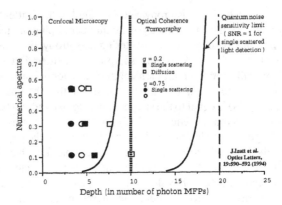

Figure 12.   Parametric plot of the theoretical limits of CM and OCM as functions of the focal point depth, scaled by the number of photon mean free paths in the sample. The plot assumes $D=7$ mm, $n_0=1.33$, $L_c=20$ mm, $l=850$ nm, $\mu_b=0.3191$, μ_s and $E=130$ mJ (see Izatt[13]). The solid symbols indicate the end of the single scattering regime, while the open symbols correspond to the beginning of the light diffusion regime as determined from the DLCI measurements. The dotted and the dashed lines in the graph indicate the imaging depths at which diffuse light is being detected with DLCI in turbid media with $g=0.9$ and 0.95 respectively.

suit the imaging parameters of the DLCI system. The data points in Fig. 12 indicate the boundaries between the single scattering, multiply scattering and diffuse regimes as defined from the linewidth data (see arrows in Fig. 7a) for different cases of sample scattering anisotropy and imaging lens NA. (The solid symbols (square and circle) indicate the depths at which multiply scattered light detection causes significant broadening of the spectrum linewidth measured in turbid samples with scattering anisotropy $g=0.2$ and 0.75. The hollow symbols indicate the experimentally determined imaging depths at which diffuse light becomes the dominant component in the light detected by DLCI. The dotted and the dashed lines in the graph indicate the imaging depths at which DLCI is expected to detect diffuse light in turbid media with $g=0.9$ and 0.95 respectively.

   It is obvious from the graph that though OCM has the dynamic range to provide single scattered light detection in homogeneous turbid media at depths up to 15 photon MFPs, the detected light becomes predominantly diffuse at much shallower imaging depths (except for the case of $g=0.95$, see dashed line). Better rejection of multiply scattered and diffuse light is observed in cases of almost isotropically scattering media when low NA imaging objectives are used. Another interesting feature observed in the graph is the non-monotonic dependence of the position of the multiply scattered and diffuse light boundaries on the media scattering anisotropy. Due to the confocal and coherence gate rejection of background light utilized in DLCI, the diffuse light boundary for media with g=0.2 appears at depths of 6-10 photon MFPs. As the scattering anisotropy of the suspension increases to g=0.75, the boundary moves to shallower depths (toward the sample surface) and appears at the position calculated from the DWS theory. Further increase in the scattering anisotropy factor results in a shift of the boundary position deeper into the turbid sample (dotted and dashed lines) since highly forward scattered light requires many more scattering events before randomization of the photon direction of propagation and polarization occurs. As mentioned above, the g dependence of the diffuse boundary position can be explained with the fact that DLCI utilizes confocal and coherence photon gating techniques that very effectively reject multiply scattered background light in turbid media with low scattering anisotropy.

Figure 12 reminds us of a question addressed above: how significant is the effect of multiply scattered background light on the OCM point-spread function (image contrast and resolution). To answer this question, the OCM axial and lateral resolution was measured in turbid media. For experimentally difficult cases, a Monte Carlo model was used to simulate the measurements. The results from the OCM lateral resolution measurement presented in Fig. 10a demonstrate a clear loss of resolution for imaging depths greater than 3 photon MFPs. Comparison with DLCI linewidth data (Fig. 9a) reveals that an almost two-fold loss of OCM lateral resolution corresponds to only a few percent change in the spectrum linewidth at the same imaging depth. Considering that the presented lateral resolution data is derived from a single measurement and therefore the chances for an experimental error cannot be dismissed, a conclusion whether the OCM lateral resolution measurement is very sensitive to detection of multiply scattered background light or the data is a measurement artifact cannot be reached unless measurements are repeated. OCM axial resolution data (Fig. 10b) demonstrates no significant deterioration for the range of imaging depths that was investigated. Considering that the confocal parameter for the lens used (NA=0.32) was 17 μm in water, the OCM axial resolution in this case is defined by the width of the coherence gate ($L_c = 13$μm) which agrees well with the experimental data. No loss of axial resolution is observed for the range of imaging depths explored. This result compares well with the small linewidth broadening measured with DLCI for the same imaging lens NA (0.32)in samples with the same scattering anisotropy (g=0.2), leading to the conclusion that in this case the OCM axial resolution is not as sensitive to detection of multiply scattered light as OCM lateral resolution. Since no experimental data is currently available for turbid samples with different scattering properties therefore no definitive conclusions can be drawn for the effect of media scattering anisotropy on the OCM resolution degradation.

Summarized results from the Monte Carlo simulations of OCM lateral and axial resolution measurements in turbid media are presented in Fig. 11a and

Fig. 11b correspondingly. The results in Fig. 11a demonstrate no significant change in the lateral OCM resolution for the case of high NA imaging lens for depths in the scattering medium smaller than 4 photon MFPs (hollow squares and circles). This behavior appears to be independent of the sample scattering anisotropy. For depths greater than 4 MFPs, a rapid degradation of the OCM lateral resolution is observed and it is more pronounced for suspensions with low g factor. For the case of small NA of the imaging objective, simulation results also show deterioration of the OCM lateral resolution, though not as rapid as in the case of high NA (solid squares and circles). The resolution degradation appears to be more pronounced and to occur at shallower depths for media with high scattering anisotropy. This result compares very well with the DLCI experimental data where the effect of the multiply scattered light on the OCM PSF is measured as a total change in spectrum linewidth. Changes in OCM lateral resolution (Fig. 11a) are apparent for photons that had traveled between 6 and 16 mean paths in the scattering suspension (corresponding to focal depths of 3 to 8 photon MFPs) and for the same range of photon paths broadening of the Lorentzian linewidth is also observed (Fig. 9a and Fig. 9b).

Summarized results from the OCM axial resolution simulations are presented in Fig. 11b. As we have seen experimentally, for the case of an almost isotropically scattering background, there is no change in the axial resolution for small NA and depths less than 7 photon MFPs, where the resolution is determined by the coherence length Lc of the light source. A small change in the axial resolution is observed for an NA=0.5 for the same background sample properties for shallow depths where there is a transition from the value predicted for non-scattering media to almost the full width of Lc. In the case of background with high anisotropy factor (g=0.9) there is a pronounced loss of OCM axial resolution with imaging depth for simulations with both small and large NA. Again, loss of axial resolution is observed for a region of photon MFPs in turbid media corresponding to a pronounced spectrum broadening measured with DLCI (Fig. 9a,b). A possible explanation for the observed OCM resolution dependence on sample optical properties and instrument imaging geometry is that in a medium with high scattering anisotropy, light scatters preferentially in the forward direction resulting in a relatively small lateral spread of the light beam. The small-angle scattering also causes a snake-like propagation of the scattered photons, causing small increase of the total photon pathlength compared to the case of photon ballistic travel from the sample surface to the lens focal point position and back. Therefore, the confocal filter and the coherence gate do not provide good rejection of multiply scattered light thus causing rapid and pronounced loss of both axial and lateral OCM resolution at shallower imaging depths. For the case of measurements in almost isotropically scattering media the combination of small NA lens and a coherence gate results in preferential detection of single scattered light at shallow imaging depths, thus causing no apparent change in the measured OCM axial and lateral resolution.

Though the simulation result presented in Fig. 11a demonstrate loss of OCM lateral resolution in scattering media as the imaging depth increases, the deterioration effect is not as dramatic as observed in the experimental results (Fig. 10a). It is imperative to point out again that due to the fact that only one experimental measurement was made and due to the high chance for experimental error, further investigation is necessary to reveal whether the discrepancy between the experimental and the simulation data is real.

Summary

In summary, a number of experimental and computational methods were used to study the process of multiply scattered light detection in OCM and its effect on image resolution. A new method (DLCI), sensitive to changes in the photon total momentum transfer occurring in the process of scattering, was employed to investigate the transition from single scattered to diffuse light in an imaging system applying confocal and coherence photon gating techniques. Experimental results demonstrated the ability of DLCI to discriminate between single scattered, multiply scattered and diffuse light and allowed for determination of the imaging depth (as a function of sample scattering anisotropy and imaging lens NA) at which a dominant portion of the detected light becomes diffuse. In addition other experimental and computational results revealed that the detection of multiply scattered light, manifesting itself as a spectrum linewidth broadening (DLCI), causes significant lateral and axial OCM resolution degradation and that the deterioration effect is dependent on the media scattering anisotropy and imaging lens NA. Though the work presented in this chapter provides answers to some of the questions raised in the introduction of this paper, some of them are left unanswered and even accompanied with a few more. For example, it is still unclear what the OCM image penetration depth is (as a function of sample g and lens NA) and whether this depth is limited by the measurement signal-to-noise ratio or the signal-to-background ratio. Furthermore, is image resolution degradation exclusively due to background light detection (light that does not carry any information about the imaged object buried in the turbid media) or is light reflected off the object and multiply scattered traversing the turbid layer above also responsible for loss of image resolution? These questions and more will be addressed in our future studies.

References

[1] J. Powley, *Handbook of biological confocal microscopy* (Plenum Press, New York, 1995).

[2] R. Webb, Rep. Prog. Phys. **59**, 427-471 (1996).

[3] M. Rajadhyaksha, Appl. Opt. **38**, 2105-2115 (1999).

[4] D. Huang, E. A. Swanson, C. P. Lin, J. S. Schuman, W. G. Stinson, W. Chang, M. R. Hee, T. Flotte, K. Gregory, C. A. Puliafito and J. G. Fujimoto, Science **254**, 1178-1181 (1991).

[5] J. Schmitt, A. Knuttel, A. H. Gandjbakhche and R. F. Bonner, Proc. SPIE **1889**, 197-211 (1993).

[6] B. Bouma, Opt.Lett. **20**, 1486-1488 (1995).

[7] A. Yodh and B. Chance, Phys. Today, **48**, 34-40 (1995).

[8] S. Arridge, Inverse Problems, **15**, R41-R94 (1999).

[9] M. Kempe, A. Z. Genack, W. R. Dorn and P. Dorn, J. Opt. Soc. Am. A **14**, 216-223 (1997).

[10] M. Kempe, J. Opt. Soc. Am. A **13**, 46-52 (1996).

[11] M. Yadlowsky, J. M. Schmitt and R. F. Bonner, Appl. Opt. **34**, 5699-5707 (1995).

[12] J. Izatt, M. R. Hee, G. M. Owen, E. A. Swanson and J. G. Fujimoto, Opt. Lett. **19**, 590-592 (1994).

[13] Y. Pan, R. Birngruber and R. Engelhardt, Appl. Opt. **36**, 2979-2983 (1997).

[14] D. A. Boas, K. K. Bizheva and A. M. Siegel, Opt. Lett. **23** (1998).

[15] K. Bizheva, D. A. Boas, Phys. Rev. E, **58**, 7664-7667 (1998).

[16] P.J. Berne and R. Pecora, (Wiley, New York, 1976).

[17] R. Pecora, (Plenum Press, New York, 1985).

[18] D. J. Pine, D. A. Weitz, P. M. Chaikin and E. Herbolzheimer, Phys. Rev. Lett. **60**, 1134-1137 (1988).

[19] G. Maret and P. E. Wolf, Z. Phys. B **65**, 409-413 (1987).

[20] D. Durian, Phys. Rev. E **51**, 3350-3358 (1995).

[21] P. Kaplan, M. H. Kao, A. G. Yodh and D. J. Pine, Appl. Opt. **32**, 3828-3836 (1994).

[22] A. Yodh, P. D. Kaplan and D. J. Pine, Phys. Rev. B **42**, 4744-4747 (1990).

[23] S. L. Wang, S. L. Jacques, and L. Zheng, Computer Methods and Programs in Biomedicine **47**, 131-146 (1995).

[24] A.K. Dunn, C. Smithpeter, A.J. Welch and R. Richards-Kortum, Appl. Opt. **35**, 3441-344 (1996).

Chapter 2

SCATTERING BY A THIN SLAB: COMPARISON BETWEEN RADIATIVE TRANSFER AND ELECTROMAGNETIC SIMULATION

J.-J. Greffet and J.B. Thibaud

Laboratoire EM2C, U.P.R. 288 CNRS, Ecole Centrale Paris
92295 Châtenay-Malabry Cedex, France
greffet@em2c.ecp.fr

L. Roux, P. Mareschal and N. Vukadinovic

Dassault-Aviation
92552 Saint-Cloud Cedex 300, France

Abstract   The paper is devoted to the study of scattering of waves by a slab containing randomly located cylinders. For the first time, the complete transmission problem has been solved numerically. We have compared the radiative transfer theory with a numerical solution of the wave equation. We discuss the coherent effects such as forward scattering dip and backscattering enhancement. It is seen that the radiative transfer equation can be used with great accuracy even for optically thick systems whose geometric thickness is comparable to the wavelength. We have also shown the presence of dependent scattering.

Keywords: scattering, radiative transfer, propagation

Introduction

Propagation of waves in random media can be described by two different theories. The simplest approach is based on a phenomenological description of radiative transfer. The relevant equation is the radiative transfer equation [1], which describes the propagation of the specific intensity. In this model, the properties of the medium are described through coefficients defined per unit volume in order to account for absorption and scattering. It is important to realize that the phase of the wave is lost in this description. One deals only

P. Sebbah (ed.), Waves and Imaging through Complex Media, 299–305.

with a specific intensity or in other words with a density of energy in phase space. A completely different approach is based on solving the propagation equation for a field in a random medium. This approach fully accounts for the phase of the wave. It is obviously necessary in order to represent phenomena such as backscattering enhancement or speckle correlations. The question of the relationship between the two approaches has been examined recently. An expression of the specific intensity in terms of fields in the framework of statistical optics (or coherence theory) has been given in the late sixties [2]. Yet, the definition yields a specific intensity that can be negative. This problem was finally fully understood in the beginning of the eighties [3]. Once the link between the basic concepts is made, the remaining problem is to establish a link between the equations for these quantities. So far, all the derivations of the radiative transfer equation rely on sufficient approximations (e.g. ladder approximation). Therefore, we do not know what are the necessary conditions for the radiative transfer equation to be valid.

The problem that we address in this paper is the validity of the radiative transfer equation for highly scattering media with a thickness of the order of the wavelength. An interesting heuristic discussion of the conditions of validity of the radiative transfer equation can be found in [6]. The radiative transfer equation is based on the assumption that for a given length scale L one can replace the actual heterogeneous medium by using homogeneous coefficients that describes the amount of scattering and absorption per unit volume. A first condition is that a volume L^3 should contain many particles so that its properties are averaged. The second condition given by the authors is that the wavelength should be smaller than the decay length. This condition expresses the idea that radiative transfer is incoherent. Indeed, it has been shown [7] that the coherence length in a random medium is the wavelength (or the decay length if it is smaller than the wavelength). The third condition is that L must be smaller than the decay length. These conditions seem to imply that one cannot use the radiative transfer equation to deal with a slab with a thickness of the order of the wavelength. Yet, we have found that the validity of the radiative transfer equation is larger than expected a priori.

1.   OUTLINE OF THE TWO MODELS

To be able to assess the validity of a solution of the radiative transfer equation, it is necessary to be able to compare it with a complete numerical solution of the problem. Due to numerical limitations, this is possible only for a bidimensional (2D) problem. Thus, we have developed a solution of the radiative transfer equation for a 2D problem. We consider radiation scattering in a plane parallel medium composed of perfectly aligned, randomly distributed, infinite circular cylinders. The medium is illuminated by a monochromatic directional plane wave normally incident on the cylinders axis. The scattering direction always lies in the plane of incidence, perpendicular to the cylinders axis. In this particular case, the two linear polarizations can be treated independently. In what follows, we only consider the case of an electric field in the plane of incidence (V polarization). We start by describing briefly how the problem can be solved within the framework of radiative transfer (RT). The method used is the discrete ordinate technique first described by Chandrasekhar. The integro-differential radiative transfer equation is converted into a first order differential equation by using a Gauss quadrature for the integral over all the scattering angles. Details can be found in [8, 9].

We now consider the same problem of reflection and transmission by a 2D scattering slab with an electromagnetic method (EM). Space is divided in homogeneous regions with permittivity ϵ. For numerical reasons, the width of the system is finite. In order to avoid edge effects, we have considered a periodic system. Thus, on each side of the slab the electric field can be described using two Rayleigh expansions. A detailed description of the numerical method we use is presented by Moore in [10] and by Mareschal in [11]. It is based on a surface integral approach with periodic boundary conditions. The discretized domain is limited to a single period L of the medium. We checked the quality of the results by making sure that the energy was conserved within 2%. Convergence of the results as a function of the discretization and the period was also checked. The radiometric quantities are obtained by averaging over speckle patterns produced by 200 random realizations.

2. RESULTS AND DISCUSSION

The comparisons are restricted to cases where the particles are in a vacuum in order to avoid refraction effects. Thus, we only deal with scattering effects. If reflection or refraction effects are seen, they are only due to the effective index of the random medium. We are interested in the effect of scatterers concentration on radiometric quantities. We have used the same particles in all calculations. They are monodisperse cylinders of refractive index $n = 2.3$ randomly distributed in a slab of thickness d. In all calculations, we have used a radius $a = \lambda/6$ because there is a Mie resonance for polarization V. All calculations have been done with a mesh definition of $\lambda/15$. The specular reflectances and transmittances are compared in Fig. 1. The transmission is essentially driven by the exponential decay. The agreement between radiative transfer and the electromagnetic simulation is quite good for low concentrations but there is a large discrepancy for a volume fraction of 15%. The RT model overestimates scattering extinction because independent scattering overestimates the attenuation coefficient in dense media. This is a well-known behaviour of dependent scattering with particles smaller than the wavelength [12]. Note also that co-

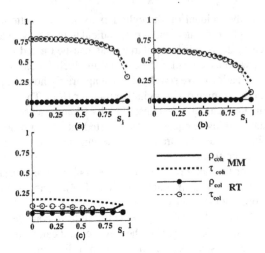

Figure 1.   Comparison of the reflected and transmitted specular components. Radiative transfer results are labelled *col* for collimated whereas electromagnetic results are labelled *coh* for coherent. a) c=1.5%, b) c=3%, c) c=15%.

herent reflection is no longer negligible at this concentration. Especially at grazing incidence, the random slab *reflects specularly* a significant part of the energy. This clearly shows that the coherent field "sees" an effective medium with an effective refractive index. Obviously, no specular reflection is present in the radiative transfer equation because no interface has been introduced in the model. The two effects that we have mentioned, anomalous decay of the coherent and existence of a specular reflection could be accounted for by introducing a complex effective index in the radiative transfer model.

We show the comparison for the case of a concentration of 3% in Fig. 2. the thickness of the slab is 6λ and the scattering mean free path is 3.1λ. The upper part of the figure displays the reflection and the lower part the transmission. The abscissa gives the angle of incidence and the ordinate provides the scattering angle. Thus a vertical line in the graphs represents a scattering pattern for a fixed angle of incidence. Although the numerical simulation is noisy, it is seen that the general shape and the scattering levels are in good agreement apart for two directions : the specular direction in transmission (first diagonal) and the backscattering direction in reflection (second diagonal).

In the forward direction, or transmitted specular direction, we observe a well defined dip of the incoherent flux obtained from the electromagnetic simulation Fig. 2. By contrast, this effect is not obtained when using the radiative transfer equation. This effect can be understood by noting that in the forward direction,

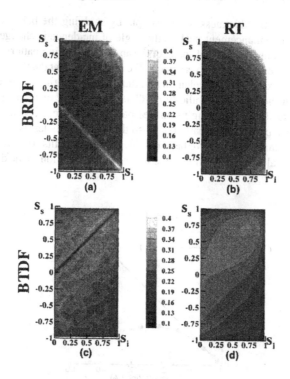

Figure 2. Comparison of the reflected and transmitted diffuse components. Radiative transfer results are labelled RT, electromagnetic results are labelled EM.

the difference of phase due to single scattering is independent of the positions of the particles. In other words, single scattering in the forward direction is coherent. This property is no longer true for other directions. Thus, the incoherent energy in the forward direction is only produced by multiple scattering. Since a medium with a low optical thickness does not produce much multiple scattering, the forward scattering is essentially coherent. The second difference between the two approaches is the presence of a peak in the backscattering direction. This peak is the well-known enhanced backscattering peak. It is interesting to observe that the peak changes as the angle of incidence increases. It is narrower and stronger for large angles of incidence, a behaviour not observed so far to our knowledge. We attribute this effect to the presence of long scattering paths when the illumination is at large angles of incidence. Thus, one might extract information on the thickness of the sample by studying the behaviour of the backscattering enhancement peak as the angle of incidence is changed.

We now discuss the general shape of the angular scattering pattern. It is seen in Fig. 2 that the agreement is very good. We emphasize that the thickness of the slab is of the order of the wavelength and that there is multiple scattering. This suggests that the domain of validity of RT is much larger than was previously thought. Indeed, the scattering mean free path in this case is $l = 3.1\lambda$ and the mean distance between two particles 2λ so that many particles cannot fit into a length scale smaller than l. We have obtained similar results for smaller thicknesses and smaller concentrations. We will now pay attention to the differences

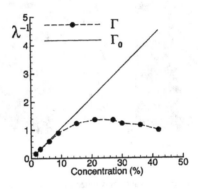

Figure 3.   Scattering mean free path

between radiative transfer calculations and the electromagnetic simulations that reveal the presence of dependent scattering effects at large volume fractions. For a concentration $c = 15\%$, the optical thickness is 2.41 so that multiple scattering is important. The scattering patterns obtained by RT and EM differ markedly (see [9]). This is due to the fact that the parameters used in the RT calculation do not account for dependent effects. In order to quantitatively show the effect of dependent scattering, we can extract the scattering mean free path from the numerical simulation. For the specular components, the random slab can be replaced by a slab with an effective index as shown by Dyson equation for the mean field. A fit yields the effective index whose imaginary part gives the scattering mean free path. Its inverse denoted by Γ is shown in Fig. 3 in units of $1/\lambda$. It is clearly seen that the linear variation with the number of particles per unit volume is no longer valid. This effect was observed experimentally [12].

Conclusion

The main result obtained is that the radiative transfer equation can be used even if the system size is comparable to the wavelength. Besides, we have found that the width of the backscattering enhancement peak depends on the sample thickness. We have also observed a dip of the diffuse transmissivity in the forward scattering direction. This is a coherent phenomenon. Finally, we have discussed the dependent scattering properties. It has been shown that the radiative transfer equation fails to describe systems with a high volume fraction. These results suggest that the limits of validity of the radiative transfer equation are far beyond what is usually accepted. Thus, there is a need for further theoretical developments on the foundations of radiometry.

References

[1] S. Chandrasekhar, *Radiative Transfer* (Dover Publications, 1960).

[2] A. Walther, J. Opt. Soc. Am. **58**, 1256 (1968).

[3] L. Mandel and E. Wolf, *Optical coherence and quantum optics* (Cambridge University Press, 1995).

[4] J.-J. Greffet and M. Nieto-Vesperinas, J. Opt. Soc. Am. A **15**, 2735 (1997).

[5] J.-B. Thibaud, *Propagation de la lumière en milieu aléatoire. Fondements et limites de la description radiométrique. Application à l'imagerie*, Ph.D. thesis, Ecole Centrale Paris, 2000.

[6] R. West, D. Gibbs, L. Tsang, and A. K. Fung, J. Opt. Soc. Am. A **11**, 1854 (1994).

[7] P. Sheng, *Introduction to Wave Scattering, Localization, and Mesoscopic Phenomena* (Academic Press, San Diego, 1995).

[8] Z. Jin and K. Stamnes, Appl. Opt. **33**, 431 (1994).

[9] L. Roux, P. Mareschal, N. Vukadinovic, J.-B. Thibaud, and J.-J. Greffet, to be published in J. Opt. Soc. Am. A

[10] J. Moore, H. Ling, and C. S. Liang, IEEE Trans. Antennas Propagat. **41**, pp 1281 (1993).

[11] P. Mareschal, Ph.D. thesis, Ecole Centrale Paris, 1996.

[12] A. Ishimaru and Y. Kuga, J. Opt. Soc. Am. **72**, 1317 (1982).

Chapter 3

METHODS FOR THE INVERSE PROBLEM IN OPTICAL TOMOGRAPHY

S. R. Arridge

Dept. Computer Science, University College London
Gower Street, London, UK
S.Arridge@cs.ucl.ac.uk

Abstract   In this paper we give a brief overview of the image reconstruction problem in Optical Tomography together with some examples of simulation reconstructions using a numerical optimisation scheme. We discuss first a Diffraction Tomography approach, wherein it is assumed that the measurement is of a scattered wave representing the difference in fields between an unknown and a known state. Both the Born and Rytov approximations are presented and lead to a linear reconstruction problem. Secondly we discuss image reconstruction as an Optimization problem, wherein we develop a model capable of predicting the total field and minimize a least-squares error functional. We introduce the Finite Element Method as a tool for calculating photon density fields in general complex geometries. By means of this method we simulate a number of images and their reconstructions using both a Born and Rytov approximation. The Rytov approximation appears superior in all cases.

Keywords: optical tomography, Green's functions, Born approximation, Rytov approximation, finite element method, optimisation

Introduction

Optical Tomography may be defined as the problem of recovering spatially varying maps of optical parameters, principally absorption and scattering coefficients, from discrete sets of measurements of the transmitted and/or reflected radiation when light is delivered to a heterogeneous body. When considering medical applications, which have received the most attention, the frequency spectrum of interest is in the near infra-red region, 700-1000nm, wherein tissue is comparatively transparent [1-3].

P. Sebbah (ed.), Waves and Imaging through Complex Media, 307–329.
© *2001 Kluwer Academic Publishers. Printed in the Netherlands.*

It is a subject that has accelerated in interest over the last decade, and given rise to many novel mathematical, physical, and instrumental issues. The most prevalent approach involves the so-called "Frequency-domain" measurement systems, wherein the light delivered is modulated at frequencies of the order of 1 GHz, and the detected radiation can be treated as a wave, with measurable amplitude and phase[4-6].

In this paper we consider the image reconstruction problem for these types of systems.

1.   PHOTON TRANSPORT MODELS

We will assume a closed finite domain Ω with boundary $\partial\Omega$. Let r be a point in the interior of Ω and m a point on $\partial\Omega$. Let $\hat{\nu}(m)$ be the outward directed normal at boundary position m. We will assume that we are dealing with a highly scattering medium such that photon density, $\Phi(r;\omega)$ is described by the diffusion equation:

$$\mathcal{L}_{\mu_a,\kappa}\Phi := -\nabla \cdot \kappa(r)\nabla\Phi(r;\omega) + \left(\mu_a(r) + \frac{i\omega}{c}\right)\Phi(r;\omega) = q(r;\omega), \quad (1)$$

where ω is the modulation frequency of the source, $c = c_0/n$ is the speed of light in a medium with refractive index n, μ_a is the spatially varying absorption coefficient, and the diffusion coefficient κ is related to μ_a, the scattering coefficient μ_s, and the phase function $\Theta(\cos\vartheta)$ (assumed angularly independent), of the Boltzmann equation through the relations:

$$\text{average cosine:} \quad \Theta_1 \quad = \int_{-1}^{1} \tau\,\Theta(\tau)d\tau, \quad (2)$$

$$\text{reduced scattering coefficient:} \quad \mu_s'(r) \quad = (1 - \Theta_1)\mu_s(r), \quad (3)$$

$$\text{diffusion coefficient:} \quad \kappa(r) \quad = \frac{1}{3(\mu_a(r) + \mu_s'(r))}. \quad (4)$$

When boundaries are considered, the most usual boundary condition utilised is the Robin condition, or partial current boundary condition [7, 8]:

$$\Phi(m;\omega) + 2\zeta\,\kappa(m)\frac{\partial\Phi(m;\omega)}{\partial\nu} = 0, \quad m \in \partial\Omega, \quad (5)$$

where the coefficient ζ is related to the refractive index mismatch at the boundary. Under these circumstances the measured data is given by Ficks law

$$g(m;\omega) = -\kappa(m)\frac{\partial\Phi(m;\omega)}{\partial\nu} = \frac{\Phi(m;\omega)}{2\zeta}. \quad (6)$$

2. DIFFRACTION TOMOGRAPHY

2.1. HELMHOLTZ EQUATION

One approach to the imaging problem is to relate optical tomography to inverse scattering problems that have well established solution methods.

In a homogeneous medium, with constant optical parameters (μ_{a_0}, κ_0), Eq. (1) is equivalent to a Helmholtz equation

$$\nabla^2 \Phi(r; \omega) + k^2 \Phi(r; \omega) = -S(r; \omega), \tag{7}$$

with

$$k^2 = -\left(\frac{\mu_{a_0} c + i\omega}{c\kappa_0}\right) \quad ; \quad S = \frac{q(r; \omega)}{\kappa_0}. \tag{8}$$

The solution in various geometries is easily derived, of which the most useful in Optical Tomography is that due to a δ-function. In an infinite medium this is simply a spherical wave

$$\Phi(r; \omega) \equiv G(r, r_s; \omega) = \frac{e^{ik|r - r_s|}}{|r - r_s|}, \tag{9}$$

where the notation $G(r, r_s; \omega)$ defines the Green's function for a source at position r_s. Due to the imaginary part of the wave number k this wave is damped. This fact is the main reason that results from Diffraction Tomography are not always straightforwardly applicable in Optical Tomography. In particular, one cannot easily make use of the far-field results which approximate a spherical wave by a plane wave. Instead, OT is concerned with *near-field* effects.

In fact, the heterogeneous equation can also be expressed as a Helmholtz equation under a transformation of variables [3]. However, we will not make use of this scheme here.

2.2. BORN APPROXIMATION

For the Born Approximation, we assume that we have a reference state $x_0 = (\mu_{a_0}, \kappa_0)$, with a corresponding wave Φ_0, and that we want to find the *scattered* field Φ^{sca} due to a change in state $\Delta x = (\alpha, \beta)$. Note that it is not necessary to assume that the initial state is homogeneous. Assuming that the source terms are unchanged we will have

$$-\nabla \cdot (\kappa_0 + \beta)\nabla\Phi + \left(\mu_{a_0} + \alpha + \frac{i\omega}{c}\right)\Phi = q, \tag{10}$$

with

$$\Phi = \Phi_0 + \Phi^{sca}. \tag{11}$$

Equation (10) can be solved by introducing the Green's operator for the reference state, $\mathcal{G}_0$

$$\Phi = \mathcal{G}_0 \left[q + \nabla \cdot \beta \nabla \Phi - \alpha \Phi \right] \tag{12}$$

$$= \Phi_0 - \int_\Omega \begin{array}{c} \beta(r') \nabla G_0(r, r') \cdot \nabla \Phi(r') \\ + \quad \alpha(r') G_0(r, r') \Phi(r') \end{array}, \, \mathrm{d}^n r' \tag{13}$$

where we used the divergence theorem and assumed $\beta(m) = 0; m \in \partial\Omega$.

In analogy to quantum mechanics, we may define a "potential" as the differential operator

$$\mathcal{V}(\alpha, \beta) := \nabla \cdot \beta \nabla - \alpha \tag{14}$$

and using Eq. (11) in Eq. (13) we obtain the familiar Born Series

$$\Phi = \Phi^{(0)} + \Phi^{(1)} + \Phi^{(2)} + \cdots , \tag{15}$$

where

$$\Phi^{(0)} = \Phi_0 \tag{16}$$

$$\Phi^{(1)} = \mathcal{G}_0 \mathcal{V}(\alpha, \beta) \Phi_0 \tag{17}$$

$$\Phi^{(2)} = \mathcal{G}_0 \mathcal{V}(\alpha, \beta) \mathcal{G}_0 \mathcal{V}(\alpha, \beta) \Phi_0 \tag{18}$$

$$\vdots$$

The first order approximation is then simply

$$\Phi^{sca}(r; \omega) = \Phi^{(1)}(r; \omega)$$

$$= \int_\Omega \begin{array}{c} \beta(r') \nabla G_0(r, r'; \omega) \cdot \nabla \Phi_0(r'; \omega) \\ + \quad \alpha(r') G_0(r, r'; \omega) \Phi_0(r'; \omega) \end{array} \mathrm{d}^n r' . \tag{19}$$

2.3.   RYTOV APPROXIMATION

The Rytov Approximation is derived by considering logarithm of the field as a complex phase [9, 10]:

$$\Phi(r; \omega) = e^{\phi(r; \omega)} , \tag{20}$$

so that, in place of Eq. (11) we have

$$\ln \Phi = \ln \Phi_0 + \phi^{sca} . \tag{21}$$

Following a standard derivation, the first order approximation becomes

$$\phi^{sca}(r; \omega) = \frac{\Phi^{(1)}(r; \omega)}{\Phi_0(r; \omega)}$$

$$= \frac{1}{\Phi_0(r; \omega)} \int_\Omega \begin{array}{c} \beta(r') \nabla G_0(r, r'; \omega) \cdot \nabla \Phi_0(r'; \omega) \\ + \quad \alpha(r') G_0(r, r'; \omega) \Phi_0(r'; \omega) \end{array} \mathrm{d}^n r' . \tag{22}$$

The Rytov Approximation is thought to have a wider range of applicability than the Born Approximation.

2.4. ADJOINT FIELD FORMULATION

A key point in the ensuing is the definition of an *adjoint* field as the solution to the adjoint problem

$$\mathcal{L}^+_{\mu_a,\kappa}\Phi^+ := -\nabla \cdot \kappa(r)\nabla\Phi^+(r;\omega) + \left(\mu_a(r) - \frac{i\omega}{c}\right)\Phi^+(r;\omega) = q^+(r;\omega).$$
(23)

In particular we define the *adjoint Green's function* as the solution to

$$\mathcal{L}^+_0 G^+_0 = \delta,$$
(24)

whence the first Born Approximation, Eq. (19), can be written

$$\Phi^{sca}(r;\omega) = \int_\Omega \begin{array}{l} \beta(r')\nabla G^+_0(r',r;\omega) \cdot \nabla\Phi_0(r';\omega) \\ + \quad \alpha(r')G^+_0(r',r;\omega)\Phi_0(r';\omega) \end{array} d^n r'$$
(25)

and the first Rytov Approximation, Eq. (22), as

$$\phi^{sca}(r;\omega) = \frac{1}{\Phi_0(r;\omega)} \int_\Omega \begin{array}{l} \beta(r')\nabla G^+_0(r',r;\omega) \cdot \nabla\Phi_0(r';\omega) \\ + \quad \alpha(r')G^+_0(r',r;\omega)\Phi_0(r';\omega) \end{array} d^n r'.$$
(26)

Note the reversal of the parameter order in the argument to G^+; to be specific we have :

$$G^+(r',r;\omega) = G(r,r';\omega),$$
(27)

indicating that the adjoint Green's function is a propagator in the opposite direction to the normal.

3. FORWARD PROBLEM

The preceding has given the standard physical treatment of the propagation of waves in random media, and the approximation of scattered waves due to small perturbations. We now want to relate this approach to the more general one of Inverse Problems. Our mechanism will be one of forward and inverse mappings, which are essentially non-linear, and the linearisation of the former through the Fréchet derivative.

Up to now we have considered the description of a scattered wave as a field inside the medium. For an imaging system, what is measured is simply the total field Φ at some detector positions. In *dynamic* systems we indeed have measurements of the reference state Φ_0, in which case we are seeking only to image the state changes (α, β) that give rise to the scattered field. However, in *absolute* imaging, we have no reference measurement and our task is to derive the distribution of optical parameters that give rise to the total field.

3.1.   SINGLE FREQUENCY, SINGLE SOURCE

Let us define a solution state as

$$\boldsymbol{x} \equiv \boldsymbol{x}(\boldsymbol{r}) := (\mu_a(\boldsymbol{r}), \kappa(\boldsymbol{r}))^{\mathrm{T}} \ \boldsymbol{r} \in \Omega. \tag{28}$$

Let us assume that the detectors are on the boundary of the medium, then the forward problem is properly considered as a linear mapping form the space of solutions to a complex scalar on the boundary:

$$y_j(\boldsymbol{m}; \omega) = \mathcal{P}_j(\boldsymbol{x}), \tag{29}$$

where the subscript j indicates a particular initial wave, at modulation frequency ω. In the case of plane or spherical waves, j will index the wave number, although in general we will consider more complex initial waves.

We will also define a residual operator

$$\mathcal{R}_j(\boldsymbol{x}) := (g_j - \mathcal{P}_j(\boldsymbol{x})) =: b_j(\boldsymbol{m}; \omega), \tag{30}$$

where g_j is the measured data, c.f. Eq. (6). In terms of the previous analysis, the residual operator determines Φ^{sca} at the boundary. For dynamic imaging we will seek (α, β) so that the approximations of section 2 agree with this scattered field at the discrete measurement sites. For absolute imaging, the "scattered field" is an artificial concept, since we are considering only one field, and the inverse problem seek a state $\boldsymbol{x}$ such that Φ agrees with the data at the measurement sites.

3.2.   SINGLE FREQUENCY, MULTIPLE SOURCES

For boundary value problems involving the Helmholtz equation, it is not possible to obtain a unique solution from only one source field. Strictly an infinite set of sources functions $\{q_j(\boldsymbol{m}')\}$ is required, which form a basis for any function on $\partial\Omega$. In practice, this cannot be achieved, and so we assume a finite set $\{q_1, \ldots, q_S\}$ wherein each source q_j is of finite support.

Using this notation we consider the forward mapping as a parallel set of projections

$$\boldsymbol{y} = \mathcal{P}(\boldsymbol{x}), \tag{31}$$

where

$$\mathcal{P} := (\mathcal{P}_1, \ldots, \mathcal{P}_S)^{\mathrm{T}} \tag{32}$$

$$\mathcal{R} := (\mathcal{R}_1, \ldots, \mathcal{R}_S)^{\mathrm{T}} \tag{33}$$

$$\boldsymbol{y} := (y_1, \ldots, y_S)^{\mathrm{T}}. \tag{34}$$

We define similarly, the total given data, and total residual data by the stacked functions :

$$g := (g_1, \dots, g_S)^{\mathrm{T}} \tag{35}$$
$$b := (b_1, \dots, b_S)^{\mathrm{T}} . \tag{36}$$

4. INVERSE PROBLEM

4.1. ERROR FUNCTIONALS

We now define the inverse problem as an optimisation problem. Specifically, we require

$$\hat{x} = \arg\min_x \left(\|\mathcal{R}(x)\|_Y \right) \tag{37}$$

where $\|.\|_Y$ is a suitable norm in the space of the data. In this paper we will assume a weighted $L2$-norm of the form

$$\|b\|_Y = \frac{1}{2} b^{\mathrm{T}} R^2 b, \tag{38}$$

with R^2 the inverse of the covariance matrix of the data. This choice of norm is identical to assuming that the detected photons obey Gaussian statistics with zero mean multivariate Gaussian noise. Such an assumption can be justified using the Central Limit Theorem that proposes that any probability distribution approaches a multivariate Gaussian in the limit of sufficiently large numbers of photons[11]. If we make the further assumption that individual photons are uncorrelated, then the covariance matrix can be taken to be diagonal, and we can rewrite Eq. (38) as

$$\|b\|_Y = \frac{1}{2} \sum_j \int_{\partial\Omega} \left(\frac{b_j(m)}{\sigma_j(m)} \right)^2 \mathrm{d}^{n-1}m . \tag{39}$$

Note that up to this point we have considered *continuous* data, i.e. functions on the boundary of the domain $\partial\Omega$. In practice this data will be sampled at discrete points $\{m_1, \dots, m_M\}$, with a sampling function of finite support. For simplicity we take these to be δ-functions. Thus we arrive at a fully discrete data set, and the corresponding error function :

$$\|b\|_Y = \frac{1}{2} \sum_{i,j} \frac{b_{i,j}^2}{\sigma_{i,j}^2} = \frac{1}{2} \sum_j \left\| \frac{b_j}{\sigma_j} \right\|^2 , \tag{40}$$

where j indexes the sources, as before, and i indexes the detectors.

4.2.   LINEARISATION

Born Approximation.   The solution of the reconstruction problem requires the linearisation of the forward operator Eq. (31). For the diffusion problem the Fréchet derivative of the forward operator is given by :

$$\rho_{i,j}(r) := \frac{\partial y_{i,j}}{\partial x(r)} = \begin{pmatrix} G_i^+(r;\omega)\,G_j(r;\omega) \\ \nabla G_i^+(r;\omega)\cdot\nabla G_j(r;\omega) \end{pmatrix}, \tag{41}$$

where G_i^+, G_j are the forward and adjoint Green's functions for source j and detector i. Using this expression, a linear integral equation is obtained by assuming a known background state x_0 and solving the linear problem

$$b = \int_\Omega K_0(r)\,\Delta x(r)\,\mathrm{d}^n r\,, \tag{42}$$

where the kernel of the integral equation, K_0, is given by

$$K_0(r) = \begin{pmatrix} \rho_{1,1}^T(r) \\ \vdots \\ \rho_{M,S}^T(r) \end{pmatrix}. \tag{43}$$

This formulation is called *continuous-discrete* since the solution is still expressed in terms of functions, whereas the data is discrete. To obtain a discrete matrix formulation we need to define a basis for the solution :

$$\kappa(r) = \sum_k \kappa_k v_k^{(\kappa)}(r) \qquad \mu(r) = \sum_k \mu_{a_k} v_k^{(\mu)}(r)\,, \tag{44}$$

whence Eq. (42) becomes

$$b = \left(A^{(\mu)}, A^{(\kappa)} \right) \Delta x\,, \tag{45}$$

where the components of matrix A are given by

$$A_{ij,k}^{(\mu)} = \int_\Omega v_k^{(\mu)}(r)\,G_i^+(r;\omega)\,G_j(r;\omega)\,\mathrm{d}^n r \tag{46}$$

$$A_{ij,k}^{(\kappa)} = \int_\Omega v_k^{(\kappa)}(r)\,\nabla G_i^+(r;\omega)\cdot\nabla G_j(r;\omega)\,\mathrm{d}^n r\,. \tag{47}$$

It should be noted that under a Helmholtz Approximation, the mapping defined by Eq. (45) is $\mathbb{C}(\Omega) \to \mathbb{C}(\partial\Omega)$ i.e. from a complex wave number to a complex scalar field. When considering the Diffusion Approximation, we have a mapping from $(\mathbb{R}(\Omega) \times \mathbb{R}(\Omega)) \to \mathbb{C}(\partial\Omega)$. However, when applying the adjoint operator, we have by definition an operator $\mathbb{C}(\partial\Omega) \to \mathbb{C}(\Omega)$ in both cases. If this were used for the diffusion approximation it would result in imaginary terms for the update in absorption and scattering. The resolution to this problem is to separate the real and imaginary parts of the measurement operator, and consider the data itself as $(\mathbb{R}(\partial\Omega) \times \mathbb{R}(\partial\Omega))$. Thus the linear inverse problem that we will consider is

$$
\begin{pmatrix} \mathrm{Re}(b) \\ \mathrm{Im}(b) \end{pmatrix} = \begin{pmatrix} \Delta\mathrm{Re}(y) \\ \Delta\mathrm{Im}(y) \end{pmatrix} = \begin{pmatrix} \mathrm{Re}(A^{(\mu)}) & \mathrm{Re}(A^{(\kappa)}) \\ \mathrm{Im}(A^{(\mu)}) & \mathrm{Im}(A^{(\kappa)}) \end{pmatrix} \begin{pmatrix} \alpha \\ \beta \end{pmatrix}. \tag{48}
$$

Rytov Approximation.   The Rytov Approximation was derived by considering the change in the logarithm of the field. Thus the norm that we are considering is

$$
\|\mathcal{R}(x)\|_{Y} = \frac{1}{2} \sum_{i,j} \frac{(\ln g_{i,j} - \ln y_{i,j})^2}{\sigma_{i,j}^2}. \tag{49}
$$

Fréchet differentiation obeys the chain rule and we have in place of Eq. (41)

$$
\rho_{i,j}^{Ryt}(r) := \frac{\partial \ln y_{i,j}}{\partial x(r)} = \begin{pmatrix} \frac{G_i^+(r;\omega)\, G_j(r;\omega)}{y_{i,j}(\omega)} \\ \frac{\nabla G_i^+(r;\omega)\cdot\nabla G_j(r;\omega)}{y_{i,j}(\omega)} \end{pmatrix}, \tag{50}
$$

and Eq. (48) becomes

$$
\begin{pmatrix} \mathrm{Re}(b) \\ \mathrm{Im}(b) \end{pmatrix} = \begin{pmatrix} \Delta\mathrm{Mod}(y) \\ \Delta\mathrm{Arg}(y) \end{pmatrix} = \begin{pmatrix} \mathrm{Re}(Y^{-1}A^{(\mu)}) & \mathrm{Re}(Y^{-1}A^{(\kappa)}) \\ \mathrm{Im}(Y^{-1}A^{(\mu)}) & \mathrm{Im}(Y^{-1}A^{(\kappa)}) \end{pmatrix} \begin{pmatrix} \alpha \\ \beta \end{pmatrix}
$$
$$\tag{51}$$

where

$$
Y = \mathrm{diag}\{y\}. \tag{52}
$$

5.   FINITE ELEMENT METHODS

The preceding discussion is quite general, in that we have described how to derive the linearisation of a nonlinear forward operator in terms of forward and adjoint fields. When developing a practical inversion scheme we need a mechanism to compute these fields. One possibility is to use analytical expressions such as the infinite space Green's function, Eq. (9), or the derived forms for certain well-defined geometries [12]. However, for general geometries, and heterogeneous parameter distributions, a numerical method is required. In this paper we will use the Finite Element Method (FEM) which allows a straightforward extension of our previous results. We will give a brief summary here and refer the reader to [13, 7, 14] where we presented detailed schemes for solving the steady-state, time-dependent, and frequency-domain versions of the Diffusion Approximation.

The assumption in the FEM approach is that Φ is approximated by the piecewise polynomial function $\Phi^h(r) = \sum_m^D \Phi_m u_m(r) \in \mathcal{U}^h$, where $\mathcal{U}^h$ is a finite

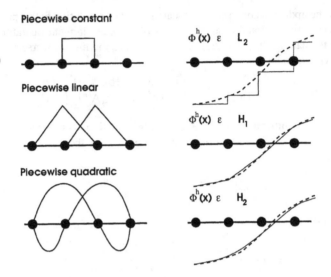

Figure 1.   Different FEM basis functions. Left : different polynomial shape functions u_m. Right : the fields that they can represent, Φ^h, (solid line) represent different approximations to the true field Φ (dashed line). Increasing orders of approximation represent functions in spaces with Sobolev norms of higher order ($H_0 \equiv L_2$).

dimensional subspace spanned by basis functions $\{u_m(r); m = 1 \ldots D\}$ chosen to have limited support. Examples of FEM basis functions in one dimension are shown in Fig. 1.

In this framework, the Diffusion Equation becomes

$$K(\omega)\Phi(\omega) = q(\omega), \tag{53}$$

where Φ and q are D-dimensional complex vectors and K is a $D \times D$ complex matrix with entries given by :

$$
\begin{aligned}
K_{lm}(\omega) &= \int_\Omega \left(\kappa(r)\nabla u_l(r) \cdot \nabla u_m(r) + \left(\mu_a(r) + \frac{i\omega}{c} \right) u_l(r)u_m(r) \right) \mathrm{d}^n r \\
&\quad + \int_{\partial\Omega} \frac{1}{2\zeta} u_l(r)u_m(r) \, \mathrm{d}^{n-1} m \, .
\end{aligned} \tag{54}
$$

Equation (53) is formally solved by matrix inversion

$$\Phi = K^{-1}(\omega)q = G(\omega)q \tag{55}$$

where $G(\omega)$ is the discrete representation of the Green's operator $\mathcal{G}$.

5.1. BORN APPROXIMATION

The Born Approximation in the discrete FEM framework proceeds analogously to the continuous case. We first define the *basis system matrices* whose entries are given by

$$V^{(\kappa)}_{k,ij} = \int_\Omega v^{(\kappa)}_k(r)\nabla u_i(r)\cdot\nabla u_j(r)\,d^n r \tag{56}$$

$$V^{(\mu)}_{k,ij} = \int_\Omega v^{(\mu)}_k(r)u_i(r)u_j(r)\,d^n r\,. \tag{57}$$

The "potential" of a perturbation can be expressed

$$V(\alpha,\beta) = \sum_k \alpha_k V^{(\mu)}_k + \beta_k V^{(\kappa)}_k\,, \tag{58}$$

whence the scattered field is obtained exactly via

$$K(\omega)\Phi^{sca}(\omega) = V(\alpha,\beta)\Phi_0(\omega) \tag{59}$$

or in the first approximation by

$$\Phi^{sca}(\omega) \simeq \Phi^{(1)}(\omega) = G_0(\omega)V(\alpha,\beta)\Phi_0(\omega)\,. \tag{60}$$

The Rytov Approximation is obtained in analogous fashion.

5.2. LINEARISATION

The derivation of the linearised form using the direct or adjoint method is again analogous to the continuous case. We give a brief summary here, and refer the reader to [3] for more details. The equivalent of the matrix elements Eq. (46) and Eq. (47) are

$$A^{(\mu)}_{ij,k} = \left\langle G^+_i(\omega), V^{(\mu)}_k G_j(\omega)\right\rangle \tag{61}$$

$$A^{(\kappa)}_{ij,k} = \left\langle G^+_i(\omega), V^{(\kappa)}_k G_j(\omega)\right\rangle\,, \tag{62}$$

where $G^+_i(\omega)$ is the discrete adjoint Green's function given as the solution to

$$K^+(\omega)G^+_i(\omega) = q^+_i(\omega)\,. \tag{63}$$

We can now state the analogy to Eq. (41)

$$
\rho^{\mathrm{T}}(i,j) = \begin{pmatrix} \left\langle G_i^+(\omega), V_1^{(\mu)} G_j(\omega) \right\rangle \\ \vdots \\ \left\langle G_i^+(\omega), V_N^{(\mu)} G_j(\omega) \right\rangle \\ \left\langle G_i^+(\omega), V_1^{(\kappa)} G_j(\omega) \right\rangle \\ \vdots \\ \left\langle G_i^+(\omega), V_N^{(\kappa)} G_j(\omega) \right\rangle \end{pmatrix}. \tag{64}
$$

6.   RECONSTRUCTION ALGORITHMS

We consider here only one type of reconstruction scheme, based on direct optimisation of Eq. (40) using its gradient. Under a Born Approximation we obtain the gradient as

$$
z := \sum_{i,j} \frac{b_{i,j}}{\sigma_{i,j}^2} \, \rho^{\mathrm{T}}(i,j) = \begin{pmatrix} \sum_j \mathrm{Re} \left\langle \Psi_j(\omega), V_1^{(\mu)} G_j(\omega) \right\rangle \\ \vdots \\ \sum_j \mathrm{Re} \left\langle \Psi_j(\omega), V_N^{(\mu)} G_j(\omega) \right\rangle \\ \sum_j \mathrm{Re} \left\langle \Psi_j(\omega), V_1^{(\kappa)} G_j(\omega) \right\rangle \\ \vdots \\ \sum_j \mathrm{Re} \left\langle \Psi_j(\omega), V_N^{(\kappa)} G_j(\omega) \right\rangle \end{pmatrix}, \tag{65}
$$

where Ψ_j is the solution to

$$
\mathsf{K}^+(\omega)\Psi_j(\omega) = \left(\frac{b_j(\omega)}{\sigma_j^2(\omega)} \right)^*, \tag{66}
$$

where $*$ indicates complex conjugate. For the Rytov Approximation, Eq. (66) is replaced by

$$
\mathsf{K}^+(\omega)\Psi_j(\omega) = \left(\frac{b_j(\omega)}{\sigma_j^2(\omega)y_j(\omega)} \right)^*. \tag{67}
$$

In nonlinear conjugate gradient methods a set of conjugate search directions is generated to find the minimum of the objective function. At each iteration step a one-dimensional line minimization along the current search direction is performed. Conjugate gradient methods are well-established in nonlinear optimisation. See for example [15]. A variety of line search methods can be used, either utilising gradient information, such as the secant method, or using only function evaluations such as the quadratic fit method. Often an exact line search is too computationally expensive due to the large number of function or derivative computations. Inexact line search methods such as Armijo's Rule define the bounds for acceptable step lengths which guarantee convergence. There the line search is terminated when the step length is within the valid range.

The conjugate gradient algorithm is only one example of a gradient-based optimisation algorithm. For a discussion of other algorithms and their relative computational complexity, see [16].

7. RESULTS

In the following examples, several different forward models are used to generate data. In each case a two dimensional circular model of radius 35 mm is used, with 16 sources and detectors, placed equidistant on the outer boundary. The mesh used is shown in Fig. 2. The background optical parameters were $(\mu_a = 0.01 \text{ mm}^{-1}, \mu_s' = 1 \text{ mm}^{-1}, n = 1.4)$, and the modulation frequency was 100 MHz in all cases. The sources were modelled as an isotropic point source, placed one scattering distance (1 mm) inside the outer boundary. In each case the object was a Gaussian blob in absorption only,

$$\alpha(r) = A \exp\left(-\frac{||r - r_0||^2}{2s^2}\right), \tag{68}$$

of height A, halfwidth s centered at position r_0. There was no perturbation in scattering. Images were reconstructed from this data using both the Born Approximation (i.e. real and imaginary complex data) and the Rytov Approximation (i.e. using amplitude and phase of the complex data). Although the object was an absorber only, we simultaneously reconstructed for absorption and scatter, in order to illustrate the effect of crosstalk.

Reconstructions were run without noise and iterated until the change in error norm was below a fraction of the initial value. The number of iterations achieved varied between the different cases.

No penalty terms were added to the cost functional for regularisation, but each solution was filtered with 3 iterations of a local neighbourhood median filter, that replaces the value of a node of the mesh with the median of that nodes values and those with which it shares an edge in the mesh graph.

The first case considered a blob of constant amplitude and half width ($A = 0.04 \text{ mm}^{-1}$, $s = 5$ mm) placed at varying distances from the mesh centre. Images from the Born Approximation method are shown for each position of the anomaly in Fig. 3(i)-(xii) (μ_a image) and Fig. 4(i)-(xii) (μ_s' image). The latter is purely artefact, since there the original μ_s' profile was homogeneous.

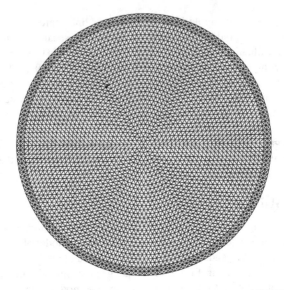

Figure 2.   Mesh used for the results in this section. It has 3511 nodes and 6840 linear triangular elements. Radius 35 mm.

The same cases for the Rytov Approximation are shown in Fig. 5(i)-(xii) (μ_a image) and Fig. 6(i)-(xii). In each image, the range is scaled to the maximum and minimum found, so although the qualitative result is apparent, the quantitative results are not. Thus the bottom figure in each case shows the radial profile through the blob extremum, for each radial position superimposed. From these we see the typical behaviour that the resolution and contrast is decreased with distance from the boundary. The contrast and resolution achieved with the Born Approximation is much worse than with the Rytov Approximation, and the cross-talk between absorption and scattering is also much worse in the former case.

As a second example, we get the position and amplitude of the blob fixed and varied its width. The results are in Fig. 7 (μ_a image) and Fig. 8 (μ'_s image). Again the results indicate that the Rytov Approximation is better able to recover quantitatively accurate images, and to reduce cross-talk.

Finally we considered keeping position and width constant and varying amplitude. The results are summarised in Fig. 9(i)-(iv) where the profile cross-sections are shown for the μ_a and μ'_s reconstructions, for both Born and Rytov approximations. The results are also consistent with the improved performance of the Rytov Approximation.

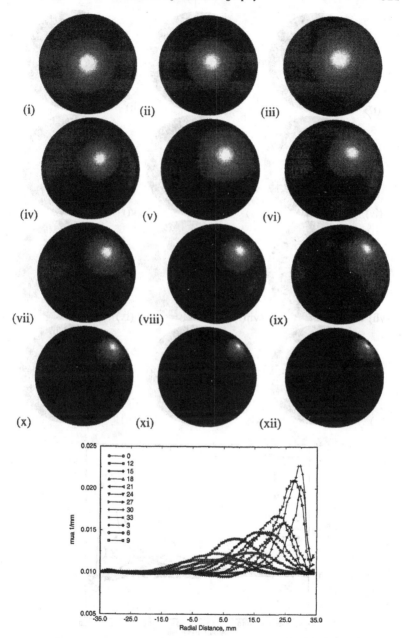

Figure 3. Reconstructed images of μ_a, using the Born Approximation, where the anomaly is displaced radially from the centre of the model (i) to a radius of 33 mm (xii) in 12 equal steps.

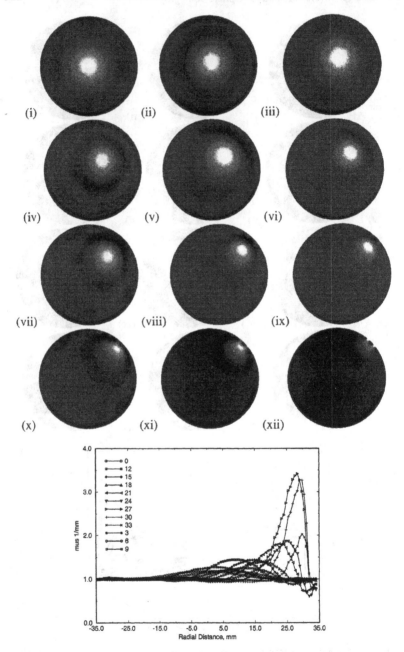

Figure 4.   Reconstructed images of μ_s' for the same case as Fig. 3

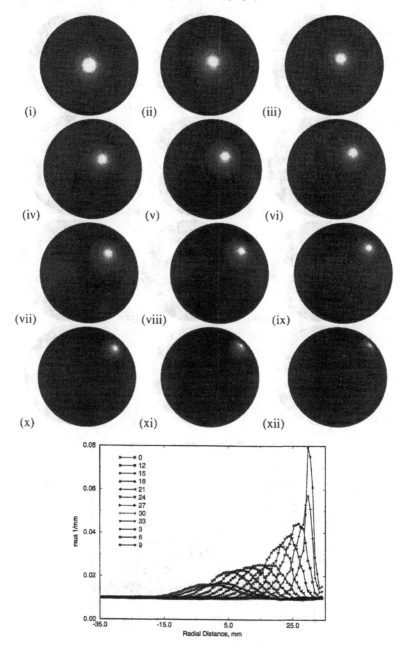

Figure 5. as Fig. 3 but using the Rytov Approximation.

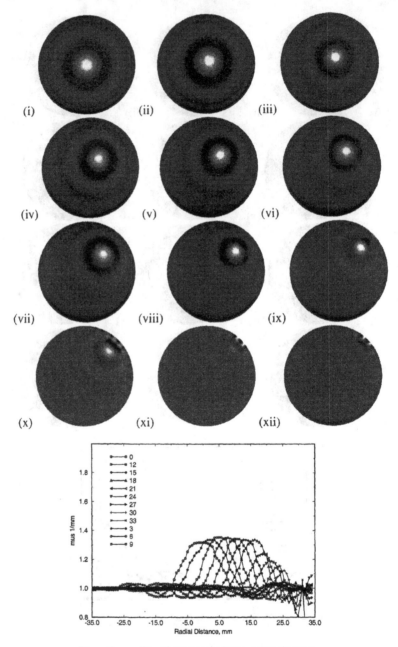

Figure 6.   as Fig. 4 but using the Rytov Approximation.

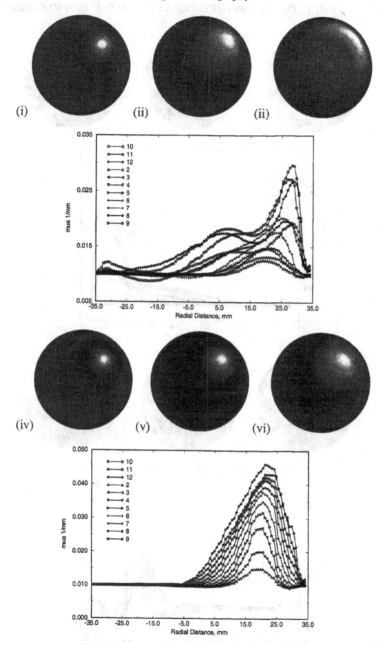

Figure 7. Reconstructed images of μ_a where the anomaly is centred at (20,20) and increased in size from a halfwidth of $s = 2$ mm to $s = 12$ mm in 1 mm steps. Top row : Born Approximation, (i) $s = 2$ mm, (ii) $s = 7$ mm, (iii) $s = 12$ mm. Second Row : Profiles of cross-sections for all Born approximation cases. Third row : Rytov approximation, (iv) $s = 2$ mm, (v) $s = 7$ mm, (vi) $s = 12$ mm. Bottom Row : Profiles of cross-sections for all Rytov Approximation cases.

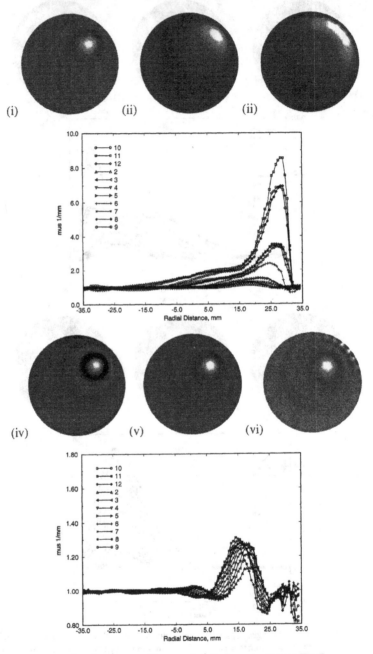

Figure 8.   Reconstructed images of μ_s' for the same cases as Fig. 7.

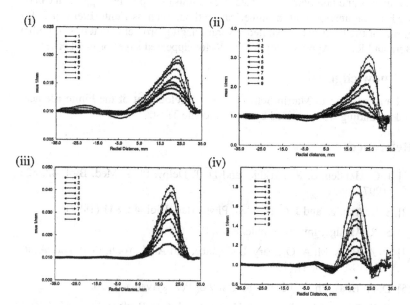

Figure 9.   Profiles through reconstructed images where the anomaly is centred at (20,20) with halfwidth of 4mm and is increased in amplitude from 0.01 mm^{-1} to 0.09 mm^{-1} in 0.01 mm^{-1} steps. Top row : μ_a (i) and μ_s' (ii) using the Born Approximation. Bottom row : μ_a (iii) and μ_s' (iv) using the Rytov Approximation.

Conclusion

In this paper we have examined the image reconstruction problem in Optical Tomography, both from the point of view of a Diffraction Tomography approach, wherein we argue that the measurement is of a scattered wave representing the difference in fields between an unknown and a known state, and the point of view of an Optimization problem, wherein we develop a model capable of predicting the total field.

Whereas the Diffraction Tomography approach is generally limited to simple geometries and homogeneous states, the optimisation problem approach can be used in conjunction with a numerical method, such as Finite Elements. By means of the latter, we illustrated several imaging problems in terms of both a Born and Rytov Approximation. The Rytov appeared superior in all cases.

Acknowledgments

I am indebted to Martin Schweiger for development of the Finite Element code. Funding was provided by the Wellcome Trust.

References

[1] J. C. Hebden, S. R. Arridge, and D. T. Delpy, Phys. Med. Biol. **42**, 825 (1997).

[2] S. R. Arridge and J. C. Hebden, Phys. Med. Biol. **42**, 841 (1997).

[3] S. R. Arridge, Inverse Problems **15**, R41 (1999).

[4] D. A. Boas, M. A. O'Leary, B. Chance, and A. G. Yodh, Proc. Nat. Acad. Sci. USA **91**, 4887 (1994).

[5] D. A. Boas, Ph.D. thesis, University of Pennsylvania, 1996.

[6] B. W. Pogue, M. S. Patterson, H. Jiang, and K. D. Paulsen, Phys. Med. Biol. **40**, 1709 (1995).

[7] M. Schweiger, S. R. Arridge, M. Hiraoka, and D. T. Delpy, Med. Phys. **22**, 1779 (1995).

[8] R. Aronson, J. Opt. Soc. Am. A **12**, 2532 (1995).

[9] A. Ishimaru, *Wave Propagation and Scattering in Random Media* (Academic, New York, 1978), Vol. 1.

[10] A. C. Kak and M. Slaney, *Principles of Computerized Tomographic Imaging* (IEEE Press, New York, 1987).

[11] S. R. Arridge, M. Hiraoka, and M. Schweiger, Phys. Med. Biol. **40**, 1539 (1995).

[12] S. R. Arridge, M. Cope, and D. T. Delpy, Phys. Med. Biol. **37**, 1531 (1992).

[13] S. R. Arridge, M. Schweiger, M. Hiraoka, and D. T. Delpy, Med. Phys. **20**, 299 (1993).

[14] M. Schweiger and S. R. Arridge, Med. Phys. **24**, 895 (1997).

[15] M. S. Bazaraa, H. D. Sherali, and C. M. Shetty, *Nonlinear Programming: Theory and Algorithms*, 2nd ed. (Wiley, New York, 1993).

[16] S. R. Arridge and M. Schweiger, in *Computational Radiology and Imaging: therapy and Diagnosis*, IMA Volumes in Mathematics and its Applications, IMA, edited by C. Borgers and F. Natterer (Springer-Verlag, New York, 1998), 45–70.

Chapter 4

INVERSE PROBLEM
FOR STRATIFIED SCATTERING MEDIA

J.-M. Tualle, J. Prat, E. Tinet and S. Avrillier
Laboratoire de Physique des Lasers, CNRS UMR 7538, Université Paris 13
99, av. J.-B. Clément, 93 430 Villetaneuse
tualle@galilee.univ-paris13.fr

Abstract   The investigation of layered media is one of the first steps for the resolution of
the inverse problem in more complex situations. Furthermore, this problem is
of great interest in biomedical optics as it can be closely linked to real biological situations like the skin-fat-muscle or the skin-skull-brain successions. A
new analytical model based on the diffusion approximation is used to achieve
ultra fast calculation of the direct problem in layered media. The possibility of
completely solve the inverse problem in two-layer turbid media is investigated
through theoretical results.

Keywords: Diffusion approximation, Inverse problem, Layered media

A first step toward the resolution of the inverse problem in complex turbid
media can be the study of layered media. This problem is very interesting in
biomedical optics as it can be closely linked to real biological situations. For
instance, in the non invasive measurement of the optical coefficients of a muscle,
the main difficulty is the perturbation induced by the skin and fat layers. The
evaluation of the thickness and the determination of the optical coefficients of
each layers by the means of an inversion procedure can therefore find concrete
applications. This ill-posed problem implies at least time and space-resolved
measurement of the back-scattered light. Such measurements can be obtained
for instance by sending at $t = 0$ a short light pulse at the surface of the medium,
and by measuring the reflectance as a function of the distance r from the source,
and time t (Fig. 1).

To show that the use of both time and space-resolved measurements can
be very significant, we have performed Monte-Carlo simulations [1] in a two-

P. Sebbah (ed.), Waves and Imaging through Complex Media, 331–337.
© 2001 *Kluwer Academic Publishers. Printed in the Netherlands.*

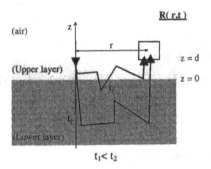

Figure 1.   The inverse problem consist in the determination of the thickness and the optical coefficients of each layer from R(r,t). Time and Space-Resolved measurements can be really useful on that purpose, as the the explored region is directly linked to the detection time.

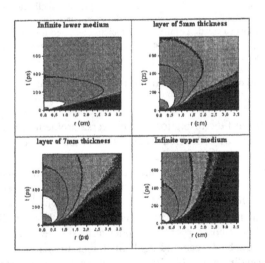

Figure 2.   These time and space-resolved map have been obtained from Monte Carlo simulations, with $\mu'_{s1} = 20$ cm^{-1}, $\mu_{a1} = 0.026$ cm^{-1}, $\mu'_{s2} = 1$ cm^{-1} and $\mu_{a2} = 0.05$ cm^{-1}. The back-scattered intensity is represented on a logarithmic scale. There is a clear signature of the layered structure.

layered medium with highly contrasted optical coefficients : the reduced scattering coefficient is $\mu'_{s1} = 20$ cm^{-1} for the upper layer and $\mu'_{s2} = 1$ cm^{-1} for the

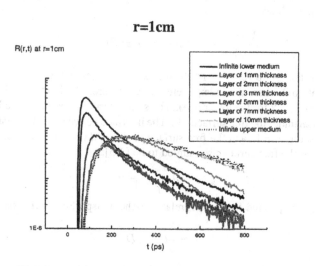

r=1cm

Figure 3. Time-resolved curves obtained from Monte Carlo simulations (logarithmic scale). There is a dependence on the first layer thickness, but not the clear signature visible on the whole map.

lower one. The results of the simulation for four different thickness of the first layer are presented on Fig. 2, which exhibits a clear signature of the layered structure. Figure 3 presents the same results at one fixed distance from the source, and again for different thickness of the first layer. A dependence on the first layer thickness can be seen, but the signature is not any more clearly visible as it is on the whole map. In fact, each of these curves in Fig. 3 could have been interpreted as the reflectance from a semi-infinite media. We can therefore conclude that the whole time and space resolved reflectance map is needed for the realization of the inverse problem. The inverse problem is usually treated through the use of an optimisation procedure concerning the unknown parameters which are here the optical coefficients and the layers thickness. To perform such calculations, a fast algorithm is needed for the evaluation of the direct problem. Moreover, a better understanding of the relationship between the reflectance signal and the layered structure can be useful for the design of an inversion procedure [2]. On that purpose we investigated the direct problem in two-layered media using the diffusion approximation.

For the mathematical formulation of this problem, we used classical assumptions. The fluence density function φ is assumed to satisfy the diffusion equation in each layer:

$$\frac{1}{c_i}\frac{\partial}{\partial t}\varphi - D_i\Delta\varphi + \mu_{ai}\varphi = S(\overrightarrow{r},t),\tag{1}$$

where c_i is the group velocity, D_i the diffusion constant and μ_{ai} the absorption coefficient, and where the index i refers to considered medium ($i = 1$ for the upper layer, $i = 2$ for the lower one). The source term is a point source located at a depth z_0 under the free surface [3]. The interface between the two scattering media is set at $z = 0$, and the free surface at $z = +d$ (Fig. 1). The boundary condition at the free surface is assumed to be [3-5]:

$$\varphi(z = d + z_b) = 0,\tag{2}$$

where the extrapolated length z_b is related to the transport mean free path l_t by:

$$z_b \approx \frac{2}{3}l_t \approx 2D.\tag{3}$$

The boundary conditions at the interface between the two scattering media are [6-8]:

$$\begin{cases} D_1\frac{\partial}{\partial z}\varphi(z = 0^+) = D_2\frac{\partial}{\partial z}\varphi(z = 0^-) \\ \\ n_2^2\varphi(z = 0^+) = n_1^2\varphi(z = 0^-). \end{cases}\tag{4}$$

This problem can easily be solved after a time and transverse-space Fourier transform [6]. But, due to the high dynamic of the reflectance function, the numerical inverse Fourier transform is numerically difficult to perform and is time consuming. We therefore decided to work directly in the real space. The case of a semi-infinite medium satisfying boundary condition (2) can be solved with the method of images: One can replace the semi-infinite medium by an infinite one, with the same optical properties, but containing an additional negative image source [9, 10], linked to $S(\overrightarrow{r},t)$ by:

$$S_{im}(\overrightarrow{r},t) = \widehat{R}_f S(\overrightarrow{r},t) = -S(\overrightarrow{\rho},2(d + z_b) - z,t),\tag{5}$$

where $\overrightarrow{\rho}$ is the transverse space coordinates. The operator $\widehat{R}_f$ allows us to associate to any source distribution the added image source distribution. We in fact extend this method to the case of an interface between two scattering media, where the boundary conditions are given by Eq. (4).

We indeed proved [11] that to solve such a problem for $z > 0$ (with an interface at $z = 0$ and a source in the half plane $z > 0$), one can replace the two semi-infinite media by an infinite one having the same properties as the medium 1, but containing, in addition to S, an image source distribution defined by (Fig. 4):

$$S_{im}(\overrightarrow{r}, t) = \widehat{R}_m S(\overrightarrow{r}, t) =$$

$$\int d^2\rho' dz' d\tau S_R(\overrightarrow{\rho} - \overrightarrow{\rho}', z, z', t - \tau) S(\overrightarrow{\rho}', z', \tau),$$

(6)

where

$$S_R(\overrightarrow{\rho}, z, z', t) =$$

$$\delta(\overrightarrow{\rho})\delta(z + z')\delta(t) - 4D_1 \frac{n_1^2}{n_2^2}\theta(-z - z')\frac{\partial^2}{\partial z^2}G_2(\overrightarrow{\rho}, \frac{D_2 n_2^2}{D_1 n_1^2}(z + z'), t)$$

(7)

and G_2 is the Green's function for the medium 2 considered as infinite.

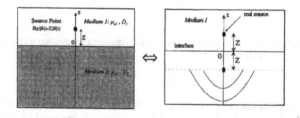

Figure 4.   The interface between two scattering media can be replaced by adding a specific source distribution.

This property can be used in the case of multiple layers through the iteration of such operators. For example, in the case of the medium described in Fig. 1, we replace the two layers and the air by an infinite medium with the following source distribution S_{equ}:

$$S_{equ} = [1 + (1 + \widehat{R}_f)\sum_{n=1}^{\infty}(\widehat{R}_f\widehat{R}_m)^n](1 + \widehat{R}_f)S.$$

(8)

In Eq. (8), the integer n corresponds to the number of light round-trips between the two interfaces. This source distribution is represented on Fig. 5, where the contributions of each boundary to the signal can be isolated. In a way, the inverse problem, which consists in the determination of the geometry of the medium, can here be solved through the determination of a source distribution

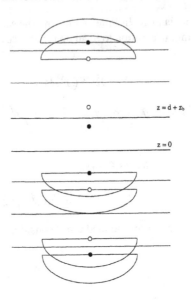

Figure 5.   In the case of a two-layered media, one can consider an equivalent source distribution in an infinite scattering medium.

in an infinite medium. The advantage of this method is to allow very efficient numerical processing. For instance, a 256×256 pixels time and space-resolved reflectance map can be obtained in less than one second computation time with a standard Pentium II $300MHz$. We have used this direct problem algorithm to test an optimization procedure for the inverse problem. We present here a result obtained from a Monte Carlo simulation of a two-layered media. The parameters of the simulation are given in Table 4.1. We used a Henyey-Greenstein phase function with an anisotropy factor $g = 0.8$ and index matching. We used a standard optimization procedure, where the contributions of each spatial position to the χ^2 error function were normalized. The fitting zone was ranging from 1 cm to 3.5 cm in the space domain, and from 0 to 2.5 ns in the time domain. As an initial guess we started with a semi-infinite medium ($\mu_{a0} = 0.1$ cm^{-1}, $\mu'_{s0} = 15$ cm^{-1}) and a layer thickness of 10 mm. We fitted 5 parameters, plus the absolute amplitude of the map. The optimization took about 700 steps in less than 2 minutes. The results are given in Table 4.1.

In conclusion, this generalization of the method of images to interfaces between scattering media leads to a fast and accurate algorithm for the direct problem calculation; it is then possible to solve the complete inverse problem for a two-layered medium, by using the whole reflectance map. The potential of

Table 4.1 . Parameters used for the simulation versus fitted ones.

Parameter	Simulation	Fit
	Upper layer	
μ_s' (cm^{-1})	10	9.9
μ_a (cm^{-1})	0.026	0.03
Thickness (mm)	6	6.01
	Lower layer	
μ_s' (cm^{-1})	30	30.75
μ_a (cm^{-1})	0.05	0.051

this tool in the case of multi-layered media has now to be explored, as well as the possibility of a more efficient inversion procedure based on the reconstruction of the equivalent source distribution.

References

[1] E. Tinet, S. Avrillier, J.-M. Tualle and J.-P. Ollivier, J. Opt. Soc. Am. A **13**, 1903–1915 (1996).

[2] M. Mochi, G. Pacelli, M.C. Recchioni and F. Zirilli, Journal of Optimization Theory and Applications, **100**, 29–57 (1999).

[3] M.S. Patterson, B. Chance, B.C Wilson, Appl. Opt. **28**, 2331–2336 (1989).

[4] R.C. Haskell, L.O. Svaasand, T.-T. Tsay ,T.-C. Feng, M.S. McAdams and B.J. Tromberg, J. Opt. Soc. Am. A **11**, 2727–2741 (1994).

[5] E. Amic, J.M. Luck, Th.M. Nieuwenhuizen, J. Phys. A **29**, 4915–55 (1996).

[6] A. Kienle, T. Glanzmann, G. Wagnières and H. van den Bergh, Applied Optics, **37**, 6852–6862 (1998).

[7] R. Aronson, J. Opt. Soc. Am. A **12**, 2532–2539 (1995).

[8] S.E. Skipetrov and R. Maynard, Phys. Lett. A **217**, 181–185 (1996).

[9] E. Akkermans, P.E. Wolf, and R. Maynard, Phys. Rev. Lett., **56**, 1471–1474 (1986).

[10] B. Davison and J.B. Sykes, *Neutron transport theory* (New York, 1957).

[11] J.-M. Tualle, E. Tinet, J. Prat, B. Gélébart and S. Avrillier, Proceeding of CLEO/Europe-EQEC, Adv. in Opt. Imaging ..., 192–194 (1999).

Chapter 5

SCATTERING
ON MULTI-SCALE ROUGH SURFACES

C. A. Guérin

Laboratoire d'Optique Electromagnétique, UPRESA 6079,
Faculté de Saint-Jérôme, case 262,
F-13397 Marseille cedex 20, FRANCE
caguerin@loe.u-3mrs.fr

M. Saillard

Laboratoire d'Optique Electromagnétique, UPRESA 6079,
Faculté de Saint-Jérôme, case 262,
F-13397 Marseille cedex 20, FRANCE
marc@loe.u-3mrs.fr

M. Holschneider

Geosciences Rennes,
Campus de Beaulieu, Bat. 15,
F-35042 Rennes, FRANCE
Matthias.Holschneider@univ-rennes1.fr

Abstract   We present a method to recover a fractal dimension of a multi-scale rough surface, the so-called correlation dimension, from the knowledge of the far-field scattered intensity. The results are validated by numerical experiments on Weierstrass-like surfaces.

Keywords: multi-scale surface, correlation dimension, Weierstrass profile.

P. Sebbah (ed.), Waves and Imaging through Complex Media, 339–346.
© 2001 *Kluwer Academic Publishers. Printed in the Netherlands.*

Introduction

Although simple, fractal models have shown to be of particular relevance for the description of natural rough surfaces. An important issue in remote sensing is therefore to understand the impact of fractal characteristics on electromagnetic wave scattering. Recently, several simple models such as Weierstrass functions of fractional Brownian motion have been proposed to describe multiscale rough surfaces and the diffracted field has been studied by means of the usual approximations (Kirchhoff approach [1-7], the Extended Boundary Condition Method [8],[9] or Integral Equation Method [10]). Some qualitative results relating the scattering amplitude to fractal dimensions of the surface have been exhibited but failed to give a precise and general way of computing the latter quantities.

In a previous paper, the authors proposed a method to compute at least one fractal dimension (the so-called correlation dimension) of a rough surface by the sole knowledge of the far-field intensity. The technique relies mainly on the small-perturbation method and its simple connection to Fourier analysis. We will here give an outline of this method and provide an illustration on the simple example of band-limited Weierstrass functions.

1.   DESCRIPTION OF THE SURFACE

Very often the rough surfaces are modeled by stationary random processes with given covariance and distribution heights. We will, however, work in a deterministic setting for the sake of simplicity. The spirit of the method is essentially the same in the random case, but one has to cope with additional complications pertaining to estimation theory. Although the deterministic description might be less realistic, it provides a good starting point to experiment the method.

Let us consider an infinite surface, invariant along the z direction, whose height about some reference plane is given by $y = f(x)$. The profile will be set to zero outside some interval of interest, say $f(x) = 0$ for $|x| > L$, and of zero mean inside, $\int f = 0$. Note that there is no restriction in supposing the range of interest of finite extent in the scattering problem as long as the latter remains significantly greater than the illuminated region.

The two-points correlation function

$$C(\tau) = \frac{1}{2L} \int_{-L}^{+L} dx f(x) f(\tau + x)$$

describes the spatial variation of heights. The root mean square (RMS) height $\sigma = \sqrt{C(0)}$ quantifies the roughness of the surface.

Usually, the fractal nature of a profile is expressed in terms of its Hausdorff-Besicovitch dimension (e.g. [11]) or simply box counting dimension, which is obtained in the following way. Given a function $f(x)$, count how many boxes or balls of radius ϵ are necessary to cover its graph. Calling $N(\epsilon)$ the number thus obtained for the optimal covering, the Hausdorff-Besicovitch dimension D_{HB} is the exponent governing the growth of $N(\epsilon)$ at small scales:

$$N(\epsilon) \sim \epsilon^{-D_{HB}}, \ \epsilon \to 0.$$

The dimension D_{HB} can be seen as an indicator of the roughness of the surface. At the two extremes we have $D_{HB} = 1$ for smooth curves and $D_{HB} = 2$ for curves that fill the plane. A curve with $D_{HB} > 1$ is non-differentiable. If it is, however, Hölder continuous with some Hölder exponent β (i.e if $|f(x + \epsilon) - f(x)| \leq C|\epsilon|^{\beta}$ for some positive constants $C, \beta > 0$), then $D_{HB} \leq 2 - \beta$. Now there is a close relationship between the fractality of a graph as described by the Hausdorff-Besicovitch dimension and the small-scale behavior of its correlation function. Suppose the following asymptotic behavior holds for all fixed τ and for some $0 < d < 2$:

$$C(\tau + h) - C(\tau) \sim h^{d}, \ h \to 0.$$

Then it is easy to show that $D_{HB} \leq 2 - \frac{d}{2}$ (in the random case where $C(\tau) = \ <f(x + \tau)f(x)> \ $ is a "true" correlation function we have even the stronger statement $D_{HB} = 2 - \frac{d}{2}$; see [12]). Hence the fractality of a profile can be quantified by the regularity of its correlation function at small scale. In fact, the exponent d has been shown [13] to coincide with one of the so-called generalized wavelet dimensions [14] constructed by means of the continuous wavelet transform: the wavelet correlation dimension, to which we will simply refer as the correlation dimension.

The advantage of the correlation dimension is that it can be computed very easily using quadratic averages of the Fourier transform. If an asymptotic scaling of the type

$$\int_{\alpha}^{\lambda \alpha} |\hat{f}(\alpha')|^2 d\alpha' \sim \alpha^{-\gamma}, \ \alpha \to \infty \tag{1}$$

is observed for some fixed $\lambda > 1$, then

$$\gamma = d, \tag{2}$$

as was shown in [13]. In practice the surfaces of interest are never true fractals, since they must be at least differentiable for the electromagnetic boundary conditions to be well-defined. However, the scaling (1) will hold in some range of scales and the governing exponent will be identified with the correlation dimension.

2.   THE ELECTROMAGNETIC SCATTERING PROBLEM

Now let us come to the physical problem. Suppose the surface described above separates vacuum from a perfectly conducting metal. The variable parameter here will be the maximum amplitude of the surface, so it is convenient to introduce the normalized function $h(x)$ and the RMS height σ defined by $f(x) = \sigma\, h(x)$. An s polarized beam, with frequency ω and harmonic time dependence $e^{-i\omega t}$ is impinging from the top. In the low frequency limit, a rough surface scattering problem can be solved with the help of the small perturbation method. Up to first order in σ, it is shown that the scattering amplitude at infinity can be expressed in terms of the Fourier transform of the profile [15]. Indeed, denoting by α and α' the x components of the wave-vectors of the incident and scattered plane waves, respectively, the scattering amplitude $r(\alpha, \alpha')$ writes

$$r(\alpha, \alpha') = -\delta(\alpha' - \alpha) + i\,\frac{\beta(\alpha)}{\pi}\,\sigma\,\hat{h}(\alpha' - \alpha) + o(\sigma), \qquad (3)$$

with $\beta = \sqrt{k^2 - \alpha^2}$ if $|\alpha| \le k$ or $\beta = i\sqrt{\alpha^2 - k^2}$ if $|\alpha| > k$, and where δ denotes Dirac's distribution.
The problem and the notations are depicted on Fig. 1.

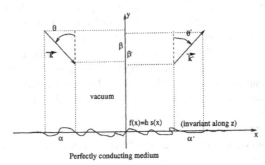

Figure 1.   Description of the problem and notations

We will use formulae (1) and (2) to retrieve the correlation dimension of the surface from the scattering pattern. Numerical application will be performed on the so-called band-limited Weierstrass function:

$$h_N(x) = \frac{1}{\sigma_N} \sum_{n=1}^{N} a^n \cos\left(2\pi b^n x + \varphi_n\right). \qquad (4)$$

Here a, b are two real positive numbers such that $a < 1$, $b > 1$ and $ab > 1$, and σ_N is chosen in such a way that $< h_N^2 > = 1$. The set $(\varphi_n)_n$ is a set of specified (arbitrary) phases. In some frequency range limited by the highest spatial frequency of h_N, the asymptotic of the Fourier transform of h_N is governed by the correlation dimension d of the true Weierstrass function h_∞ (that is the untroncated series). A simple computation gives

$$d = -2\frac{\log a}{\log b}. \tag{5}$$

Let us now consider an isotropic source, the so-called wire source in bidimensional problems, located at $S(0, y_S)$, $y_S > 0$. The incident electric field writes, for $y < y_S$,

$$E^i(x, y) = E_0 \, H_0^{(1)}(k\sqrt{x^2 + (y - y_S)^2}) = E_0 \int_{-\infty}^{\infty} \frac{d\alpha}{\beta} \, e^{i\alpha x - i\beta(y - y_S)}, \tag{6}$$

E_0 being a complex constant, and $H_0^{(1)}$ the zero order Hankel function of the first kind. Assuming the source is far from the surface ($ky_S \gg 1$), the integral can be restricted to propagating waves. Therefore, the incident field writes as a superposition of plane waves where each incidence angle θ is present, with amplitude $1/k \cos \theta$. Focusing on the backscattered field $E^b(S)$, we derive in the frame of the small perturbation method:

$$\begin{aligned} E^b(S) &\simeq E_0 \int_{-k}^{k} d\alpha \, \frac{r(\alpha, -\alpha)}{\beta} \, e^{2i\beta y_S} \\ &\simeq E_0 \left(-e^{2iky_S} + \frac{i}{\pi} \int_{-k}^{k} d\alpha \, \sigma \hat{h}(-2\alpha) \, e^{2i\beta y_S} \right). \end{aligned} \tag{7}$$

The first term results from the specular reflection under normal incidence. Generally, remote sensing experiments avoid normal or near-normal incidences, because of the lack of information contained in the echo. In the same way, mathematically, such a term with constant modulus prevents any informative asymptotic behavior of the backscattered field. Therefore, in the following, we consider a wire source with some mask which stops the waves radiated around normal incidence. Denoting by $[-\theta_m, \theta_m]$ the darkened angular interval, and assuming that the contributions from the various incidence angles are not correlated in the far field, we obtain

$$I^b(k) = |E^b(S)|^2 = 2 \left| \frac{E_0}{\pi} \right|^2 \int_{k \sin \theta_m}^{k} d\alpha \, |\sigma \hat{h}(-2\alpha)|^2. \tag{8}$$

According to (2), $I^b(k)$ behaves as k^{-d}. Consequently, the correlation dimension can be deduced from a few measurements.

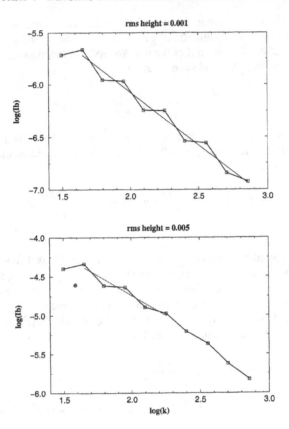

Figure 2.   Back-scattered intensity from a Weierstrass surface for increasing RMS.

2.1.   THE RIGOROUS SOLUTION

The rigorous solution of the scattering problem is achieved thanks to a classical boundary integral method, described in [16], coupled with a boundary finite-element method. We consider a band-limited Weierstrass function with $b = \sqrt{\pi}$, $a = 1/\sqrt{b}$ and $N = 9$. The ratio between the highest and the lowest spatial frequency is thus close to 100, and the correlation dimension $\kappa_2 = 1$. As a check, we have first computed the backscattered efficiencies for $\sigma = 10^{-3}$. With such a RMS height, when the wavenumber k increases from 2π to $2\pi b^8$, the ratio λ/σ decreases from 1000 to 20. In this range, the perturbation theory is supposed to give reliable results. As expected, a linear regression obtained from a $\log - \log$ plot with nine values of the frequency gives a slope $\gamma = -1.01$ with a correlation coefficient $\rho = 0.99$, whereas the theoretical value is $\gamma = -\kappa_2 = -1$ [13]. Then, we have multiplied the RMS height by a factor of five, $\sigma = 5.10^{-3}$. Performing the linear regression on increasing ranges $\lambda/\sigma < 10$, $\lambda/\sigma < 4$ and $\lambda/\sigma < 1.2$, respectively, we obtain $\gamma = -1.03$, $\gamma = -1.08$ and $\gamma = -1.19$ respectively. This shows that the estimated dimension departs only slowly from the real one while increasing the ratio λ/σ, that is while leaving the domain of validity of the small perturbation method. The results are displayed on Fig. 2.

Conclusion

Relying upon the small-perturbation approximation, we have been able to recover in a simple way a fractal dimension of a rough surface. In addition, numerical experiments performed on the simple example of Weierstrass functions suggest that the method remains satisfactory far beyond the domain of validity of the small-perturbation. This is, however, only a first step. The correlation dimension alone is not sufficient to identify completely fractal surfaces, and a challenging problem for the future is to be able to recover the whole set of generalized (wavelet) fractal dimensions.

References

[1] M. Berry, J.Phys. A: Math. Gen. **14**, 3101 (1981).

[2] Y. Agnon and M. Stiassnie, J. Geophys. Res. **96**, 12773 (1991).

[3] P. Rouvier, S. Borderies and I. Chenerie, Radio Science **32**, 285 (1997).

[4] D. L. Jaggard and X. Sun, J. Opt. Soc. of Am. **7**, 1131 (1990).

[5] G. Franceschetti, A. Iodice, M. Migliaccio, and D. Riccio, IEEE Trans. Antennas and Propagation **47**, 1405 (1999).

[6] G. Franceschetti, A. Iodice, M. Migliaccio, and D. Riccio, Radio Sci. **31**, 1749 (1996).

[7] F. Berizzi and E. Dalle-Mese, IEEE Trans. Antennas and Propagation **47**, 324 (1999).

[8] P. Savaidis, S. Frangos, D. L. Jaggard, and K. Hizanidis, J. Opt. Soc. of Am. **14**, 475 (1997).

[9] S. Savaidis, P. Frangos, D. L. Jaggard, and K. Hizanidis, Opt. Lett. **20**, 2357 (1995).

[10] F. Mattia, Journal of Electromagnetic waves and applications **13**, 493 (1999).

[11] K. Falconer, *The geometry of fractal sets*, Cambridge tracts in Mathematics 85 (Cambridge University Press, Cambridge, 1985).

[12] R. Adler, *The geometry of random fields* (Wiley, New-York, 1981).

[13] C. Guérin, M. Holschneider, and M. Saillard, Waves in Random Media **7**, 331 (1997).

[14] M. Holschneider, Comm. Math. Phys. **160**, 457 (1994).

[15] D. Maystre, O. M. Mendez, and A. Roger, Optica Acta **30**, 1707 (1983).

[16] M. Saillard and D. Maystre, J. Opt. Soc. of Am. **7**, 331 (1990).

VI

DYNAMIC MULTIPLE LIGHT SCATTERING

DYNAMIC MULTIPATH CH sca TTERING

Chapter 1

IMAGING OF DYNAMIC HETEROGENEITIES IN MULTIPLE LIGHT SCATTERING

G. Maret

Fachbereich Physik, Universität Konstanz
Postfach 5560, D-78457 Konstanz, Germany, phone ++49 7531 884151
Georg.Maret@uni-konstanz.de

M. Heckmeier

Merck KGaA, Liquid Crystals Division
D-64271 Darmstadt, Germany, phone ++49 6151 727687
Michael.Heckmeier@Merck.de

Abstract   Illumination of strongly scattering media with coherent light generates speckle patterns. This article provides a tutorial overview how from measurements of time and position dependent laser speckle intensity fluctuations information about inhomogeneous motions of scatterers inside the medium can be obtained. This principle of low resolution imaging of dynamic heterogeneities is described and illustrated by a series of experiments on turbidity matched inclusions buried inside media with different dynamics. This reveals both the potential and the limits of the method. Areas of possible applications are briefly discussed.

Keywords: multiple light scattering, speckle fluctuations, diffusing wave spectroscopy, flow imaging, turbid suspensions

Introduction

Most objects in nature strongly scatter visible light: white paints, milk, many dairy products, snow or clouds are examples where the scattering largely dominates over the absorption of light, while many biological tissues, wood, rocks, mud etc ... , have substantial absorption in addition to scattering. In the former class of materials almost all scattered light escapes back into free space and, therefore, the angular integrated reflectivity is close to unity, these

P. Sebbah (ed.), Waves and Imaging through Complex Media, 349–367.
© 2001 *Kluwer Academic Publishers. Printed in the Netherlands.*

materials look white. The latter have of course (much) smaller reflectivity since a significant amount of light is absorbed at least in some parts of the visible spectrum. Consequently these substances look colored or grey. The strong scattering of light is caused, usually, by a high concentration of inhomogeneities such as colloidal particles suspended in a background medium having a rather different refractive index. The fat droplets in milk, the water droplets in clouds or the grains in sand are typical examples. Light scattering is strongest for large index mismatch combined with particles sizes of order of the wavelength of light, but the most essential requirement for the turbid appearance is that the piece of material has dimensions L larger than the typical distance over which light propagates inside it without significant change in direction due to scattering. This distance is usually called the photon transport mean free path l^*. The criterion for strong multiple scattering is thus $L \gg l^*$. In this limit the transport of light is well described by diffusive spreading of the light intensity, or photon diffusion, or photon random walks. Individual photons are scattered along many random scattering paths and their distribution in space as well as in length depend on the size and shape of the multiple scattering object and, in addition, on the distribution of the light sources at the surface of the object.

Although the photon diffusion picture was invented almost 100 years ago to describe light propagation in the interior of the sun [1] - and successfully used ever since -, its most severe deficiency is to neglect wave interference effects. Indeed, inspection of the light of a laser pointer scattered from a sheet of paper reveals strong intensity fluctuations when the angle of observation is varied, in sharp contrast to the very smooth angular variation expected from the photon diffusion model. This irregular pattern called 'speckle' is the result of interferences between the light waves scattered along the various multiple scattering paths. One may think of the many scattering particles along a given scattering path as a series of tiny mirrors which built up a tortuous branch of an interferometer, and many of these branches are brought to interference outside the sample. This makes clear the granular appearance of the speckle pattern as well as the fact that speckles appear as long as the lengths of essentially all scattering paths are shorter than the longitudinal coherence length of the light source. In fact, the rapid development of the field of optical multiple scattering has set in with the discovery of distinct interference effects such as weak localization [2], coherent backscattering [3, 4] and dynamic speckle fluctuations [5] of light. We concentrate here on the *temporal* fluctuations of the multiply scattered speckle patterns which are caused by motion of scatterers. While parts of the underlying physics was discussed already in 1983/84 [6, 7], dynamic multiple light scattering has been introduced and made a quantitative tool by the work on calibrated colloidal latex particles in aqueous suspensions

[5, 8]. It has rapidly evolved into a powerful technique called Diffusing Wave Spectroscopy, (DWS) {9, 10].

In order to understand the principle of DWS, realize that a speckle pattern is a (complex) interferometric fingerprint of the positions of all scattering particles illuminated by the diffusing light. A change of the scatterers relative positions goes along with a change in the relative phases of the different light waves and, consequently, the intensity of the speckle spots changes, pretty much like in standard quasi-elastic single light scattering (QELS)[11]. However, in contrast to QELS, because the light is scattered *many* times on average before leaving the sample, the light waves traveling along multiple scattering paths pick up much larger phase shifts in a given time interval, as the phase shifts induced by all particle motions add up along the paths. Therefore dynamic speckles generated by multiple scattering fluctuate much faster than corresponding single scattering speckles.

Like in any diffusion problem, the spatial distribution of the diffusing flux density depends on the distribution of sources and of scatterers. Therefore, a non-homogeneous distribution of scatterers over space inside the scattering material - in the sense of a position-dependent transport mean free path - results in a different light flux distribution as compared to the homogeneous case. Thus, the spatial distribution of light intensity at the surface of the medium will be correspondingly modified by the presence of the inhomogeneity. Imagine, for example, a turbid medium surrounding a spatially localized region with smaller/larger transport mean free path where the multiple scattered light intensity is observed in transmission. The surface area of the medium closest to this region will emit, respectively, less/more light than surface areas further away. This means the light intensity profile at the surface contains information about the location of the object, even if the object is located many mean free paths inside the sample. The same holds, if the localized region (the 'object') contains more/less absorption, respectively, than its environment (the 'medium'). Obviously, when the distance from the observed surface of the medium to the object becomes larger, or the contrast of the object smaller - both in absorption or scattering -, the change in the surface intensity becomes smaller. The larger the distance of the object from the observed surface, the larger the area of modified luminosity of the surface will be since the object may be roughly considered as a localized sink or source, respectively, of diffusing light intensity. The spatial distribution of the light intensity emerging from the surface of the medium thus contains not only information about the location but also about the contrast of the object. This principle of low resolution imaging has been quantitatively studied by measuring static surface intensity profiles on samples consisting of absorbing and transparent rods embedded inside a multiple scattering medium [12].

In this article, we utilize the very same physical idea to localize and in principle image objects buried inside a multiple scattering medium, but instead of using contrast in the average diffusing intensity we exploit *dynamical* contrast in the speckle pattern. Dynamic contrast is generated by a motion of the scattering particles inside the object which differs from the motion of particles outside the object. In DWS, as discussed above, the time scale of fluctuations of the speckle intensity depends both on the path length and on the velocity of scattering particles. By using 'turbidity matched' objects, i.e. objects which have the same transport mean free path than the medium, we make sure that the spatial path length distribution is the same than in the homogeneous medium (without the object)[1]. To understand the principle of "dynamic imaging" consider a multiple scattering path that crosses the object in comparison to another path *of the same length* that does not. The phases of the fields of both paths will fluctuate at different time scales despite of their identical length, since the scattering particles inside and outside induce different phase shifts. Thus, in their time dependence, the speckle fluctuations outside the medium contain information about the object. Like in the static imaging above, the larger the fraction of the diffusing photons crossing the object or, equivalently, the larger the number of paths intersecting the inclusion, the larger will be the (dynamic) signature of the object in the speckle fluctuations. Measuring the *time*-dependence of the speckle fluctuations as a function of the position at the surface of the medium thus provides a dynamic low resolution image.

1.   DIFFUSING WAVE SPECTROSCOPY OF HOMOGENEOUS MEDIA

The theory of DWS as originally developed [5] in the framework of the photon diffusion picture can be found in tutorial reviews (e.g. [9, 10]). We therefore only briefly resume here the essential steps.

In a dynamic light scattering experiment the time dependent intensity $I(\mathbf{r}, t)$ scattered into one speckle spot is measured. Experimentally this is essentially done by imaging a small area (at position r, diameter d) of the sample surface either onto a pinhole or onto a mono-mode optical fiber positioned in front of a photomultiplier tube. The pinhole size is matched to the angular size λ/d of a speckle spot. An electronic correlator evaluates the time autocorrelation function $G(\mathbf{r}, t)$ of $I(\mathbf{r}, t)$

$$G(\mathbf{r},t) = \frac{< I(\mathbf{r},0)I(\mathbf{r},t) >}{< I(\mathbf{r},t) >^2} = \frac{< E(\mathbf{r},0)E^*(\mathbf{r},0)E(\mathbf{r},t)E^*(\mathbf{r},t) >}{< (E(\mathbf{r},t)E^*(\mathbf{r},t))^2 >}. \quad (1)$$

---

[1]This is not a necessary restriction but it simplifies the analysis and makes the physics of dynamic imaging particularly simple.

$E(\mathbf{r}, t)$ denotes the total scattered light field and $< \,..\, >$ averaging over many configurations of the positions of the scatterers, which - in an ergodic system such as a colloidal suspension undergoing for instance Brownian motion - is equivalent to a t-average. Since $E(\mathbf{r}, t)$ is given as the sum over the fields $E_i(\mathbf{r}, t)$ scattered along the different multiple scattering paths $E(\mathbf{r}, t) = \sum_i^\infty E_i(\mathbf{r}, t)$ the intensity correlation function $G(\mathbf{r}, t)$ is a four field correlation function containing mixed contributions from all paths. In the case of random and non-correlated scattering paths, the $E_i(\mathbf{r}, t)$ are independent Gaussian variables, which tremendously simplifies $G(\mathbf{r}, t)$, since all mixed terms $E_i E_j$ vanish on average. Only the auto-correlations of fields $E_i(\mathbf{r}, 0) E_i^*(\mathbf{r}, t)$ scattered along individual paths (index i) contribute to $G(\mathbf{r}, t)$. One obtains

$$G(\mathbf{r}, t) = 1 + |g(\mathbf{r}, t)|^2 , \tag{2}$$

$$g(\mathbf{r}, t) = \frac{< E(\mathbf{r}, 0) E^*(\mathbf{r}, t) >}{< |E(\mathbf{r}, t)|^2 >} = \sum_{n=1}^{\infty} P(\mathbf{r}, n) < e^{i\Delta\phi_n(t)} > , \tag{3}$$

where $P(\mathbf{r}, n)$ is the fraction of the light intensity scattered into paths containing n scattering events and $\Delta\phi_n(t)$ is the phase difference between $E_n(\mathbf{r}, 0)$ and $E_n(\mathbf{r}, t)$. $\Delta\phi_n(t)$ is the sum of all phase shifts $\delta\phi_\nu(t)$ due to the displacements of all n scatterers in a given path, $\Delta\phi_n(t) = \sum_{\nu=1}^{n} \delta\phi_\nu(t) = \mathbf{k}_0\mathbf{r}_1 + \mathbf{k}_1\mathbf{r}_2 - \mathbf{k}_1\mathbf{r}_1 + \mathbf{k}_2\mathbf{r}_3 - \mathbf{k}_2\mathbf{r}_2 + ... + \mathbf{k}_{n-1}\mathbf{r}_n - \mathbf{k}_{n-1}\mathbf{r}_{n-1}$. Here $\mathbf{r}_i$ denotes the position of scatterer i and $\mathbf{k}_i$ the wave vector of light propagating from scatterer i to scatterer $i+1$. For random scattering paths $\Delta\phi_n(t)$ has a Gaussian distribution and one obtains

$$g(\mathbf{r}, t) = \sum_{n=1}^{\infty} P(\mathbf{r}, n) e^{-<\Delta\phi_n^2(t)/6>}. \tag{4}$$

If the phase shifts along the scattering paths are uncorrelated[2] and n is large we have - as a result of the central limit theorem - simply $< \Delta\phi_n^2(t) > = n < \delta\phi_\nu^2(t) >$. Finally, because under typical circumstances of strong multiple scattering $(L \gg l^*)$ many paths are much longer than l^* or, equivalently, $n \gg 1$ we can replace the above sum by an integral:

$$g(\mathbf{r}, t) = \int_{n=1}^{\infty} dn \, P(\mathbf{r}, n) \, e^{-n<\delta\phi_\nu^2(t)>/6}. \tag{5}$$

---

[2] for example because of uncorrelated motions of the scatterers as in the case of simple Brownian motion.

This is the central equation of DWS, which provides a quantitative description of all features of DWS discussed in the introduction. $\delta\phi_\nu^2(t)$ measures how fast individual scatterers move, it sets the time scale of the average single scattering contribution to $g(\mathbf{r}, t)$. The factor n in the exponential - which is proportional to the path length[3] - expresses the fact that the time scale of the speckle fluctuations decreases with the order n of multiple scattering; Multiple scattering accelerates speckle fluctuations and, hence, DWS is sensitive to much smaller displacements of particles than QELS. $P(\mathbf{r}, n)$ contains the sensitivity of $g(\mathbf{r}, t)$ to the sample dimensions and geometry. Explicit formulas for $P(\mathbf{r}, n)$ have been worked out for various geometries such as backscattering and transmission from slabs, pairs of optical fibers dipping into a turbid sample and others by solving the diffusion equation under appropriate boundary conditions.

1.1.   EXAMPLE: BROWNIAN MOTION

In order to use DWS to study the displacements of the scatterers, $< \delta\phi_\nu^2(t) >$ has to be evaluated for different types of motion. Let us first briefly discuss Brownian motion of non-interacting particles, which is characterized by a mean square displacement $< \Delta r(t)^2 >= 6Dt$ of the particles corresponding to a 3 dimensional random walk. Here D is the diffusion constant of the particles. In a single scattering QELS-experiment, motion by $\Delta\mathbf{r}$ gives rise to speckle fluctuations which are described by the time autocorrelation function $< E(0)E^*(t) >=< \exp(-i\mathbf{q}\Delta\mathbf{r}(t)) >$ of the scattered field [11]. $\mathbf{q}$ denotes the scattering vector given by the difference between the wave vectors of the scattered and incident waves $\mathbf{q} = \mathbf{k}_s - \mathbf{k}_0$. We therefore have an exponential decay of the field autocorrelation function $< E(0)E^*(t) >= \exp(-q^2 Dt)$ with a characteristic time scale $\tau = 1/(Dq^2)$. q is a well controlled quantity in a single scattering experiment, while in multiple scattering it is not, because of the random tortuous configurations of the multiple scattering paths. $\Delta\phi_n(t)$ is now written as a sum over the phase factors appearing at each scattering event, which fluctuate independently since the scatterers positions $\mathbf{r}_\nu(t)$ are independent. Therefore

$$< e^{i\Delta\phi_n(t)} >=< e^{i\sum_{\nu=0}^n \mathbf{q}_\nu \mathbf{r}_\nu(t)} >= \Pi_{\nu=0}^n < e^{i\mathbf{q}_\nu \mathbf{r}_\nu(t)} > . \qquad (6)$$

Evaluation of $< \delta\phi_\nu^2(t) >=< q^2 6Dt >$ reduces to an evaluation of $< q^2 >$, where the average has to be made by properly weighting how much of the light is scattered into a given direction in each single scattering event. In the case of non-interacting scatterers considered here, $< .. >$ denotes an average over

---

[3] $n = s/l$ with s the contour length of the path and l the average distance between scattering events usually called the scattering mean free path.

the angular (or q) dependent scattering form factor. The form factor strongly depends on the particle size; While small particles essentially scatter equally into all directions (Rayleigh scattering), larger particles scatter more into small scattering angles θ (Mie scattering). As a result, the transport mean free path $l^* = l/(1- < \cos\theta >)$ becomes larger than the scattering mean free path l for larger particles. From the relation between l and l^* and the definition of $q = 2k_0 \sin(\theta/2)$ one immediately obtains $< q^2 > = 2k_0^2 l/l^*$.

Thus, we have $< \delta\phi_\nu^2(t) > = 12\,(t/\tau_0)\,(l/l^*)$ where the characteristic time $\tau_0 = 1/(Dk_0^2)$ was introduced. Inserting this result into Eq. (5) and replacing n by s/l, s being the contour length of the path, gives the DWS autocorrelation function for Brownian motion.

$$g(\mathbf{r}, t) = \int_{s=l^*}^{\infty} P(\mathbf{r}, s)\, e^{-2t/\tau_0\, s/l^*}\, ds. \tag{7}$$

One now sees very clearly two important points: (1) the time scale $\tau_0 l^*/s$ associated with paths of length s in the DWS correlation function is inversely proportional to s. (2) The path length distribution $P(\mathbf{r}, s)$ is controlled by the sample geometry: In backscattering from a very thick slab $P(\mathbf{r}, s)$ is very broad and decays at large s asymptotically like $s^{-3/2}$ [9, 10]. This results in a nearly exponential decay of $g(\mathbf{r}, t)$ with $\sqrt{t/\tau_0}$

$$g(\mathbf{r}, t) = e^{-\gamma\sqrt{6t/\tau_0}}, \tag{8}$$

with γ being a coefficient of order two related to the description of the light propagation near the surface of the sample. In transmission through a slab of thickness L the path length distribution $P(\mathbf{r}, s)$ is peaked at $s \approx L^2/l^*$ giving a nearly exponential decay of $g(\mathbf{r}, t)$ at time scale $\tau_0 l^{*2}/L^2$. These relations were critically tested experimentally on suspensions of well characterized colloidal suspensions, see e.g. [9, 10]. Points (1) and (2) above are also essential to understand how dynamic multiple scattering imaging works, as discussed below.

1.2.   EXAMPLE: SHEAR FLOW

Let us consider for a moment a system under homogeneous shear where the scattering particles are displaced relative to each other in a completely deterministic way given by the homogeneous gradient Γ of the velocity field. Note that this is not very realistic since in a real liquid under shear flow particles will undergo also Brownian motion in addition to the shear. In the case for pure shear it is useful to lump together the terms in $\Delta\phi_n(t)$ in a different way

$$< e^{i\Delta\phi_n(t)} > = < e^{i\sum_{\nu=1}^{n-1} \mathbf{k}_\nu(\mathbf{r}_{\nu+1}(t)-\mathbf{r}_\nu(t))} > = \Pi_{\nu=1}^{n-1} < e^{i\mathbf{k}_\nu(\mathbf{r}_{\nu+1}(t)-\mathbf{r}_\nu(t))} > .$$
$$(9)$$

While the distance vector between consecutive scattering events $\mathbf{r}_{\nu+1}(t) - \mathbf{r}_\nu(t)$ now evolves deterministically in time, the phase factors are still random variables because of the randomness of $\mathbf{k}_\nu$ (photon random walk)[4]. One obtains $< \delta\phi_\nu^2(t) > = (2t/\tau_s)^2$ with $\tau_s = \sqrt{30}/(k_0 l\Gamma)$ where the factor $\sqrt{30}$ originates from isotropic angular averaging [15]. τ_s is the characteristic time needed by a pair of scatterers initially separated by a distance l to move a relative distance of one wavelength λ due to the velocity gradient Γ. The quadratic scaling of $< \delta\phi_\nu^2(t) >$ with time is the signature of the deterministic motion under shear.

It was shown [15] that for the case of slowly[5] varying shear gradients the theory is only slightly modified in that τ_s then becomes $\tau_s = \sqrt{30}/(k_0 l\Gamma_{eff})$ where Γ_{eff} is the effective shear rate appropriately averaged over the spatial distribution of the density of diffusing photons. The DWS correlation function can thus be used to determine the shear rate in turbid flowing media [15]. Spatial resolution of the measurement of Γ can be obtained by a restriction of the extension of the photon cloud, for example by putting the locations of the entrance and exit of detected light closely together.

Under the assumption[6] that shear motion and Brownian motion do not affect each other, both contributions to the phase fluctuations discussed above simply add. The superposition of motions immediately implies that in the DWS autocorrelation functions the dimensionless time scale becomes $6t/\tau_0 + 6(t/\tau_s)^2$ instead of $6t/\tau_0$ for Brownian motion alone. Thus, under circumstances that Brownian motion and shear are homogeneous over the extension of the photon cloud, the decay of $g(\mathbf{r}, t)$ is always dominated by Brownian motion at times smaller than the cross-over time $\tau_c = \tau_s^2/\tau_0$, while at $t \gg \tau_c$ it merely probes the shear flow.

The above expressions have been quantitatively tested by measurements of $g(\mathbf{r}, t)$ in transmission through a slab containing planar Poiseuille flow [13] and in backscattering on planar shear flow [15]. These experiments demonstrated that DWS is a useful new tool to measure shear in a remote and non-invasive way. The planar shear was generated by a Couette cell consisting of two concentric cylinders with the inner one rotating. In the Couette-geometry the laminar planar shear flow becomes unstable above a critical rotation speed of the inner cylinder and breaks up into a flow pattern consisting of a set of pairs of counter rotating toroidal rolls (Taylor vortex flow). While in this flow pattern velocity

---

[4]The argument is given here for isotropic scattering ($l = l^*$) only!

[5]on the scale of l.

[6]which is usually well fulfilled in colloidal suspensions.

and shear are strongly inhomogeneous - reflecting the periodic axial arrangement of the Taylor rolls - the roll pattern in a concentrated turbid colloidal suspension cannot be observed by eye. Nevertheless when the impinging laser spot and the detection area were both reduced to a size substantially smaller than the diameter of the rolls the Γ-values extracted from the DWS signal showed large oscillations as a function of the distance along the cylindrical axis of the Couette cell. They directly reflect the Taylor roll pattern illustrating the imaging capability of space resolved DWS.

1.3.   EXAMPLE: PERIODIC LONGITUDINAL SHEAR

From the above discussion it becomes clear that any type of motion, where the relative positions of scatterers are changed (stochastically or deterministically) in time, gives rise to a signature in the DWS autocorrelation function. As a further example related to dynamic multiple scattering imaging let us mention longitudinal periodic shear as generated by ultrasonic waves propagating in a turbid medium. This motion can be quantitatively described along the lines outlined above for transverse shear [17]. In a solid material $g(\mathbf{r}, t)$ exhibits a periodic modulation at the ultrasound frequency with amplitude related to the acoustic amplitude. In a colloidal liquid this modulation is superimposed on the relaxation of $g(\mathbf{r}, t)$ due to Brownian motion and, hence, can be detected as long as the ultrasound frequency is higher than the inverse of the characteristic relaxation time. Experiments to test the theory were performed in transmission through slabs of colloidal latex suspensions using a correlator or Fabry-Perot Interferometer to detect ultrasound at about 2 MHz or 20 MHz, respectively [17].

1.4.   CORRELATION DIFFUSION MODEL

An alternative way - which is particularly powerful for inhomogeneous flow - to describe the field autocorrelation function $g(\mathbf{r}, t)$ in a strongly multiple scattering medium has been introduced by Boas *et al.* [18]. Like the intensity $E(\mathbf{r})E^*(\mathbf{r})$ in the photon diffusion model the unnormalized field correlation function $< E(\mathbf{r}, 0)E^*(\mathbf{r}, t) >$ obeys a diffusion equation inside the turbid medium. In other words, inside a stationary medium (no motion of scatterers) the field correlation diffuses in just the same way as the light intensity, while correlation is absorbed when scatterers undergo motion. For the case of homogeneous Brownian motion combined with homogeneous shear one easily sees [18, 19] that the absorption rate $\kappa^2(t)$ in the diffusion equation of $< E(\mathbf{r}, 0)E^*(\mathbf{r}, t >$ is given by

$$\kappa^2(t) = 6t/(\tau_0 l^2) + 6t^2/(\tau_s l)^2 \,. \tag{10}$$

Therefore the calculation of $g(\mathbf{r}, t)$ maps onto the calculation of the diffusing light intensity in presence of absorption, which means that our problem to image a region containing distinct motions of scatterers maps onto the problem of calculating the effects of absorption of intensity (absorption coefficient κ) in that region onto the light intensity at the surface of the multiple scattering medium. In the following sections this will be illustrated by a series of experiments in comparison with this type of theory.

2.   DYNAMIC IMAGING

2.1.   EXPERIMENTS ON LOCALIZED FLOW EMBEDDED IN COLLOIDAL SUSPENSIONS

The above mentioned visualization of Taylor rolls, in fact, monitors a sample-intrinsic dynamic behavior where the dynamic regions (the rolls) are distributed periodically over a macroscopic volume fraction of the sample. In the following, we report DWS experiments on samples which contain macroscopic inclusions of different scatterer dynamics compared to the environment scatterers. As discussed, there is a well-defined localization of this inclusion (the object) which is embedded in a multiple scattering environment (the medium). A typical experimental setup, where the object is a thin X-ray capillary filled with a colloidal suspension and the medium a large cell filled with the very same suspension is shown in figure 1. Since the photon transport mean free path is identical inside and outside the capillary, the contrast which enables its localization with DWS is generated solely by the different types of motion of scatterers inside and outside. Inside the capillary the particles move by their Brownian motion as well as by laminar Poiseuille flow while outside the motion is purely Brownian. Therefore, photons crossing the capillary on their way through the sample pick up characteristic phase shifts of this flowing region and contribute in a specific manner to the decay of the autocorrelation function $g(\mathbf{r}, t)$. The relative contribution of these photons to the decay of $g(\mathbf{r}, t)$ strongly depends on the distance between the capillary and the sample front surface where the laser beam impinges upon. This is demonstrated in figure 2. The solid curves are theoretical predictions based on the model of diffusion of correlation for $G(\mathbf{r}, t)$. A detailed description of this theory which is found in perfect qualitative agreement with the experimental data can be found in [19]. Here we discuss key features of the data presented in figure 2 in more qualitative physical terms: Considering scatterers motion inside the capillary alone results in the above defined cross-over time $\tau_c = \tau_S^2/\tau_0$ between the stochastic Brownian motion and the deterministic flow motion. For correlation times $t < \tau_c$, the decay of $g(\mathbf{r}, t)$ is governed by Brownian motion of the scatterers alone and the flow effect is not visible. This is in agreement with the data in figure 2 where all curves tend to merge for short correlation times

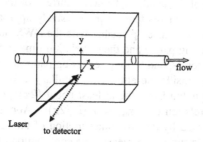

Figure 1. Schematic view of the experimental setup. A large rectangular cell (5 cm x 5 cm x 2 cm) is filled with a colloidal suspension. A cylindrical capillary (diameter 1.5mm) oriented along the z-axis and containing the same or a different suspension can be placed at variable positions (x) inside the cell. The laser impinges at position $y = 0$ which is centered with respect to the capillary. $x = 0$ corresponds to the capillary touching the front wall of the cell. For the case of a liquid inclusion inside a non-moving (solid) medium, the cell is replaced by a block of Teflon.

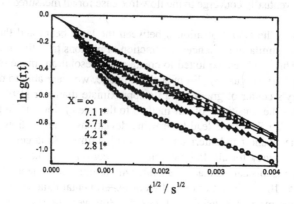

Figure 2. Measured time auto-correlation functions $g(\mathbf{r}, t)$ for various in-depth positions x of the capillary, centered at $y = 0$ with respect to the point of illumination ($y = 0$) for identical colloidal suspensions inside and outside the capillary, flow rate $Q = 0.5$ml/s. Sample: Aqueous suspension of polystyrene latex particles, volume fraction of particles 0.058, diameter of particles 0.12 μm, $l^* = 69$ μm, $\tau_0 = 1.0610^{-3}$ s. Full lines are theoretical results from ref. [19].

of about 10^{-7} s. Regarding longer correlation times, the curves significantly deviate from the flow-free case (straight line) and the deviation clearly depends on the distance x of the capillary from the surface of the cell. At even larger correlation times, all curves tend to progressively approach the data without flow. This effect can be made plausible with the DWS-intrinsic relation [5, 8] between the scattering path s and the correlation time t: $t = \tau_0 l^*/s$, i.e. short correlation times correspond to long scattering paths and vice versa. To probe the flow dynamics inside the embedded capillary at position x, the average path length of a photon has to be at least about $(2x)^2/l^*$ corresponding to a characteristic correlation time $\tau_{Path} = \tau_0(l^*/2x)^2$. Above this time the decay of $g(\mathbf{r}, t)$ is governed by the Brownian motion of the scatterers between the tube and the sample surface. A measurable contribution of flow to the decay of $g(\mathbf{r}, t)$ is therefore expected for $t < \tau_{Path}$. An increasing depth of the flow leads to a shorter time of convergence of the flow influenced curves and the flow free case. Taking together both effects leads to a defined interval of correlation times $\tau_c \leq t \leq \tau_{Path}$ for the applicability of the described method of flow imaging. This is nicely confirmed by the data in figure 2. Within the discussed window of correlation times, the flow induced reduction of the autocorrelation function below the flow free case in figure 2 decreases with increasing position x of the capillary. The deeper its position inside the cell, the smaller the contribution of photons which have crossed the capillary. For values of x larger than about $11l^*$ the curves eventually converge to the flow free case for all measured correlation times t.

Changes in the lateral position y between the laser beam and the embedded capillary similarly influences the fraction of photons traveling through the capillary. This has been exploited to create a low resolution image of the embedded capillary (Figure 3). From such an image, we were able to determine the capillary's center of gravity and its approximate diameter [19]. Figure 4 demonstrates the sensitivity of our method to the rate of flow inside the capillary. Increasing flow rates lead to growing deviations from the flow free case (straight line) at short and intermediate correlation times. At larger times, there is a clear tendency for all flow rate curves to converge to the curve without flow. As above this can be made plausible with intuitive arguments: The range of $t < \tau_c \approx 10^{-7}$ s is not covered by our experimental data so that the pure Brownian regime is not visible in figure 4. However, the shortest measured correlation times are close to τ_c which leads to the pronounced sensitivity to the flow-rate in the short time decay of $g(\mathbf{r}, t)$. The absence of the flow effect for large correlation times accordingly reflects the described position dependence due to the DWS time-path relation. Large correlation times correspond to short photon path lengths. For short photon path length, however, the probability that a photon has passed the capillary is quite low and only a minor effect is seen in the corresponding correlation time range. From the presented data it is evident

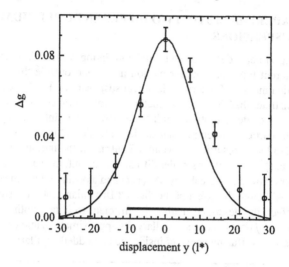

Figure 3. Maximum difference Δg of the auto-correlation functions $g(\mathbf{r}, t)$ with and without flow (at rate $Q = 0.5$ ml/s) for the sample specified in fig.2, at $x = 7.1l^*$, as a function of the difference in y between the capillary and the point of detection. The horizontal line indicates the location of the capillary and the continuous line the theory [19].

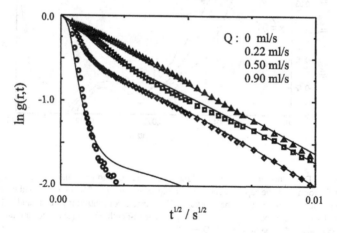

Figure 4. Auto-correlation functions $g(\mathbf{r}, t)$ for different flow rates Q for the sample described in fig.2, $y = 0$, $x = 2.8l^*$, in comparison with theory [19] (cont. lines).

that if either the depth of the capillary or the flow rate is known the respective other parameter can be determined by our developed imaging method.

2.2.   EXPERIMENTS ON TWO DIFFERENT COLLOIDAL SUSPENSIONS

In the experiments described so far the correlation contrast has been created by two different types of scatterer motion inside and outside the object, flow and Brownian motion of otherwise identical suspensions. Further experiments [19] demonstrate, that this imaging method is not restricted to the case of different types of scatterer motion. Filling larger scatterers into the capillary and matching the particle concentration to achieve identical l^* inside and outside the capillary also leads to sufficient dynamic contrast in the measured correlation function $g(\mathbf{r}, t)$. Due to their smaller diffusion constant, the particles inside the capillary now slow down the decay of $g(\mathbf{r}, t)$ and a positive dynamic contrast is generated. Although the effect of the distinct Brownian motion is smaller than the flow effect, the described method is sufficient to obtain both information about the x and y position of the capillary, its approximate diameter and to estimate the particle diffusion constant inside the embedded capillary. Combining

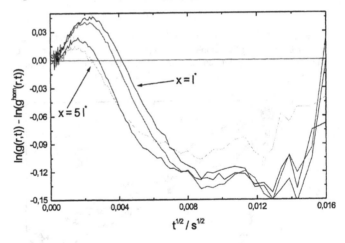

Figure 5.   Difference of the correlation functions with large particles (diameter 1.16 μm) under Poiseuille flow inside the capillary and the correlation function $g^{hom}(\mathbf{r}, t)$ for Brownian particles with diameter 0.12 μm outside the capillary. The particle concentrations inside and outside of the capillary are adjusted to match $l^* = 100$ μm. The in-depth position x of the capillary varies between l^* and $5l^*$. The flow rate is $Q = 0.01$ ml/s.

the Brownian motion of larger scatterers inside the capillary with additional flow of this particles nicely confirms our developed qualitative picture. Figure 5 shows the difference of the measured correlation function with large particles (diameter 1.16 μm) under Poiseuille flow in the capillary as compared to the homogeneous correlation function $g^{hom}(\mathbf{r}, t)$. For short correlation times, the slower Brownian motion of the scatterers inside the capillary reduce the decay of $g(\mathbf{r}, t)$ leading to a positive difference to the homogeneous case. After passing a maximum value the difference becomes zero and driven by the flow, the correlation functions decay faster than $g^{hom}(\mathbf{r}, t)$. The resulting crossover time indicates an equivalent contribution to the induced photon phase shift by the slower Brownian motion and the faster flow motion of the scatterers inside the capillary, compared to the homogeneous case. In this range, the dynamic heterogeneity hardly can be distinguished from a homogeneous sample; this has to be taken into account for possible applications. For larger correlation times, the difference of the correlation functions becomes smaller and tends to converge to zero[7], which can be explained as above by the DWS intrinsic time-path relation.

2.3.   EXPERIMENTS ON COLLOIDAL SUSPENSIONS EMBEDDED IN A TURBID SOLID

A natural extension of the idea of imaging dynamic heterogeneities is to substitute the filled cell for a solid environment and the capillary for a cylindrical cavity inside the solid block [22]. Experimentally, this is realized by a Teflon block containing a cylindrical cavity which is filled with a colloidal suspension adjusted to provide only dynamic but no static scattering contrast. Since microscopic motion of scatterers is now confined inside the cavity, photons that did not cross the cavity generate a static speckle pattern, while photon paths which have crossed the suspension create a temporal fluctuating speckle pattern. This situation corresponds to a non-ergodic system, i.e. time average and ensemble average of the speckle patterns are different. Usually, both averages are made equal to recover ergodicity and herewith reliable normalization of $g(\mathbf{r}, t)$ by averaging the static speckle pattern in translating or rotating the sample.

Following these ideas, we slightly rotated the Teflon block during the measurement in order to average its static contribution to the speckle patterns. Selecting the corresponding angular velocity appropriately two well separated decays in the measured intensity auto-correlation function $G(\mathbf{r}, t)$ are found. The fast decay results from the Brownian motion of the scatterers, while the slow decay reflects the external rotation of the Teflon block. Analyzing the fast decay allows, like in the pure liquid case, to create low resolution images of the embedded cavity and to extract information about the particle motion inside the cavity. The achieved results are quite similar to those described above for the pure liquid case. In particular, the maximum depth x up to which the embedded

---

[7]In this time range the corresponding values of the correlation functions are of order e^{-4} times their initial value, which leads to a certain noise due to the restricted measurement time.

cylinder can be detected is in the same range as the diameter of the cylinder
itself. Similar in-depth resolution was obtained in other experiments [23].

However, this limitation in the in-depth resolution can be overcome with
the help of the so-called dark speckle technique [22]: Measuring $G(\mathbf{r}, t)$ at
a fixed angular position where the static scattering intensity exhibits a local
minimum puts a strong relative weight on photon paths that have crossed the
embedded suspension since the paths through the static medium are essentially
in destructive interference thus not contributing to the static speckle pattern.
In ref.[22] we demonstrate that there is a well-defined relation between the
static scattering intensity and corresponding intercept of $G(\mathbf{r}, t)$. The lower
the static intensity at a certain angular position, the higher the intercept of
$G(\mathbf{r}, t)$. This can be exploited to localize and image the dynamic object from
deeper inside the sample. Taking advantage of this technique allows to extract
a measurable dynamic contrast even in the case where the cavity is positioned
at $x = 37l^*$ which is more than five(!) times the diameter of the embedded
cylinder. Thus exploiting dark speckle spots significantly increases the range
of possible applications of the dynamic imaging method. The quantitative
results are in agreement with a theory made to localize absorbing or transmitting
objects inside a multiple scattering environment by measuring static scattering
intensities [12]. This correspondence leads to the interpretation, that in our
case correlation is absorbed inside the embedded cavity, since every photon
that penetrates the colloidal suspension contributes to the decay of the measured
correlation function.

2.4. IMAGING USING THE SPECKLE INTENSITY DISTRIBUTION

The outlined possibilities of imaging dynamic heterogeneities are based on
a position dependent and time resolved analysis of intensity fluctuations as
generated by the dynamics of scatterers inside the object. An electronic auto
correlator as well as a fast and sensitive photomultiplier including photon counting are used as detection system, which may be considered too sophisticated
equipment for certain applications. It is known, on the other hand, that static
speckle patterns generated by a multiple light scattering object have a negative
exponential intensity distribution $P(I)$, just like in the single scattering case.
The speckle statistics can be used to visualize a dynamic object inside a turbid environment without a direct time-resolved measurement of the scatterer's
dynamics inside. Hence, surprisingly at first, an object without any static scattering contrast can still be visualized in principle in a static light scattering
experiment [24].

Under the assumption of uncorrelated scatterers and uncorrelated scattering
paths, the distribution of the scattered intensity of a multiple scattering sample

has the following form

$$P(I) = \frac{1}{<I>} \exp(-\frac{I}{<I>}),$$ (11)

where $<I>$ stands for the average static scattering intensity. Taking the above mentioned Teflon block with the embedded suspension, all photons that crossed the cavity lead to a rapidly fluctuating speckle pattern. The time scale of this fluctuation created by the Brownian dynamics of the scatterers is adjusted to be in the range of 10^{-5} s. Furthermore, the above mentioned rotation of the Teflon block also leads to a temporal fluctuating speckle pattern generated by the photons that did not cross the embedded suspension. The rotational velocity of the Teflon block is selected to achieve a typical time scale of this intensity fluctuation in the order of 10^{-2} s. Hence, detection of $P(I)$ with a temporal resolution (viz. sampling time) of 10^{-3} s averages the fast fluctuations but sensitively probes the speckle statistics of the rotated Teflon block. Physically, the surface region near the dynamic object has less speckle contrast in this measurement than other regions since the motion of scatterers inside the object washes out contrast on a non resolved time scale. In the data analysis we empirically take this into account by adding the averaged intensity of the photons that passed the suspension as a incoherent shift factor ΔI in Eq. (11). Fitting experimental $P(I)$ data reveals ΔI as unique adjustable parameter[8]. Its value directly reflects the amount of photons that crossed the embedded cavity, therefore it is position dependent and usable to localize and image the embedded suspension. As a result, we showed [24] that a static intensity measurement (on the time scale of 10^{-3} s), is sensitive to visualize a dynamic heterogeneity with underlying dynamics in the time range of 10^{-5} s. In addition, we illustrated [24] that the applicability of position dependent measurements of $P(I)$ for localizing and imaging dynamic objects is not restricted to the case of a solid environment. An extension to liquids embedded in liquids is possible provided that the time scales of the generated speckle patterns from photons that passed the embedded liquid and photon paths that did not pass the embedded liquid are sufficiently separated.

Perspectives and applications

The principle of dynamic speckle imaging discussed here and illustrated by experiments on calibrated colloidal samples with well-known parameters can be applied to a vast variety of problems ranging from fundamental science to daily-live applications. For example, flow patterns can be studied well beyond the laminar regimes investigated so far, with the particularly interesting feature that

---

[8] apart from an independently determined experimental constant.

the mean free path of light sets an additional - often widely variable - length scale on which velocity gradients are probed; this might be useful to study turbulence in complex fluids [15]. Convective motions driven by heat gradients or light radiation pressure [25] can be imaged as well. Applications to combustion are discussed by Snabre *et al.* in this volume. Many medical applications are beginning to appear [20]; evidently blood flow in superficial vessels can be measured, and given the rather large values of both the photon mean free path and the light absorption length in tissues in the near infrared spectral range the technique will find useful diagnostic applications exploiting penetration depths of many cm. Another promising way to enhance the image resolution is to combine the acoustic modulation of the optical speckles [17] to generate dynamic contrast with the dynamic imaging principle; first experiments [26] show that by using a well focused beam of ultrasound the region of dynamic optical contrast (viz. modulation response) can be substantially reduced below the extension of the photon cloud and can be scanned inside the tissues with high precision.

Further instrumental developments can be foreseen: The concept of dynamic cross-correlation imaging [27] will have to be experimentally explored. The development of (dedicated) multichannel electronic correlators will substantially reduce the image acquisition time which nowadays are still far too long for many practical applications. Another route will be numerical calculations of the correlation functions from (high speed) digital video images. Finally, the principle of dynamic multiple scattering imaging is expected to be useful for various types of radiation (e.g. ultrasound, radio waves, x-rays ...) other than visible light.

Acknowledgments

We kindly acknowledge many discussions with S.Skipetrov and R.Maynard, as well as financial support by the Deutscher Akademischer Austauschdienst (DAAD) through the HSPII/AUFE program.

References

[1] A.Schuster, Astrophys. J. **21**, 1 (1905).

[2] P. W. Anderson, Phil. Mag. **52**, 505 (1985).

[3] M. P. van Albada, A. Lagendjik, Phys. Rev. Lett. **55**, 2692 (1985).

[4] P. E. Wolf, G. Maret, Phys. Rev. Lett. **55**, 2696 (1985).

[5] G. Maret, P. E. Wolf, Z. Phys. **B65**, 409 (1987).

[6] D. Y. Ivanov, A. F. Kostko, Opt. Spectrosk. (USSR) **55(5)**, 950 (1983).

[7] A. A. Golubentsev, Sov. Phys. JETP **59**, 26 (1984).

[8] D. J. Pine, D. A. Weitz, P. M. Chaikin, E. Herbolzheimer, Phys. Rev. Lett **60**, 1134 (1988).

[9] D. J. Pine, D. A. Weitz, G. Maret, P. E. Wolf, E. Herbolzheimer, P. M. Chaikin, in *Scattering and localisation of classical waves in random media*, edited by P. Sheng (World Scientific, Singapore, 1990), 312-372.

[10] D. A. Weitz, D. J. Pine, *Diffusing wave spectroscopy* in *Dynamic light scattering*, Edited by W. Brown (Oxford U. Press, New York, 1993), Chap. 16, 652-720.

[11] B. J. Berne, R. Pecora, *Dynamic light scattering*, (John Wiley & Sons Inc., New York, 1976).

[12] P. N. DenOuter, Th. M. Nieuwenhuizen, A. Lagendijk, J. Opt. Soc. Am. A **10**, 1209 (1993).

[13] X. L. Wu, D. J. Pine, P. M. Chaikin, J. S. Huang, D. A. Weitz, J. Opt. Soc. Am. B **7**, 15 (1990).

[14] D. Bicout, E. Akkermans, R. Maynard, J. Phys. France I **1**, 471 (1991).

[15] D. Bicout, G. Maret, Physica A **210**, 87 (1994).

[16] D. Bicout, R. Maynard, Physica B **204**, 20 (1994).

[17] W. Leutz, G. Maret, Physica B **204**, 14 (1995).

[18] D. A. Boas, L. E. Campbell, A. G. Yodh, Phys. Rev. Lett. **75**, 1855 (1995).

[19] M. Heckmeier, S. E. Skipetrov, G. Maret, R. Maynard, J. Opt. Soc. Am. A **14**, 185 (1997).

[20] D. A. Boas, A. G. Yodh, J. Opt. Soc. Am. A **14**, 192 (1997).

[21] M. Heckmeier, G. Maret, Europhys. Lett. **34**, 257 (1996).

[22] M. Heckmeier, G. Maret, Prog. Coll. Pol. Sci. **104**, 12 (1997).

[23] S. E. Skipetrov, I. V. Meglinskii, JETP **86**, 661 (1998).

[24] M. Heckmeier, G. Maret, Optics communications **148**, 1 (1998).

[25] S. E. Skipetrov, M. A. Kazaryan, N. P. Korotkov, S. D. Zakharov, J. Moscow Phys. Soc. **7**, 411 (1997); S. E. Skipetrov, S. S. Chesnokov, S. D. Zakharov, M. A. Kazaryan, V. A. Shcheglov, JETP Letters **67**, 635 (1998); S. E. Skipetrov, S. S. Chesnokov, S. D. Zakharov, M. A. Kazaryan, N. P. Korotkov, V. A. Shcheglov, Quantum Electronics **28**, 434 (1998).

[26] M. Kempe, M. Larionov, D. Zaslavsky, A. Z. Genack, J. Opt. Soc. Am. A. **14**, 1151, (1997).

[27] S. E. Skipetrov, Europhys. Lett., **40**, 381 (1997).

Chapter 2

DIFFUSE LASER DOPPLER VELOCIMETRY FROM MULTIPLE SCATTERING MEDIA AND FLOWING SUSPENSIONS

P. Snabre

Institut de Sciences et de génie des matériaux et procédés UPR 8621
B.P.5, 66125 Font-Romeu, France
snabre@imp-odeillo.fr

J. Dufaux

Laboratoire de Biorhéologie et d'hydrodynamique physicochimique ESA 7057
Université Paris VII, 2 place Jussieu, 75225 Paris, France
dufaux@ccr.jussieu.fr

L. Brunel

Institut de Sciences et de génie des matériaux et procédés UPR 8621
B.P.5, 66125 Font-Romeu, France
brunel@imp-odeillo.fr

Abstract   A study of the Doppler frequency spectrum (DFS) of the light multiply scattered from a random collection of moving anisotropic scatterers is presented. The analytical approach based on photon diffusion approximation and statistical models yields functional forms of DFS for random or simple shear flow consistent both with experiments in a backscatter geometry and predictions from the Diffusing Wave Spectroscopy theory (DWS). Homodyne and heterodyne detection modes are further analyzed and the potential applications of the Doppler heterodyne analysis in the field of suspension dynamics are investigated.

Keywords: laser Doppler velocimetry, multiple scattering, suspension dynamics

P. Sebbah (ed.), Waves and Imaging through Complex Media, 369–382.

Introduction

Laser Doppler velocimetry (LDV) also known as laser Doppler anemometry (LDA) is a well known coherent technique that measures flow induced Doppler frequency shifts by moving particles in the single scattering regime. More than two decades ago, temporal fluctuations in the laser light multiply scattered from a biological tissue were observed to be related to the microcirculatory flow [1, 2, 3]. Now, Laser Doppler flowmetry has become a competitive method of measuring blood perfusion in the superficial microcirculation [4, 5] with advantages over acoustic Doppler imaging of limited spatial resolution [6].

When illuminating a turbid medium with a coherent laser light, the loss of coherence of the scattered field arises from motions of the scattering centers with respect to each other [7]. Intensity fluctuation spectroscopy and diffuse laser Doppler velocimetry (DLDV) are then uniquely suited for measurement of temporal variations of the scattered light and analysis of particle motions over very small distances within a turbid medium. By the Wiener-khintchine theorem, the autocorrelation function and the Doppler frequency spectrum (DFS) are simple Fourier transform of each other [7]. A "wandering photon model" for DLDV in a random perfused biological tissue was proposed in the seventies by Bonner and Nossal [8]. Most of theoretical models are now carried out in terms of intensity autocorrelation function based on the formalism of diffusing wave spectroscopy (DWS) introduced by Maret *et al* [9] and Pine *et al* [10].

The present chapter concerns the extension of the classical approach for LDV to the multiple scattering regime [11]. Within the framework of photon diffusion approximation, functional forms of the DFS of the light multiply scattered from a random collection of moving anisotropic scatterers (random or simple shear flow) are derived from statistical models and compared to Monte Carlo simulations, predictions from DWS theory and experiments in a backscatter geometry. Finally, the potential applications of DLDV for studying particle dynamics in a dense suspension are illustrated and advantages of either DWS or Doppler heterodyne spectral analysis are discussed.

1.   DIFFUSE LASER DOPPLER VELOCIMETRY

1.1.   DOPPLER FREQUENCY SHIFT

One considers the interaction of a coherent scalar light wave[1] (incident wavevector $\mathbf{k}_i$) with moving particles randomly distributed in a fluid. For a sample thickness L much less than the photon mean free path l, the Doppler frequency shift ω of a single scattered wave (wavevector $\mathbf{k}_s$) is the scalar prod-

---

[1] Polarization effects are ignored.

uct $\mathbf{q} \bullet \mathbf{V}$ of the Bragg scattering vector $\mathbf{q} = \mathbf{k}_s - \mathbf{k}_i$ and the particle velocity vector $\mathbf{V}$.

In the weak and multiple scattering regime ($kl \gg 1$), interferences effects are averaged out over the length scale l and the photon diffusion approximation can be used [12]. In such conditions, the Doppler frequency shift of a multiply scattered wave can be viewed as a superposition of Doppler shifts due to individual scattering events ($\omega = (\sum_{p=1}^{p=n} \mathbf{q}_p \bullet \mathbf{V}_p = -\sum_{p=0}^{p=n} \mathbf{k}_p \bullet (\mathbf{V}_{p+1} - \mathbf{V}_p)$ with $\mathbf{q}_p = \mathbf{k}_{p+1} - \mathbf{k}_p$, Fig.1). Monte Carlo simulations are then suitable for predicting the Doppler frequency shifts of the multiply scattered light[2].

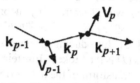

Figure 1. Particle velocity vectors $\mathbf{V}_p$ and wavevectors $\mathbf{k}_p$.

Figure 2. Transport mean path l^*.

For a random distribution of scatterers, all information about the initial direction of a wandering photon is lost over the transport mean path l^* depending on the asymmetry factor $g = \langle \cos\theta \rangle$ of scatterers (Fig.2) [9, 10, 11]. For a length scale larger than the transport mean path l^*, Doppler frequency shifts can be considered as a random gaussian variable. Directional randomization of scattering wave vectors results in a broadening of the Doppler frequency spectrum without dependence upon the scattering angle. From the central limit theorem, the DFS $S_n(\omega)$ of n diffusion paths is a gaussian:

$$S_n(\omega) \approx \frac{1}{\sqrt{(n/n^*)\sigma_n^{*2}}} \, e^{-\frac{\omega^2 - \langle\omega\rangle^2}{2(n/n^*)\sigma_n^{*2}}} \quad \text{with} \quad n \gg n^* = \frac{1}{1-g} . \quad (1)$$

Averaging Doppler frequency shifts over wave vector $\mathbf{k}$ and particle velocity vector $\mathbf{V}$ gives a gaussian width scaling either with the mean quadratic velocity $\langle V^2 \rangle^{1/2}$ (for random flows) or quadratic shear rate $\langle \gamma^2 \rangle^{1/2}$ (for laminar shear flows) averaged over the volume probed by the scattered light [12, 13]:

$$\sigma_n^{*2} \approx n^* \left\langle\left\langle (\mathbf{q} \bullet \mathbf{V})^2 \right\rangle\right\rangle_{\mathbf{q},\mathbf{V}} = \frac{2}{3} k^2 \langle V^2 \rangle , \quad (2)$$

$$\sigma_n^{*2} \approx \left\langle\left\langle (\mathbf{k} \bullet (\mathbf{V}_{n^*+1} - \mathbf{V}_1))^2 \right\rangle\right\rangle_{\mathbf{k},\mathbf{V}} = \frac{1}{15} k^2 \langle \gamma^2 \rangle \, l^{*2} . \quad (3)$$

---

[2] A Monte carlo method described in a previous publication [14] was used to calculate the Doppler frequency shift of photons.

Random flow.   In the case of random particle motion, the characteristic Doppler shift ω_o of n diffusion paths is the product of wavevector and mean quadratic velocity of scatterers (Eqs. (1) and (2) with $\langle \omega \rangle = 0$):

$$S_n(\omega) \approx n^{-1/2}\, e^{-\frac{3\omega^2}{4(n/n^*)\omega_o^2}} \quad \text{with} \quad \omega_o = k \langle V^2 \rangle^{1/2} . \tag{4}$$

For Brownian scatterers, such a frequency approach raises a problem since the mean square velocity of Brownian scatterers scales as the inverse of time scale. However, one can consider the diffusing wave spectroscopy theory that gives the electric field autocorrelation function $g_1^n(\tau) \approx n^{-1/2}\, e^{-2(n/n^*)\tau k^2 D}$ of n diffusion paths in a Brownian suspension [12]. Introducing the mean square velocity $\langle V^2 \rangle = 6D/\tau$ of Brownian scatterers (D is the Stokes - Einstein diffusion coefficient of scatterers), then the Fourier transform of the autocorrelation function $g_1^n(\tau)$ equals the Doppler frequency spectrum $S_n(\omega)$ of n diffusion paths (Eq. (4)) and decays exponentially with frequency square.

Simple shear flow.   For correlated motions of scatterers in a simple shear flow (uniform shear rate γ)[3], the characteristic Doppler frequency shift ω_s is the product of wavevector and velocity difference $\gamma\, l^*$ over the transport mean path (Eqs. (1) and (3) with $\langle \omega \rangle = 0$):

$$S_n(\omega) \approx n^{-1/2}\, e^{-\frac{15\omega^2}{2(n/n^*)\omega_s^2}} \quad \text{with} \quad \omega_s = k\gamma l^* . \tag{5}$$

The Fourier transform of the DFS is the temporal electric field autocorrelation function $g_1^n(\tau) \approx n^{-1/2} e^{-(n/n^*)k^2\gamma^2 l^{*2}\tau^2/30}$ given by DWS theory [13]. The bandwidth of the DFS scales as the square root of the number of scattering events in agreement with Monte Carlo simulations (Fig.3) [11].

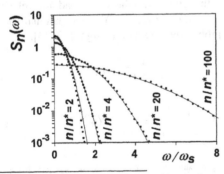

Figure 3 DFS of n diffusion paths for simple shear flow ($g = 0.75$). Monte Carlo simulations (•) and DLDV theory (—).

[3]Stationary flow with slow velocity change at the scale of scattering events.

1.2.   DOPPLER FREQUENCY SPECTRUM

The full Doppler frequency spectrum $S(\omega)$ of the scattered light emerging at the detector point is obtained by summing over all scattering paths:

$$S(\omega) = \int_1^\infty P(n)\, S_n(\omega)\, d(n)\,, \tag{6}$$

where the probability $P(n)$ of n diffusion paths is sensitive to boundary conditions (size and shape of the sample), detection geometry and optical properties of scatterers.

Conservative scattering.   Scatterers randomly distributed between two parallel infinite plates are considered. Photons are injected into the medium along the incident direction $\theta = 0$ normal to the planes limiting the scattering volume. For long path photons ($n/n^* > 4$) scattered from a non absorbing infinite medium, the probability $P(n)$ of n scattering events scales as $(n/n^*)^{-3/2}$ because of the diffusive nature of light transport. The integral (6) over the scattering numbers with $\alpha < n/n^* < \infty$ gives a functional form of the DFS close to the Fourier transform of the electric field autocorrelation function derived from DWS theory [12] and in agreement with Monte Carlo simulations for random (Fig.4a) or simple shear flow (Fig.4b) [11] [4]:

$$S(\omega) \approx \int_\alpha^\infty (n/n^*)^{-3/2}\, S_n(\omega)\, d(n/n^*) = \frac{1 - e^{-\frac{\omega^2}{\alpha\,\Omega^2}}}{\omega^2}\,, \tag{7}$$

with $\Omega^2 = 4\,\omega_o^2/3$ for random flow and $\Omega^2 = 2\,\omega_s^2/15$ for simple shear flow. The lower bound α of the integral somewhat depends on particle motion, scattering anisotropy and scattering angle since the diffusion approximation breaks down for short diffusion paths [12].

In the case of a slab of thickness L, long path photons are transmitted through the medium for a characteristic scattering number $n/n^* > (L/l^*)^2$. The less number of scattering events for the light backscattered from a finite sized turbid medium then results in a reduced DFS bandwidth:

$$S(\omega) \approx \int_\alpha^{(L/l^*)^2} (n/n^*)^{-3/2}\, S_n(\omega)\, d(n/n^*) = \frac{e^{-\frac{l^{*2}\,\omega^2}{L^2\,\Omega^2}} - e^{-\frac{\omega^2}{\alpha\,\Omega^2}}}{\omega^2}\,. \tag{8}$$

In the case of random particle motion, the average Doppler shift $\langle \omega \rangle$ rises with the optical thickness L/l^* up to a saturation level $\approx 5\omega_o$ for $L/l^* > 300$

---

[4]For correlated particle motions, the DFS is dependent upon the transport mean path l^* and the asymmetry factor g of the scattering diagram. However, second order moments of the phase function (forward or backscatter probability p) weakly influence the DFS as a result of photon randomization (Fig.4a).

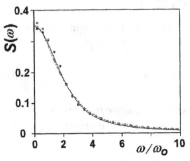

Figure 4a. Normalized DFS for random flow and backscattering geometry (non absorbing infinite medium, $g = 0.75$, backscattering probability $p = 0.07$ (•) or $p = 0$ (o)). DLDV (—) (Eq. (7) with $\alpha = 2$) and DWS theory (- - -).

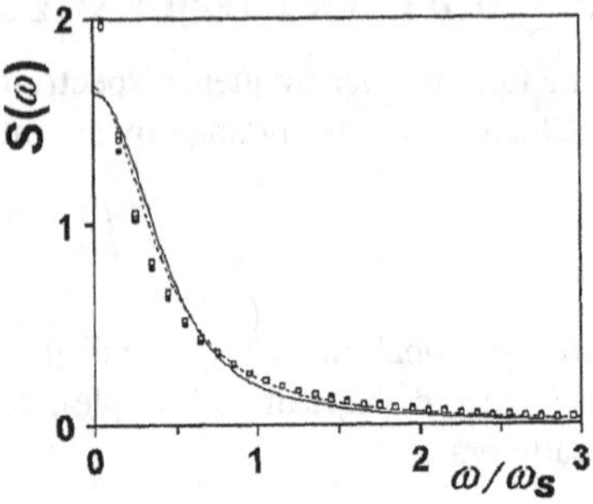

Figure 4b. Normalized DFS for simple shear flow and backscattering geometry (non absorbing infinite medium, $g = 0.5$ (•), $g = 0.75$ (o), $g = 0.9$ (□)). DLDV (—) (Eq. (7) with $\alpha = 0.9$) and DWS theory (- - -).

(Fig.4c). In the case of simple shear flow, an increase in the scatterer density reduces both the particle velocity difference γl^* along the transport mean path and the DFS bandwidth ($\langle \omega \rangle \approx \omega_S/5$ for $L/l^* > 2$, Fig.4d). Because of photon randomization, the full Doppler spectrum further becomes insensitive to the scattering angle θ in the multiple scattering regime.

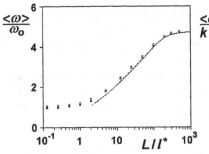

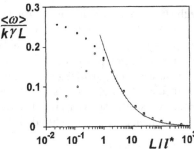

Figure 4c. Doppler frequency width *versus* optical thickness for random flow and backscatter geometry (non absorbing medium, scattering angle $\theta = 100°$ (•) or $\theta = 170°$ (o)). DLDV (—) (Eq. (8) with $\alpha = 2$) or DWS theory (- - -).

Figure 4d. Doppler frequency width *versus* optical thickness for simple shear flow and backscatter geometry (non absorbing medium, scattering angle $\theta = 100°$ (•) or $\theta = 170°$ (o)). DLDV (—) (Eq. (8) with $\alpha = 0.9$) or DWS theory (- - -).

Non conservative scattering. In the case of non conservative scattering, light absorption cancels out long diffusion paths. The maximum number of scattering events then scales as the ratio of the absorption path l_a and the transport mean path l^*. In the backscatter geometry, the DLDV theory predicts a lower DFS bandwidth in agreement with Monte Carlo simulations (Fig.5).

$$S(\omega) \approx \int_\alpha^{l_a/2l^*} (n/n^*)^{-3/2} \, S_n(\omega) \, d(n/n^*) = \frac{e^{-\frac{2l^*\omega^2}{l_a\Omega^2}} - e^{-\frac{\omega^2}{\alpha\Omega^2}}}{\omega^2}. \quad (9)$$

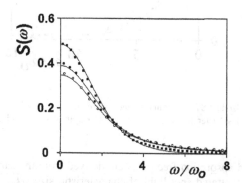

Figure 5. Normalized DFS for random flow and backscatter geometry (absorbing infinite medium: $l_a/l^* = \infty$ (o), $l_a/l^* = 250$ (•) and $l_a/l^* = 25$ (■) with $g = 0.75$). Monte Carlo simulations and DLDV theory (—) (Eq. (9) with $\alpha = 2$).

Boundary reflections. Internal reflections at interfaces influence the diffusive transport of light and introduce a discontinuity in the spatial distribution of backscattered photons [14]. Photons reinjected in the medium indeed result in a larger DFS bandwidth (Fig.6). However, the spectrum broadening remains somewhat negligible for an aqueous suspension in a glass container.

2.   DLDV EXPERIMENTS

This section concerns DLDV experiments from polystyrene spheres or red blood cells suspensions in a random or shear flow. Scattering parameters were determined independently by imagery analysis of the light flux distribution in the incoherent backscattered spot light when illuminating the turbid medium with a normally incident He-Ne laser beam [14]. For long diffusion paths, the surface flux density $F_s(\rho)$ scales as the transport free path l^* and the inverse cube of the distance ρ from the injection point [14]:

$$F_S(\rho) \approx l^* \rho^{-3} \, e^{-\rho/\sqrt{l_a l^*}} \quad \text{for} \quad \rho/l^* > 4. \quad (10)$$

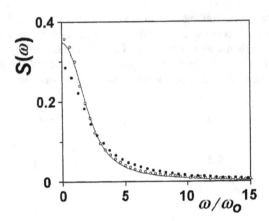

Figure 6.   Normalized DFS for random flow and backscatter geometry ($g = 0.75$, no boundary reflections (○), air/glass/water interfaces (●)). Monte Carlo simulations and DLDV theory (—)).

Transport and absorption free paths are derived from the surface flux analysis of the backscattered spot light of characteristic size $\sqrt{l_a l^*}$ ($l^* \approx 220 \mu m$ for 0.4 μm-diameter polystyrene spheres (PS) and particle volume fraction $\phi = 0.01$, Fig.7). Measurements were typically within 5% of calculations of l^* from Mie theory.

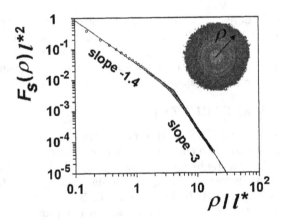

Figure 7.   Normalized surface flux density in the backscattered spot light (shown in the insert) for 0.4 μm-diameter PS in water and $\phi = 0.01$. Experimental data (○) and statistical model (—) [14].

2.1. HOMODYNE / HETERODYNE DETECTION

The Doppler frequency spectrum $S(\omega)$ was measured using either an optical fiber probe (this is the case of usual flowmeters in biomedecine) or an interferometric setup (Fig.8).

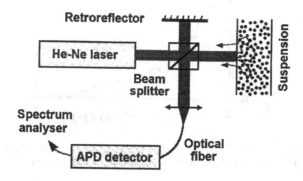

Figure 8. Schematic of the interferometric setup.

For the homodyne detection mode with an optical fiber, the Doppler shifted scattered light is mixed with itself and the mutual interferences of scattered waves increase the DFS bandwidth by a factor of about two very sensitive to the degree of coherence of the detected signal (Fig.9). In the heterodyne detection mode with an interferometer, part of the incoming beam is split-off to serve as a local oscillator that is mixed with the Doppler shifted scattered light (Fig.8). The frequency of the intensity modulation of the detector signal due to interference is the frequency spectrum $S(\omega)$ of Doppler shifted photons. In the case of heterodyne detection, the DFS becomes less sensitive to the detector geometry and further equals the Fourier transform of the temporal electric field autocorrelation function $g_1(\tau)$. The heterodyne detection mode will now be considered.

Brownian suspension. Fig.10 shows the DFS for 11.9 μm–diameter PS and $\phi = 0.1$. The Fourier transform of the temporal electric field auto-correlation function derived from DWS theory for Brownian scatterers [13] describes the experimental data and gives a reasonable characteristic diffusion time $\tau_o = 1/(4k^2 D) \approx 1.31$ ms.

$$g_1(\tau) \approx \frac{\sinh\left(\left(\frac{L}{l^*} - \alpha\right)\sqrt{\frac{3\tau}{2\tau_o}}\right)}{\sinh\left(\frac{L}{l^*}\sqrt{\frac{3\tau}{2\tau_o}}\right)}. \tag{11}$$

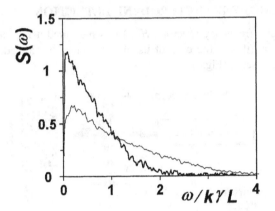

Figure 9.   Normalized DFS from a 2mm - thick plane capillary flow with 11.9 μm - diameter PS and $\phi = 0.1$ ($L/l^* = 5.5$). Homodyne (—) and heterodyne detection (—).

For non correlated Brownian scatterers, the DFS bandwidth increases linearly with the Stokes-Einstein diffusion coefficient D and scales as the inverse of particle diameter in the multiple scattering regime [11].

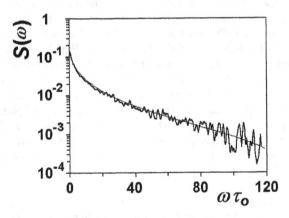

Figure 10.   Normalized DFS from 0.4 μm - diameter PS and $\phi = 0.1$ ($L/l^* = 32.2$). Experimental data (—) and DWS theory (—) (Eq. (11) with $\alpha = 2$ and $\tau_o = 1.31$ ms).

Porous flow.   DLDV experiments were performed for the flow of non Brownian PS in a porous bed made of glass particles of diameter 1 mm. Glass particles weakly influence light transport in the scattering medium. The DFS bandwidth increases linearly with flow rate and average velocity $\langle V \rangle$ of particles along the flow direction [11]. Basic assumptions proposed by Saffman [15] to describe dispersion in a statistically homogeneous and isotropic porous medium predict a linear relationship $\langle V^2 \rangle^{1/2} = \sqrt{3} \langle V \rangle$ between the mean quadratic velocity and the average velocity of particles through the medium [5]. Considering the dimensionless frequency $\omega_o = k \langle V^2 \rangle^{1/2}$, Monte Carlo simulations and DLDV theory (Eq. (8)) describe the experimental DFS (Fig.11) since velocity fluctuations of independently moving particles determine Doppler shifts of scattered photons in the multiple scattering regime.

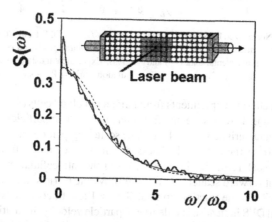

Figure 11. Normalized DFS from a porous flow (see insert) with 6.4 μm - diameter PS and $\phi = 0.05$ ($L/l^* = 32.2$). Experimental data (—), DLDV theory (—) (Eq. (8) with $\alpha = 2$) and Monte carlo simulations (- - -).

Couette flow. Scattering experiments were performed for non Brownian PS spheres in a simple shear flow. The stationary inner cylinder is black painted to absorb the transmitted photons while the rotation of the outer cylinder induces a uniform shear flow in the gap (Fig.12a). The bandwidth of experimental DFS $S(\omega/\omega_s)$ with $\omega_s = k \gamma l^*$ increases with particle volume fraction as predicted by DLDV theory (Fig.12a). However, Monte Carlo simulations underestimate the DFS bandwidth for concentrated suspensions ($\phi > 0.1$, Fig.12b). Particle collisions induce transverse velocity fluctuations that can explain the DFS broadening as indicated by Monte Carlo simulations when adding a random isotropic velocity component $\delta V = 0.1 \gamma L$ [11].

---

[5] $\langle V^2 \rangle^{1/2} = \int_0^{\pi/2} V(\theta) \sin \theta \, d\theta$ and $\langle V \rangle = \int_0^{\pi/2} V(\theta) \cos \theta \, d\theta$ with $\langle V \rangle = V_o \cos \theta$ for a 3D assemblage of randomly oriented straight uniform pores.

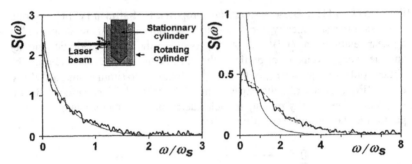

Figure 12a.   Normalized DFS from Couette flow with 6.4 μm - diameter PS and $\phi = 0.05$ ($L/l^* = 1.95$). Experimental data (—) and Monte Carlo simulations (—).

Figure 12b.   DFS from Couette flow with 6.4 μm - diameter PS and $\phi = 0.2$ ($L/l^* = 7.9$). Experimental data (—) and Monte simulations ($\delta V = 0$ (—), $\delta V = 0.1\gamma L$ (---)).

Finally, scattering experiments from human red blood cells (RBC) in Couette flow are presented. RBCs are non Brownian deformable particles of biconcave shape and characteristic size $\approx 4\mu$m. The scattering parameters of RBCs were determined from the optical thickness dependence of the diffuse reflectance [11] [6]. The weak refractive index $n \approx 1.4$ of the intracellular medium results in a dominant forward scattering ($g \approx 0.99$). Monte Carlo simulations describe DFS for RBC volume fraction up to 30% (Fig.12c). In contrast with rigid PS, experimental DFS indicate a less degree of particle velocity fluctuations because of RBC deformability. However, some deviation is observed at 40% RBC concentration as a possible result of transverse velocity fluctuations (Fig.12d).

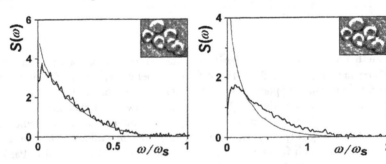

Figure 12c.   Normalized DFS from Couette flow with RBC and $\phi = 0.2$ ($L/l^* = 0.93$). Experimental data (—) and Monte Carlo simulations (—).

Figure 12d.   DFS from Couette flow with RBC and $\phi = 0.4$ ($L/l^* = 1.87$). Experimental data (—) and Monte Carlo simulations ($\delta V = 0$ (—), $\delta V = 0.1\gamma L$ (---)).

---

[6]RBC are washed and suspended in physiological saline solution to cancel aggregation phenomena.

Conclusion

When analyzing the temporal correlations of multiply scattered light, the signal intensity and the time scale of fluctuations determine data collection either in the form of autocorrelation function or frequency spectrum. The heterodyne analysis overcomes several limitations inherent in conventional photon correlation spectroscopy. Indeed, DWS intensity correlation measurements are very sensitive to the detector aperture and the scattered light has usually to be collected from a single speckle spot. In contrast, Doppler frequency spectra obtained from heterodyne spectral analysis are weakly influenced by the detector geometry or the interferometric setup when the optical path length difference is within the coherence length of the source light [11]. In addition, the Monte Carlo method directly provides the DFS derived from heterodyne Doppler analysis.

For negligible contributions of short diffusion paths, the interpretation of DWS or DLDV experiments can give useful information about the dynamics of the scattering medium. In the multiple scattering regime, the DFS bandwidth scales either as the mean quadratic velocity for random particle motions or the average velocity difference along the transport mean path l^* for shear flows. On the basis of photon diffusion approximation, functional forms of the Doppler frequency spectrum derived from statistical models reasonably describe experimental data in the backscatter geometry. The broadening of the Doppler spectrum from a flowing dense suspension of non Brownian scatterers further indicates transverse particle velocity fluctuations. Despite difficulties in the interpretation of results in terms of correlated particle motions, DLDV experiments may provide new insight into the dynamics of concentrated suspensions. Finally, the heterodyne spectral analysis has potential applications in the field of clinical medicine for measuring or imaging blood flow in biological tissues [15-18].

References

[1] M. D. Stern, Nature (Lond.), **254**, 56 (1975).

[2] T. Tanaka, and G. B. Benedek, Appl. Opt. **14**, 189 (1975).

[3] F. F. M. de Mul, J. van Spijker, D. van der Plas, J. Greve, J. G. Aarnoudse and T. M. Smits, Appl. Opt. **23**, 2970 (1984).

[4] D. Watkins and G. A. Holloway, IEEE Trans. Biomed. Eng. **25**, 28 (1978).

[5] F. F. M. de Mul , M. H. Koelink, M. L Kok, P. J. Harmsma, J. Greve, R. Graaff, Appl. Opt. **34**, 6595 (1995).

[6] Z. Chen, T. E. Milner, D. Dave and J. S. Nelson, Opt. Lett. **19**, 590 (1994).

[7] B. J. Berne and R. Pecora, Dynamic light scattering: With Applications to Chemistry, Biology, and Physics (Wiley, New York, 1976).

[8] B. Bonner and R. Nossal, Appl. Opt. **20**, 2097 (1981).

[9] G. Maret and P. E. Wolf, Z. Phys. B **65**, 409 (1987).

[10] D. J Pine, D. A. Weitz, P. M Chaikin and E. Herbolzheimer, Phys. Rev. Lett. **60**, 1134 (1988).

[11] P. Snabre, L. Brunel, E. Chazel and J. Dufaux, Eur. Phys. J. E. Submitted for publication.

[12] D. J. Pine, D. A. Weitz, G. Maret, P. E. Wolf, E. Herbolzheimer and P. M. Chaikin, in *Scattering and Localization of Classical Waves in Random Media*, ed. P. Sheng (World Scientific Pub. Co., 1990) p.312.

[13] D. Bicout, E. Akkermans and R. Maynard, J. Phys. 1 **1**, 471 (1991).

[14] P. Snabre and A.Arhaliass, Appl. Opt. **37**, 4017 (1998).

[15] P. G. Saffman, J. Fluid Mech. **6**, 321 (1959).

[16] G. Soelkner, G. Mitic and R. Lohwasser, Appl. Opt. **36**, 5647 (1997).

[17] Z. Chen, T. E. Milner, S. Srinivas, X. Wang, A. Malekafzali, M. J. C. van Gemert and J. S Nelson, Opt. Lett. **22**, 1119 (1997).

[18] J. A Izatt, M. D. Kulkarni and S. Yazdanfar, Opt. Lett. **22**, 1439 (1997).

Chapter 3

HIGH RESOLUTION ACOUSTO-OPTIC IMAGING

S. Lévêque-Fort

Laboratoire d'Optique, CNRS UPR A0005,
Ecole Supérieure de Physique et Chimie Industrielle
10 rue Vauquelin, 75005 Paris, France
fort@optique.espci.fr

Abstract   We develop an original optical imaging system based on the interaction of light with an ultrasonic field. A CCD camera followed by parallel data processing detects the modulation induced by the focused ultrasonic field in a great number of speckle grains. This acousto-optic imaging allows to reveal optical contrast through several centimeters thick biological tissues with a millimeter resolution.

Keywords: acousto-optic, 3D imaging, speckle modulation

Introduction

Non-invasive revealment of optical contrast inside organs is an important stake for new medical imaging approaches. Indeed surgeons could determine the nature of the tissues through their appearances, since optical properties are a good clue to reveal abnormal tissues. Several technique are currently implemented, and the strong multiple scattering in biological tissues imposes some selection of the photon trajectories. Many techniques take advantage of ballistic or quasi-ballistic photons, easier to extract. Nevertheless most of the light that exits the sample is multiply scattered. Two original approaches taking advantage of all photons and based on an ultrasonic field and an optical field have been developed to image optical contrast of hidden objects in tissues : photoacoustic [1, 2] and opto-acoustic [3-6] approaches.

In the photoacoustic approach, an ultrasonic transducer detects the acoustic wave induced by laser pulses in the tissues. The photoacoustic signal amplitude is determined by the local fluence in the studied zone. The depth and lateral

P. Sebbah (ed.), Waves and Imaging through Complex Media, 383–388.
© 2001 *Kluwer Academic Publishers. Printed in the Netherlands.*

resolution are reported [1] to reach 10 μm and 200 μm respectively, but limited to $\approx$ 5 mm thick samples.

We have developed an original acousto-optic imaging system which allows to reveal optical contrast with a millimeter resolution through several centimeters of tissues [7]. In this approach, we take advantage of the different photons behaviors, and rely mainly on diffused ones. We select spatially all photons that pass trough a small zone of the biological tissue by bringing the focal zone of the ultrasonic source in coincidence with the zone of interest.

1.   ACOUSTO-OPTIC SIGNAL GENERATION

The acousto-optic signal is based on the interaction of light with an ultrasonic field. A single mode laser diode at 840 nm illuminates the biological sample and generates a speckle field. An ultrasonic field, focused in a small zone, induces a periodic displacement of the scatterers (typically a few nm) at the ultrasonic frequency (2.3 MHz). The optical path of the photons that cross the ultrasonic focal zone is modulated, and so the speckle exiting the sample. By scanning the ultrasonic focus across the sample, we obtain for each voxel of the sample delimited by the ultrasonic focus zone, the amplitude of modulated light. So, we reconstruct an image of the sample related to its local optical properties. This allows one to detect a buried light-absorbing object with a millimeter resolution. When the focused zone reaches the object, we observe a decrease of the ultrasonic-modulation signal, because the modulated photons are trapped in the absorbing object, so that less modulated light reaches the CCD camera.

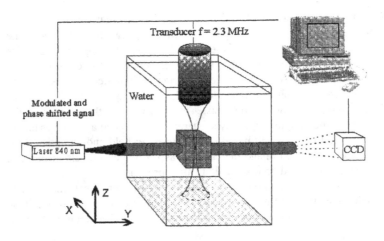

Figure 1.   Experimental set-up

The resolution is determined by the size of the focal zone of the transducer. We have estimated it by measuring the full width at half maximum of the pressure in the focal region. The resolution reaches 2 mm laterally (X,Y direction) and 25 mm longitudinally (Z direction). We could even improve the resolution by using a transducer with a larger aperture.

The detection of the modulated signal is based on a new approach which uses a CCD camera (256 × 256 pixels) and an original signal processing scheme which takes advantages of the full speckle field. This multichannel lock-in detection developed in our lab improves the signal to noise ratio. But the high modulation frequency of our ultrasonic transducer (2.3 MHz) is not compatible with the slow sampling rate of the 2D CCD camera (maximum frame: 200 Hz). To keep the benefit of the synchronous detection, we modify the conventional scheme (mixing-dc filtering) by moving the mixing to the light source. So we use a laser diode modulable at 2.3 MHz and the charge integration in the CCD pixels acts as the low-pass filter. The optical sampling of the acousto-optic signal at 2.3 MHz is obtained by integrating the light on the CCD during a quarter of the modulation period and for four phase shifts light excitation compared to the modulated signal. The first image is obtained when the ultrasonic transducer inducing the modulation and the laser diode modulated at the same frequency are in phase, the three others correspond to a phase shift of 90°, 180° and 270°. Processing the four frames, we extract the amplitude and the phase of the modulated light for each individual pixel. We can treat in parallel the modulations of a large number of speckle grains.

This method yields quickly an average modulated signal with a degree of modulation larger by two orders of magnitude than with a single detector [4-6]. A single detector records the modulation associated to a single speckle grain or to many speckle grains. The response is highly fluctuating, due to the random distribution of the speckle field. In order to get an appreciable modulation amplitude, some authors take advantage of the speckle drift associated with the evolution of the sample, but this leads to a slower acquisition.

The only drawback of our multichannel lock-in detection is that, although our 'stroboscopic' acquisition provides the correct signal at a few MHz, the noise corresponds to the frequency band associated with the acquisition frame rate (200 Hz).

2. RESULTS

Results were mainly obtained for biological tissues of thicknesses from 2 cm to 3.5 cm, and of optical and acoustic properties close to human tissues. The aim of acousto-optic imaging is to reveal optical contrast when acoustic contrast is too low for a diagnostic. Therefore, to show the efficiency of our technique we introduce into the sample an inclusion with a difference of optical properties only. The stability of our system (1%) offers good conditions for the reconstruction of the image of the sample. Each point of the reconstructed image, associated with a position of the ultrasonic focal zone in the sample volume, represents the sum of the modulation amplitudes seen by all pixels, and so reflects the local optical properties.

Figure 2 represents an experimental result with a 15 mm thick turkey breast sample: a cylinder of modelling clay (diameter 3 mm), is placed at $x = 4$ mm. Its presence is detected during the first ultrasonic scan as shown by the marked dip of the curve at this position. Then a second cylinder is placed at $x = 14$ mm.

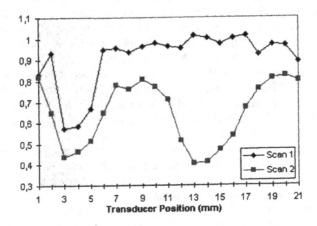

Figure 2.   1-D image of two absorbing cylinders in a 15 mm turkey breast sample

During the second scan, we find again the first cylinder with a good agreement and distinguish in addition the second one. Thus this two objects located at 7 mm from each others can be clearly resolved, and the lower measured distance between two objects was ~ 4 mm. These results are in good agreement with the 2 mm lateral resolution measured.

We obtained several reconstructed 3D images of biological samples, revealing an optical absorbing object. In Figure 3 [8] the 3 cm turkey breast sample was scanned by step of 1 mm along X direction, 2 mm along Y and 4 mm along Z. In both planes XY and XZ, the presence and the position of the absorbing object (3 mm diameter modelling clay sphere) where clearly detected. The resolution is better in XY than XZ plan as predictable by the measure of the focal zone size.

Very encouraging results were obtained regarding future application to human body imaging. On the one hand, we revealed intrinsic structures in biolog-

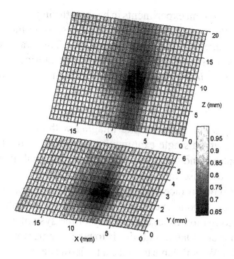

Figure 3.   3D image of a 30 mm thick turkey breast sample. The absorbing spherical inclusion (diameter 3 mm) is well located in XY and XZ maps. As mentioned in the text, due to the shape focal zone, the resolution is better along X and Y than along Z.

ical sample and in a human tissue. 2D image of a ligament inside a 2 cm thick turkey breast sample (Fig. 3) was detected, in good agreement with the real size, subsequently measured by direct examination of the dissected sample. And an

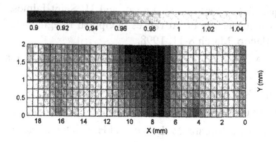

Figure 4.   Acousto-optic image of a transversal cut of a ligament inside a 2 cm turkey breast sample.

histologic cut of a womb containing a tumor, showed a clear difference of optical contrast between normal tissue and tumor, corresponding to the difference of colors visible on the surface of the cut.

On the other hand, recent experiments in backscattering geometry also yielded us good acousto-optic imaging. We detected an absorbing inclusion located at 1 cm under the surface (XZ plane) of the sample. This was cross-checked in transillumination with a good agreement. The backscattering configuration will certainly lead to easier *in vivo* examination of other human organs.

Prospects

We plan to couple our acousto-optic imaging with an echograph. This instrument will yield, in an single examination, two images related respectively to the acoustic contrast and to the optical contrast. We shall improve our setup in terms of signal to noise ratio and of measurement time for *in vivo* applications. We shall have to face various sources of noise, like the influence of blood circulation, of osmotic exchanges, and of course the patient movements. Concerning the acquisition time, the stability of the system allows a reduction of the number of pixels. One possibility is to take a camera with less pixels but a higher frame rate. We will have to make a trade-off between time acquisition, light level and signal to noise ratio. If real time is not yet possible, a first step could consist in using acousto-optic imaging as a complementary diagnostic whenever echography result reveals a suspicious zone.

References

[1] C.G.A. Hoelen, F.F.M. de Mul, R. Pongers, A. Dekker, Opt. Lett. **28**, 3 (1998).

[2] A.A. Oraevsky, Technical digest, CLEO Europe, Munich (1999).

[3] S. Lévêque, A.C. Boccara, M. Lebec, and H. Saint-Jalmes, Proc. OSA Spring Topical Meeting on Advances in Optical Imaging and Photon Migration, Orlando, March 8-11, 1998.

[4] M. Kempe, M. Larionov, D. Zaslavsky and A.Z. Genack, J. Opt. Soc. Am. A, **14**, 1151 (1997).

[5] L. Wang and X. Zhao, Appl. Opt. **36**, 7277 (1997).

[6] W. Leutz and G. Maret, Physica B **204**,14 (1995).

[7] S. Lévêque, A.C. Boccara, M. Lebec, and H. Saint-Jalmes, , Opt. Lett. **24**, 181 (1999).

[8] S. Lévêque-Fort, submitted to Appl. Opt.

VII

SPECKLE CORRELATIONS
IN RANDOM MEDIA

Chapter 1

CORRELATION OF SPECKLE
IN RANDOM MEDIA

R. Pnini

Department of Physics, Technion - Israel Institute of Technology
32000 Haifa, Israel.
pnini@physics.technion.ac.il

Abstract   The diagram technique for waves propagating in disordered media is reviewed and the short, long and infinite range correlations - known as C_1, C_2 and C_3, respectively - are calculated using the Langevin scheme. This enables one to consider several experimentally relevant effects: absorption, frequency shifts, internal reflections and the effect of geometry. Fluctuations in optical transmission through disordered slabs are described in detail. Spatial correlations and deviations from Rayleigh statistics in quasi-one-dimensional systems are discussed as well.

Keywords: speckle correlations, diagrams, diffusons

Introduction

A wave propagating in random medium produces a complicated intensity pattern in the bulk and at the boundary through which it leaves the medium [1]. As a result of the randomization of the direction of multiply scattered waves, the intensity exhibits large fluctuations with a correlation length on the scale of a wavelength [2]. The speckle pattern also exhibits weak long-range correlation, which arises from diffusion of locally large fluctuations. Such long-range correlation is the microscopic source of enhanced fluctuations in optical transmission [3] and electronic conductance [4].

In the weak disorder limit the problem has a small parameter, $1/k_0\ell$, where k_0 and ℓ are, respectively, the wavenumber and the elastic mean free path. This parameter enables one to develop a perturbative diagrammatic technique for calculating various average quantities and correlation functions. Indeed, most of the theoretical results in the field - such as universal conductance fluctuations in disordered metals [4], coherent backscattering [5], fluctuations in transmission

P. Sebbah (ed.), Waves and Imaging through Complex Media, 391–412.

coefficients, the optical memory effect, and many others [6] - were obtained
by this technique. The purpose of this paper is to review diagrams and discuss
some recent developments in the field.

Section 1 illustrates the diagram techniquefor the case of a point source embedded in an infinite medium. In this context, we introduce the diffusion ladder
and calculate the intensity-intensity correlation function in the leading order
(factorization approximation). Section 2 is devoted to the Langevin description
of long range speckle correlation in arbitrary geometries. Sections 3 and 4 deal
with two specific examples: fluctuations in optical transmission through disordered slabs, and the intensity distribution in quasi-one dimensional samples.
The conclusions are summarized in the last section.

1.   A SHORT REVIEW OF DIAGRAMS

Let us consider the scalar wave equation for a monochromatic field, of frequency ω, propagating in weakly absorbing random medium:

$$\{\nabla^2 + k_0^2[1 + \mu(\mathbf{r}) + 2i\eta)]\}\psi_\omega(\mathbf{r}) = 0 \ . \tag{1}$$

The field $\psi_\omega(\mathbf{r})$ describes a classical wave, e.g., acoustic or (if polarization
effects can be neglected) electromagnetic wave [7]. Here $0 < \eta \ll 1$ is
the attenuation index and $k_0 = \omega/c$, where c is the speed of propagation in
the average medium. The function $\mu(\mathbf{r})$ describes the fluctuating part of the
refraction index $n(\mathbf{r})$, so that $n^2(\mathbf{r}) = 1 + \mu(\mathbf{r}) + 2i\eta$. The energy current
density is defined as

$$\mathbf{J}(\mathbf{r}) = \frac{ic}{2k_0}\left[\psi(\mathbf{r})\nabla\psi^*(\mathbf{r}) - \text{c.c}\right] \ , \tag{2}$$

where c.c is the complex conjugate. To define the problem completely one
needs to specify the sources and give the statistical properties of the random
function $\mu(\mathbf{r})$. We shall assume a white-noise Gaussian statistics [8], i.e.

$$\langle\mu(\mathbf{r})\rangle = 0 \ , \quad \langle\mu(\mathbf{r})\mu(\mathbf{r}')\rangle = \epsilon\delta(\mathbf{r} - \mathbf{r}') \ , \tag{3}$$

and higher order correlations factorize into products of pairs. In Eq. (3) angular brackets denotes averaging over an ensemble of macroscopically identical
samples, ϵ is a constant satisfying $\epsilon k_0^3 \ll 1$ (weak disorder limit), and both $\mathbf{r}$
and $\mathbf{r}'$ should be within the medium [outside the medium $\mu(\mathbf{r}) \equiv 0$]. For a
given realization of disorder $\psi_\omega(\mathbf{r})$ satisfies

$$\psi_\omega(\mathbf{r}) = \psi_0(\mathbf{r}) + \int d^3r_1 G_0(\mathbf{r}, \mathbf{r}_1)[-k_0^2\mu(\mathbf{r}_1)]\psi_\omega(\mathbf{r}_1) \ , \tag{4}$$

where $\psi_0(\mathbf{r})$ is the free field in the absence of random scatterers [i.e. the solution
of Eq. (1) for $\mu(\mathbf{r}) \equiv 0$] and $G_o(\mathbf{r}, \mathbf{R})$ is the corresponding free Green's func-
tion, $\left[\nabla^2 + k_o^2\left(1 + 2i\eta\right)\right]G_o(\mathbf{r}, \mathbf{R}) = \delta(\mathbf{r} - \mathbf{R})$. We are therefore interested

in calculating quantities like the average field $\langle \psi_\omega(\mathbf{r}) \rangle$, the average intensity $\langle I_\omega(\mathbf{r}) \rangle \equiv \langle |\psi_\omega(\mathbf{r})|^2 \rangle$, the field-field correlation function $\langle \psi_\omega(\mathbf{r})\psi_{\omega'}^*(\mathbf{r}') \rangle$, or some more complicated objects involving products of fields at different points and different frequencies. This is done by expanding Eq. (4) in the standard Born series and averaging this series, or its products, term by term. Various terms are then represented by diagrams to allow for convenient partial summations. The rules for generating and calculating such diagrams are described in the literature [9]. Below we give a brief summary, using the generic example of a point source embedded in an infinite disordered medium. The field at a point $\mathbf{r}$ due to a point source at $\mathbf{R}$ is then given by the Green's function of Eq. (1), satisfying

$$G_\omega(\mathbf{r}, \mathbf{R}) = G_0(\mathbf{r}, \mathbf{R}) + \int d^3r_1 G_0(\mathbf{r}, \mathbf{r}_1)[-k_0^2\mu(\mathbf{r}_1)]G_\omega(\mathbf{r}_1, \mathbf{R}) . \quad (5)$$

The averaged Green's function.   Diagrams for the averaged Green's function, $\langle G_\omega(\mathbf{r}, \mathbf{R}) \rangle$, are generated from the Born series for $G_\omega(\mathbf{r}, \mathbf{R})$ as shown in Fig. 1. A dashed line connecting two points $\mathbf{r}_i, \mathbf{r}_j$ in Fig. 1b corresponds

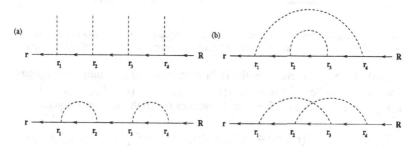

Figure 1.   (a) An example of a scattering sequence in the Born series for $G_\omega(\mathbf{r}, \mathbf{R})$ describing propagation from $\mathbf{R}$ to $\mathbf{r}$ with scattering events at 4 intermediate points. (b) Three diagrams for $\langle G_\omega(\mathbf{r}, \mathbf{R}) \rangle$ generated by the sequence in (a) upon averaging with respect to the Gaussian statistics.

to $k_0^4 \langle \mu(\mathbf{r}_i)\mu(\mathbf{r}_j) \rangle = \epsilon k_0^4 \delta(\mathbf{r}_i - \mathbf{r}_j)$, a solid line represents the free Green's function $G_0(|\mathbf{r}_i - \mathbf{r}_j|)$ and integration, over all intermediate points, is implied. Partial summation of the entire series is done with the help of Dyson's equation $\langle G \rangle = G_0 + G_0 \Sigma \langle G \rangle$. To the leading order in the self-energy, Σ, one obtains:

$$\langle G_\omega(\mathbf{r}, \mathbf{R}) \rangle = -\frac{1}{4\pi|\mathbf{r} - \mathbf{R}|} \exp\left[ik_o|\mathbf{r} - \mathbf{R}| - |\mathbf{r} - \mathbf{R}|/2\ell\right] , \quad (6)$$

where $\ell = (2\eta k_o + \epsilon k_o^4/4\pi)^{-1}$ is the mean free path, and a small correction to the speed of propagation has been neglected. It should be noted that Eq. (6) exhibits exponential decay even in the absence of absorption. This decay is related to the randomization of phases over distances $|\mathbf{r} - \mathbf{R}|$ larger than the mean free path.

The field correlation function.   The field-field correlation function consists of a trivial product of two averaged Green's functions (the disconnected part) and a connected part: $\langle GG^* \rangle = \langle G \rangle \langle G^* \rangle + \langle GG^* \rangle_c$. To the leading order in $1/k_0\ell$ the connected part is given by the sum of ladder diagrams shown in Fig. 2. Solid lines with arrows to the left represent the averaged Green's functions, given in

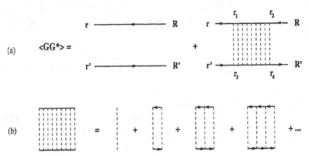

Figure 2.   (a) The field-field correlation function. Solid lines with arrows to the (left/right) represent the (retarded/advanced) averaged Green's functions. (b) The sum of ladder diagrams.

Eq. (6). Lines with arrows to the right correspond to the complex conjugate. The ladder can be written as $L_{\omega\omega'}(\mathbf{r}_1, \mathbf{r}_2, \mathbf{r}_3, \mathbf{r}_4) = (\epsilon k_0^4)P_{\omega\omega'}(\mathbf{r}_1, \mathbf{r}_2)\delta(\mathbf{r}_1 - \mathbf{r}_3)\delta(\mathbf{r}_2 - \mathbf{r}_4)$, where the function P satisfies the Bethe-Saltpeter equation:

$$P_{\omega\omega'}(\mathbf{r}_1, \mathbf{r}_2) = \delta(\mathbf{r}_1, \mathbf{r}_2) + \epsilon k_0^4 \int d^3 r \langle G_\omega(\mathbf{r}_1, \mathbf{r}) \rangle \langle G_{\omega'}^*(\mathbf{r}_1, \mathbf{r}) \rangle P_{\omega\omega'}(\mathbf{r}, \mathbf{r}_2) . \tag{7}$$

If $|\mathbf{r}_1 - \mathbf{r}_2| \gg \ell$ and $|\omega - \omega'| \equiv |\Delta\omega| \ll c/\ell$, this equation is equivalent to a diffusion equation

$$\left(\nabla^2 - \alpha^2 + i\beta^2\right) P_{\omega\omega'}(\mathbf{r}_1 - \mathbf{r}_2) = -\frac{3}{\ell^2}\delta(\mathbf{r}_1 - \mathbf{r}_2) , \tag{8}$$

where $\alpha^2 = 6\eta k_0/\ell$ is the squared inverse absorption length, $\beta^2 = \Delta\omega/D$, and $D = c\ell/3$ is the diffusion constant. Eq. (8) is solved by

$$P_{\omega\omega'}(\mathbf{r}_1 - \mathbf{r}_2) = \frac{3}{4\pi\ell^2|\mathbf{r}_1 - \mathbf{r}_2|} \exp(-\gamma|\mathbf{r}_1 - \mathbf{r}_2|) , \tag{9}$$

with $\gamma^2 = \alpha^2 - i\beta^2$ and $\mathrm{Re}\gamma \geq 0$. The function P is thus known as the diffusion propagator or "diffuson". Since $|\gamma|\ell \ll 1$, the diffuson remains almost unchanged over distances of the order of the mean free path, whereas the averaged Green's functions decay rapidly. Therefore, when $|\mathbf{r} - \mathbf{R}|, |\mathbf{r}' - \mathbf{R}'| \gg \ell$,

$$\langle G_\omega(\mathbf{r}, \mathbf{R}) G^*_{\omega'}(\mathbf{r}', \mathbf{R}') \rangle = \left(\frac{4\pi}{\ell}\right) \int d^3 r_1 d^3 r_2 \langle G_\omega(\mathbf{r}, \mathbf{r}_1) \rangle \langle G^*_{\omega'}(\mathbf{r}', \mathbf{r}_1) \rangle \quad (10)$$
$$\times P_{\omega\omega'}(\mathbf{r}_1, \mathbf{r}_2) \langle G_\omega(\mathbf{r}_2, \mathbf{R}) \rangle \langle G^*_{\omega'}(\mathbf{r}_2, \mathbf{R}') \rangle ,$$

where the exponentially small disconnected term has been neglected. Pulling the diffuson out of this integral, taking its value at $\mathbf{r}_1 = \mathbf{r}, \mathbf{r}_2 = \mathbf{R}$, one ends up with:

$$\langle G_\omega(\mathbf{r}, \mathbf{R}) G^*_{\omega'}(\mathbf{r}', \mathbf{R}') \rangle = \left(\frac{\ell}{4\pi}\right) P_{\omega\omega'}(\mathbf{r}, \mathbf{R}) f(\Delta r) f(\Delta R) , \quad (11)$$

where $\Delta r \equiv |\mathbf{r} - \mathbf{r}'|, \Delta R \equiv |\mathbf{R} - \mathbf{R}'|$ and $f(x) \equiv (\sin k_0 x / k_0 x) \exp(-x/2\ell)$. The average intensity is now obtained by setting in Eq. (11) $\Delta r = \Delta R = \Delta \omega = 0$:

$$\langle I_\omega(\mathbf{r}, \mathbf{R}) \rangle = \langle |G_\omega(\mathbf{r}, \mathbf{R})|^2 \rangle = \frac{3}{16\pi^2 \ell |\mathbf{r} - \mathbf{R}|} \exp(-\alpha|\mathbf{r} - \mathbf{R}|) . \quad (12)$$

Using Eq. (2) one also recovers Fick's law $\langle \mathbf{J}_\omega(\mathbf{r}) \rangle = -D\nabla \langle I_\omega(\mathbf{r}) \rangle$.

Intensity correlations.   Consider now the intensity-intensity correlation function,

$$C_{\omega\omega'}(\mathbf{r}, \mathbf{r}', \mathbf{R}; \mathbf{R}') \equiv \langle \delta I_\omega(\mathbf{r}, \mathbf{R}) \delta I_{\omega'}(\mathbf{r}', \mathbf{R}') \rangle , \quad (13)$$

where δI_ω is the deviation of the intensity at point $\mathbf{r}$ from its average value. This function involves averaging over a product of four Green's functions $\langle I_\omega(\mathbf{r}, \mathbf{R}) I_{\omega'}(\mathbf{r}', \mathbf{R}') \rangle$. If both $\mathbf{r}$ and $\mathbf{r}'$ are sufficiently far (i.e. at least few

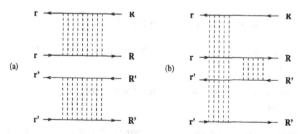

Figure 3.   Diagrams for $\langle I_\omega(\mathbf{r}, \mathbf{R}) I_{\omega'}(\mathbf{r}', \mathbf{R}') \rangle$ showing two diffusion ladders inserted between pairs of averaged Green's functions. (a) The disconnected part. (b) The short-range correlation (C_1 term).

ℓ's) from the sources, and $\Delta r \equiv |\mathbf{r} - \mathbf{r}'| \leq \ell$, then the product is given by the

diagrams in Fig. 3 (any other diagram gives an exponentially small contribution, or a contribution in high orders of perturbation theory). These diagrams
are obtained by inserting two diffusion ladders between four lines of average
Green's functions [10]. The short-range intensity-intensity correlation function
is, therefore, factorized into a product of two field-field correlation functions:

$$C_1(\mathbf{r}, \mathbf{r}'; \mathbf{R}, \mathbf{R}', \omega, \omega') = |\langle G_\omega(\mathbf{r}, \mathbf{R}) G^*_{\omega'}(\mathbf{r}', \mathbf{R}') \rangle|^2 . \qquad (14)$$

$$= \left(\frac{\ell}{4\pi} \right)^2 |P_{\omega\omega'}(\mathbf{r}, \mathbf{R})|^2 f^2(\Delta r) f^2(\Delta R) ,$$

where the subscript indicates the leading order in perturbation theory. For
$\Delta\omega = 0$ expression (14) reduces to

$$C_1(\mathbf{r}, \mathbf{r}'; \mathbf{R}, \mathbf{R}') = \langle I_\omega(\mathbf{r}-\mathbf{R}) \rangle^2 f^2(\Delta r) f^2(\Delta R) \qquad (15)$$

and if, in addition $\Delta r = \Delta R = 0$, one finds that $\langle \delta I^2 \rangle = \langle I \rangle^2$ which is the
well known large intensity fluctuations [1].

It was pointed out by Stephen and Cwilich [11] that Eq. (15) is valid only
for small distances, i.e. for $\Delta r \lesssim \ell$. Indeed, for $\Delta r \gg \ell$, expression (15) vanishes exponentially and diagrams - other than that in Fig. 3b – dominate the
intensity-intensity correlation function. The long-range correlations in the intensity speckle can be obtained from diagrams which describe the effective
interaction between two diffusons in the bulk of the random medium [12],
as shown in Figs. 4b and c. Such diagrams, or more precisely their Fourier
transforms, were identified and calculated by Feng, Kane, Lee and Stone who
studied the correlation of transmission coefficients in the waveguide geometry
[13]. Following their notations, one can distinguish between 3 terms, which
give the leading order contributions to the correlation function in a $(k_0\ell)^{-2}$
expansion:

- C_1 - short-range correlation, as in Eq. (15).

- C_2 - long-range correlation which dominates the fluctuations in total
  optical transmission [13, 14].

- C_3 - infinite-range correlation which dominates the universal conductance
  fluctuations [15].

In this context, the case of a point source in three dimensional medium
forms an exception. Since the source is "small", the intensity correlation,
$\langle \delta I_\omega(\mathbf{r}, \mathbf{R}) \delta I_{\omega'}(\mathbf{r}', \mathbf{R}) \rangle$, is built up due to interference of diffusons near the
source, as well as in the bulk. As a result, the C_2 term - which is of order
$(k_0\ell)^{-2}$ and decays under increase of Δr or $\Delta\omega$ according to a power law [11]
- is dominated by a source correlated term of order $(k_0\ell)^{-1}$, which has infinite

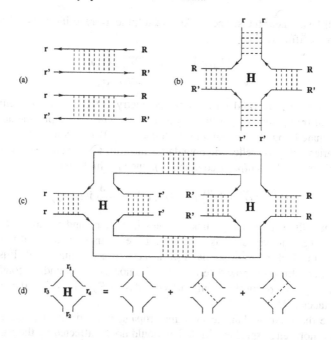

Figure 4. Diagrams for $\langle \delta I_\omega(\mathbf{r}, \mathbf{R}) \delta I_{\omega'}(\mathbf{r}', \mathbf{R}') \rangle$. (a) C_1 correlation, (b) C_2 correlation and (c) C_3 correlation. (d) The interaction vertex (Hikami box) $H(\mathbf{r}_1, \mathbf{r}_2, \mathbf{r}_3, \mathbf{r}_4) = \left(\frac{\ell^5}{48\pi k_0^2} \right) \int d^3\rho \left[(\nabla_1 + \nabla_2) \cdot (\nabla_3 + \nabla_4) + 2\nabla_1 \cdot \nabla_2 + 2\nabla_3 \cdot \nabla_4 + 2\alpha^2 \right] \prod_{i=1}^{4} \delta(\rho - \mathbf{r}_i)$.

range [16]. Being sensitive to the short-range properties of disorder this term is not universal [16, 17].

Strictly speaking, the diagrams in Figs. 4b and c are divergent. The divergencies are clearly unphysical, since they appear when the interaction points between diffusons approach either the sources or one another. To eliminate the divergencies one has to go beyond the diffusion approximation or take into account a large class of short-range diagrams as in [18]. Alternatively, one can generate C_2 and C_3 from simpler diagrams, using the principle of current conservation [19] or the Langevin scheme [20]. This is discussed in the next section.

2. SPECKLE CORRELATIONS IN THE LANGEVIN APPROACH

Speckle correlations in realistic multiply scattering samples are modified by the presence of boundaries. However, due to the exponential decay of the coherent field $\langle \psi_\omega(\mathbf{r}) \rangle$ in the sample, the boundary effects are usually restricted to small layers near the surfaces. As a result, the average intensity within the medium is diffusive, i.e.

$$\langle \mathbf{J}_\omega(\mathbf{r}) \rangle = -D\nabla\langle I_\omega(\mathbf{r}) \rangle \tag{16a}$$

$$\nabla \cdot \langle \mathbf{J}_\omega(\mathbf{r}) \rangle = -2\omega\eta\langle I_\omega(\mathbf{r}) \rangle . \tag{16b}$$

These equations are valid for arbitrary geometry in the absence of sources, a few mean free paths away from the boundaries. They are supplemented with approximated boundary conditions such that $\langle I_\omega(\mathbf{r}) \rangle$ is specified on absorbing boundaries, while on reflecting boundaries the normal derivative $\hat{n} \cdot \langle \mathbf{J}_\omega(\mathbf{r}) \rangle = 0$. Using the Bethe-Saltpeter equation (7), one also finds that

$$\left(\nabla^2 - \alpha^2 + i\beta^2\right) P_{\omega\omega'}(\mathbf{r}_1, \mathbf{r}_2) = -\frac{3}{\ell^2}\delta(\mathbf{r}_1, \mathbf{r}_2) , \tag{17}$$

which is valid for optically thick samples if both $\mathbf{r}_1$ and $\mathbf{r}_2$ are sufficiently far from the boundaries. This equation for the diffuson is identical to (8) except for the lack of translational invariance. In accordance with Eqs. (16a and b), the diffuson is now determined by homogeneous boundary conditions: $P_{\omega\omega'}(\mathbf{r}_1, \mathbf{r}_2) = 0$ on an absorbing surface and $\hat{n} \cdot \nabla P_{\omega\omega'}(\mathbf{r}_1, \mathbf{r}_2) = 0$ on reflecting surfaces [21].

Since the wave within the medium propagates by diffusion, one expects that the short range speckle correlation would not be affected by the particular geometry of the sample. Integrating out the sources in (10), one indeed finds [22]

$$\langle \psi_\omega(\mathbf{r})\psi^*_{\omega'}(\mathbf{r}') \rangle = \left(\frac{4\pi}{\ell}\right) \int d^3r_1 \langle G_\omega(\mathbf{r}, \mathbf{r}_1) \rangle\langle G^*_{\omega'}(\mathbf{r}', \mathbf{r}_1) \rangle\langle \psi_\omega(\mathbf{r}_1)\psi^*_{\omega'}(\mathbf{r}_1) \rangle ,$$

where $\langle \psi_\omega(\mathbf{r})\psi^*_{\omega'}(\mathbf{r}) \rangle \equiv \langle I_{\omega\omega'}(\mathbf{r}) \rangle$ is the field auto-correlation function at point $\mathbf{r}$. Therefore,

$$C_1(\mathbf{r}, \mathbf{r}', \omega, \omega') = |\langle \psi_\omega(\mathbf{r})\psi^*_{\omega'}(\mathbf{r}') \rangle|^2 = |\langle I_{\omega\omega'}(\mathbf{r}) \rangle|^2 f^2(\Delta r) , \tag{18}$$

which is similar to (14) apart from a trivial modification of the field auto-correlation function.

In order to calculate the C_2 and C_3 terms one has to take into account the long-range nature of external current vertices [19]. Alternatively, one can use the Langevin scheme as proposed by Zyuzin and Spivak [20]. These authors have pointed out that long-range correlation in the speckle pattern can be treated by adding suitable Langevin-type sources into the macroscopic (i.e. averaged) equations describing the diffusion of the wave in the medium. In their approach one assumes that the energy current density of the wave in the random medium can be split into two terms:

$$\mathbf{J}_\omega(\mathbf{r}) = -D\nabla I_\omega(\mathbf{r}) + \mathbf{J}_{\omega \text{ext}}(\mathbf{r}) . \tag{19}$$

The first term in this equation describes the diffusive relaxation of the intensity over distances larger than ℓ. The Langevin current, $\mathbf{J}_{\omega\text{ext}}(\mathbf{r})$, describes local spatial fluctuations which are averaged to zero over distances $\Delta r \lesssim \ell$. This equation, together with (16a and b) and the continuity condition $\nabla \cdot \delta \mathbf{J}_\omega + \alpha^2 D \delta I_\omega = 0$, enables one to express the fluctuating quantities $\delta \mathbf{J} \equiv \mathbf{J} - \langle \mathbf{J} \rangle$ and $\delta I \equiv -\langle I \rangle$, in terms of the Langevin current. Thus [23],

$$\delta I_\omega(\mathbf{r}) = \frac{\ell^2}{3D} \int_\Omega d^3 r' [\nabla' P_\omega(\mathbf{r}, \mathbf{r}')] \cdot \mathbf{J}_{\omega\text{ext}}(\mathbf{r}') \qquad (20a)$$

$$\delta \mathbf{J}_\omega(\mathbf{r}) = -D \nabla \delta I_\omega(\mathbf{r}) + \mathbf{J}_{\omega\text{ext}}(\mathbf{r}) , \qquad (20b)$$

where Ω is the volume of the scattering region and $P_\omega(\mathbf{r}, \mathbf{r}')$ is the diffuson at zero frequency shift. In this way, long range correlations in the intensity (and current density) develop from "spontaneous" fluctuations, which occur at distances smaller than ℓ and propagate by diffusion to large distances. The intensity-intensity correlation function is now obtained from Eq. (20a) in the form

$$C_{\omega\omega'}(\mathbf{r}, \mathbf{r}') = \frac{\ell^4}{9D^2} \int\!\!\int_\Omega d^3 r_1 d^3 r_2 [\nabla_{1a} P_\omega(\mathbf{r}, \mathbf{r}_1)][\nabla_{2b} P_\omega(\mathbf{r}', \mathbf{r}_2)] \Gamma^{ab}_{\omega\omega'}(\mathbf{r}_1, \mathbf{r}_2) ,$$
$$(21)$$

where $\Gamma^{ab}_{\omega\omega'}(\mathbf{r}_1, \mathbf{r}_2) \equiv \langle J^a_{\omega\text{ext}}(\mathbf{r}_1) J^b_{\omega'\text{ext}}(\mathbf{r}_2) \rangle$ denotes the correlator of the Langevin currents with components $a, b = 1, 2, 3$. The problem is therefore reduced to a diagrammatic calculation of the appropriate Langevin correlators.

The current correlator which generates the C_2 term is determined from expression (18) by identifying $\mathbf{J}_{\omega\text{ext}}(\mathbf{r}) \equiv c \Delta I_\omega(\mathbf{r}) \hat{s}$, where ΔI is the short-range intensity fluctuation and $\hat{s}$ is a unit vector in the direction of propagation. This gives

$$\langle J^a_{\omega\text{ext}}(\mathbf{r}_1) J^b_{\omega'\text{ext}}(\mathbf{r}_2) \rangle_2 = \left(\frac{2\pi \ell c^2}{3k_0^2} \right) \delta^{ab} \delta(\mathbf{r}_1 - \mathbf{r}_2) |\langle I_{\omega\omega'}(\mathbf{r}_1) \rangle|^2 , \qquad (22)$$

where the rapidly decaying function $f^2(\Delta r)$ has been replaced by a delta-function. The long-range intensity correlation can therefore be written as

$$C_2(\mathbf{r}, \mathbf{r}', \omega, \omega') = \frac{2\pi \ell^3}{3k_0^2} \int_\Omega d^3 r_1 [\nabla P_\omega(\mathbf{r}, \mathbf{r}_1)] \cdot [\nabla P_\omega(\mathbf{r}', \mathbf{r}_1)] |\langle I_{\omega\omega'}(\mathbf{r}_1) \rangle|^2 .$$
$$(23)$$

Using the explicit expression for the Hikami box, one can verify that Eq. (23) coincides with a direct calculation of the diagram shown in Fig. 4b. The current correlators which generate the C_3 term are determined in a similar way, taking

into account high order diagrams containing four diffusion propagators. Thus
[24, 25]

$$\langle J^a_{\omega\text{ext}}(\mathbf{r}_1) J^b_{\omega'\text{ext}}(\mathbf{r}_2)\rangle_3 = \left(\frac{2\pi\ell^2 c}{3k_0^2}\right)^2 [\Gamma_1 + \Gamma_2 + \Gamma_3]^{ab} , \qquad (24)$$

$$\begin{aligned}
\Gamma_1^{ab} &= \delta^{ab}\delta(\mathbf{r}_1 - \mathbf{r}_2)\int d^3r |P_{\omega\omega'}(\mathbf{r}_1,\mathbf{r})|^2 \nabla\langle I_\omega(\mathbf{r})\rangle \cdot \nabla\langle I_{\omega'}(\mathbf{r})\rangle \\
\Gamma_2^{ab} &= |P_{\omega\omega'}(\mathbf{r}_1,\mathbf{r}_2)|^2 \nabla^a\langle I_\omega(\mathbf{r}_1)\rangle\nabla^b\langle I_{\omega'}(\mathbf{r}_2)\rangle \\
\Gamma_3^{ab} &= |P_{\omega\omega'}(\mathbf{r}_1,\mathbf{r}_2)|^2 \nabla^b\langle I_\omega(\mathbf{r}_1)\rangle\nabla^a\langle I_{\omega'}(\mathbf{r}_2)\rangle
\end{aligned}$$

Eqs. (20a and b), (22) and (24) are convenient for studying various sample
geometries. Specific examples of their use are considered in the following
sections.

3.   FLUCTUATIONS IN TRANSMISSION

Fluctuations in optical transmission provide an instructive example for large
relative fluctuations, which are dominated by the slowly decaying long range
speckle correlation [26-29]. Let us consider, for instance, a disordered slab
of thickness $L \gg \ell$ in the z direction, and infinite in the transverse (x, y)
directions. Suppose now that a beam of radiation, with an arbitrary intensity
profile is impinging on the slab along the z-axis. The diffuson in this geometry
is

$$P_\omega(\mathbf{r}_1,\mathbf{r}_2) = \frac{3}{\ell^2}\int \frac{d^2p}{(2\pi)^2}\exp[i\mathbf{p}\cdot(\mathbf{R}_1 - \mathbf{R}_2)]\frac{\sinh(\gamma z_<)\sinh[\gamma(L-z_>)]}{\gamma\sinh(\gamma L)} , \qquad (25)$$

where $\mathbf{R} = (x, y)$ is a vector in the transverse direction, $\mathbf{p} = (p_x, p_y)$ the
transverse momentum, and $\gamma^2 = p^2 + \alpha^2$ with α the inverse absorption length.
The total incident flux on the $z = 0$ plane is $c\int d^2R\Lambda(\mathbf{R}) = c\tilde{\Lambda}(\mathbf{p} = 0)$, where
$\tilde{\Lambda}(\mathbf{p})$ is the Fourier transform of the incident beam profile. Therefore,

$$\langle I_\omega(\mathbf{r})\rangle = \int \frac{d^2p}{(2\pi)^2}\exp[i\mathbf{p}\cdot\mathbf{R}]\frac{\sinh\gamma(L-z)}{\sinh\gamma L}\tilde{\Lambda}(\mathbf{p}) \qquad (26a)$$

$$\langle J^z_\omega(\mathbf{r})\rangle = D\int \frac{d^2p}{(2\pi)^2}\exp[i\mathbf{p}\cdot\mathbf{R}]\gamma\frac{\cosh\gamma(L-z)}{\sinh\gamma L}\tilde{\Lambda}(\mathbf{p}) . \qquad (26b)$$

The transmission coefficient, T_ω, is defined as the ratio between the emergent
flux at $z = L$ (i.e. current density integrated over cross section) and the incident
flux:

$$T_\omega = \int d^2R J^z_\omega(\mathbf{R}, L)/c\tilde{\Lambda}(0) . \qquad (27)$$

The average transmission coefficient is thereby independent on the beam profile and equals to

$$\langle T_\omega \rangle = \frac{D}{c} \frac{\alpha}{\sinh \alpha L} = \frac{\alpha \ell}{3 \sinh \alpha L} \tag{28}$$

The relative fluctuation of T_ω, or more generally, the correlation $\langle \delta T_\omega \delta T_{\omega'} \rangle$ can then be written in the form

$$\frac{\langle \delta T_\omega \delta T_{\omega'} \rangle}{\langle T_\omega \rangle^2} = \frac{\langle \delta \tilde{J}_\omega(L) \delta \tilde{J}_{\omega'}(L) \rangle}{\langle \tilde{J}_\omega(L) \rangle^2}, \tag{29}$$

where $\tilde{J}_\omega(z) \equiv \int d^2 R J_\omega^z(\mathbf{R}, z)$ is the flux passing through z-plane (for $\alpha = 0$ the flux is obviously a z-independent random variable). Integrating the Langevin equations (20a and b) over the cross section and expressing $\delta \tilde{J}_\omega(z)$ in terms of Langevin flux $\tilde{J}_{\omega\text{ext}}(z) \equiv \int d^2 R J_{\omega\text{ext}}^z(\mathbf{R}, z)$, one obtains

$$\delta \tilde{I}_\omega = \frac{1}{D} \int_0^L dz' \left[\frac{d}{dz'} K_\alpha(z, z') \right] \tilde{J}_{\omega\text{ext}}(z') \tag{30a}$$

$$\delta \tilde{J}_\omega = - \int_0^L dz' \left[\frac{d^2}{dz \, dz'} K_\alpha(z, z') \right] \tilde{J}_{\omega\text{ext}}(z') \tag{30b}$$

where $K_\alpha(z, z') = \sinh(\alpha z_<) \sinh[\alpha(L - z_>)]/\alpha \sinh \alpha L$. Therefore,

$$\frac{\langle \delta T_\omega \delta T_{\omega'} \rangle}{\langle T_\omega \rangle^2} = \frac{1}{[D\tilde{\Lambda}(0)]^2} \int_0^L dz_1 dz_2 \cosh \alpha z_1 \cosh \alpha z_2 \langle \tilde{J}_{\omega\text{ext}}(z_1) \tilde{J}_{\omega'\text{ext}}(z_2) \rangle. \tag{31}$$

Expression (31) holds for optically thick samples, where the effects of internal reflections can be neglected [30].

Plane wave limit. In the plane wave limit $\tilde{\Lambda}(0) = I_0 A$, where I_0 is the incident intensity on a unit area. Using Eq. (22) together with (26a) and assuming $\alpha = 0$, one finds

$$\langle \tilde{J}_{\omega\text{ext}}(z_1) \tilde{J}_{\omega\text{ext}}(z_2) \rangle_2 = \frac{2}{gL} \left(\frac{I_0 AD}{L} \right)^2 \delta(z_1 - z_2)(L - z)^2 \tag{32}$$

where $g = A k_0^2 \ell / 3 \pi L \gg 1$ is the average conductance of the slab. Similarly, using (24)

$$\langle \tilde{J}_{\omega\text{ext}}(z_1) \tilde{J}_{\omega\text{ext}}(z_2) \rangle_3 = \left(\frac{2 I_0 AD}{gL^2} \right)^2 [\tilde{\Gamma}_1 + \tilde{\Gamma}_2 + \tilde{\Gamma}_3], \tag{33}$$

$$\tilde{\Gamma}_1 = \delta(z_1 - z_2) \int_0^L dz \sum_p K_p^2(z_1, z) \; ; \quad \tilde{\Gamma}_2 = \tilde{\Gamma}_3 = K_p^2(z_1, z_2) \,,$$

where the summation goes over all transverse modes. Equations (32) and (33) reproduce, respectively, the leading order contributions to the C_2 and C_3 terms. Inserting these expressions back into (31) one immediately obtains:

$$\frac{\langle \delta T_\omega^2 \rangle}{\langle T_\omega \rangle^2} = \frac{2}{3g} + \frac{3}{2g^2} \sum_p F_3(pL) \,, \tag{34}$$

with $F_3(x) = (2 + 2x^2 - 2\cosh 2x + x\sinh 2x)/x^4 \sinh^2 x$. This coincides with the much more complicated calculation of Ref. [18]. It is straightforward to generalize the calculation to include absorption and frequency shifts. The result in the presence of absorption, for example, is given by [23, 31]

$$\frac{\langle \delta T_\omega^2 \rangle}{\langle T_\omega \rangle^2} = \left(\frac{1}{8g} \right) \frac{\sinh 2\alpha L - 2\alpha L(2 - \cosh 2\alpha L)}{\alpha L \sinh^2 \alpha L} \,, \tag{35}$$

where all the sub-leading terms of order $1/g^2$ have been neglected. For strong absorption, $\alpha L \gg 1$, expression (35) assumes an absorption-independent value $\langle \delta T_\omega^2 \rangle / \langle T_\omega \rangle^2 = 1/2g$. This demonstrates the difference between absorption of classical waves and inelastic scattering of electrons [32, 33]: absorption introduces amplitude diminishment but does not necessary lead to dephasing (paths and time-reversed paths remain coherent).

Angular correlations.   The angular correlation, $\langle \delta T_\omega(\theta_1) \delta T_\omega(\theta_2) \rangle$, for a beam incident on the slab at an angle θ with respect to the normal, can be calculated by a slight modification of Eq. (32) which takes into account the phase difference of the beams at the $z = 0$ plane [24]. Substituting $\tilde{\Lambda}(\mathbf{p}) = \delta(\mathbf{p} - \Delta\mathbf{q})$ in (26a), one finds

$$\langle \tilde{J}_{\theta_1 \text{ext}}(z_1) \tilde{J}_{\theta_2 \text{ext}}(z_2) \rangle_2 = \frac{2L}{g} \left(\frac{I_0 AD}{L} \right)^2 \delta(z_1 - z_2) \frac{\sinh^2[\Delta q(L - z)]}{\sinh^2 \Delta q L} \tag{36}$$

where $q \equiv k_0 \sin\theta$. Therefore,

$$\frac{\langle \delta T_\omega(\theta_1) \delta T_\omega(\theta_2) \rangle}{\langle \delta T_\omega^2 \rangle} = \frac{3}{2} F_2(\Delta q L) + \frac{9}{4g} \left[1 - \frac{3}{2} F_2(\Delta q L) \right] \sum_p F_3(pL) \,, \tag{37}$$

where $F_2(x) = (\sinh 2x - 2x)/2x \sinh^2 x$, in agreement with [11, 13]. The F_3 term does not depend on the angle of incidence and thus controls the universal conductance fluctuations [15, 18]:

$$\langle \delta g^2 \rangle = \frac{3}{2} \sum_{\mathbf{p}} F_3(pL) = \begin{cases} A/2\pi L^2 & 3\text{D} \ (L^2 \lesssim A) \\ 2/15 & \text{quasi} - 1\text{D} \end{cases} \tag{38}$$

For strongly absorbing samples, Eqs. (31) and (33) yield $\langle \delta g^2 \rangle = (\alpha L)^2 e^{-2\alpha L}/2$ in agreement with Ref. [33].

Gaussian Beam.   The form factor F_2 also appears when one considers a slab which is illuminated with a Gaussian beam of radius R_0 along the z-axis [22, 27]. In this case $\Lambda(\mathbf{R}) = I_0 \exp[-2R^2/R_0^2]$ and the transmission fluctuations are given by

$$\frac{\langle \delta T_\omega^2 \rangle}{\langle T_\omega \rangle^2} = \frac{3\pi L}{k_0^2 \ell} \int \frac{d^2 p}{(2\pi)^2} F_2(pL) \exp[-p^2 R_0^2/4] \tag{39}$$

with the following limiting cases:

$$\frac{\langle \delta T_\omega^2 \rangle}{\langle T_\omega \rangle^2} = \begin{cases} \frac{2L}{k_0^2 \ell R_0^2} & R_0 \gg L \\ \\ \frac{3\sqrt{\pi}}{2k_0^2 \ell R_0} & R_0 \ll L \end{cases} \tag{40}$$

In the plane wave limit Eq (39) reduces to $\langle \delta T_\omega^2 \rangle / \langle T_\omega \rangle^2 = 2/3g$, where $A = \pi R_0^2$. For a narrow beam, $R_0 \ll L$, the relative fluctuations are independent of L and $\langle T_\omega \rangle^2 / \langle \delta T_\omega^2 \rangle \propto R_0$. This behaviour can be explained as a result of the spreading of the diffuse intensity in restricted geometries. It has been observed in temporal correlation experiments by Scheffold *et al.* [29] as discussed in the following chapter.

4. SPATIAL CORRELATIONS AND DEVIATIONS FROM RAYLEIGH STATISTICS

Let us concentrate now on the quasi-one-dimensional geometry, i.e. a tube of transverse dimension W and length $L \gg W$, filled with scattering medium. Let us denote by $I(\mathbf{r}, \mathbf{R}) \equiv |G_\omega(\mathbf{r}, \mathbf{R})|^2$ the intensity at $\mathbf{r}$ due to a monochromatic source at $\mathbf{R}$ and consider the spatial correlation function

$$C(\Delta r, \Delta R) \equiv \langle \delta I(\mathbf{r}, \mathbf{R}) \delta I(\mathbf{r}', \mathbf{R}') \rangle / \langle I(\mathbf{r}, \mathbf{R}) \rangle \langle I(\mathbf{r}', \mathbf{R}') \rangle , \tag{41}$$

where δI is the deviation of the intensity from its ensemble average value, $\Delta r = |\mathbf{r} - \mathbf{r}'|$, $\Delta R = |\mathbf{R} - \mathbf{R}'|$. For the quasi-one-dimensional geometry, the expression for the diffuson reads

$$P_\omega(\mathbf{r}, \mathbf{R}) = \frac{3}{\ell^2} \frac{x_0(L - x)}{AL} , \tag{42}$$

where A is the cross section of the tube, $\mathbf{r} = (x, y, z)$, $\mathbf{R} = (x_0, y_0, z_0)$ and the x-axis is directed along the sample (see Fig. 5a). Here it is assumed that $x - x_0 \gg W$ and $x_0, L - x \gtrsim W$ (both the source and the detector are placed not too close to the edges of the tube). Because of rapid decay of the average Green's functions, $\langle I(\mathbf{r}, \mathbf{R}) \rangle = (\ell/4\pi) P_\omega(\mathbf{r}, \mathbf{R})$. The diagrams contributing to $C(\Delta r, \Delta R)$ are shown in Figs. 4a-c. In these diagrams, each pair of external Green's functions $\langle G \rangle \langle G^* \rangle$ contributes a spatial form-factor as in Eq. (18). Each Hikami vertex, representing crossing diffusons, contributes a factor of the order of $1/g$. Therefore, taking into account all possible permutations of the external vertices, one obtains

$$
\begin{aligned}
C(\Delta r, \Delta R) &= C_1 + C_2 + C_3 = f^2(\Delta r) f^2(\Delta R) + \frac{2}{3g}[f^2(\Delta r) + f^2(\Delta R)] \\
&+ \frac{2}{15g^2}[1 + f^2(\Delta r) + f^2(\Delta R) + f^2(\Delta r) f^2(\Delta R)] \quad (43)
\end{aligned}
$$

where $f(x) = (\sin k_0 x / k_0 x) \exp(-x/2\ell)$. The structure of $C(\Delta r, \Delta R)$ is similar to that of the correlation function of transmission coefficients, obtained in the multi-channel formalism [13, 14, 25]. In the present case, however, all terms are described by a single spatial form-factor. Expression (43) provides a direct probe for identifying the different terms in the speckle correlation function by measuring $C(\Delta r, \Delta R)$ as a function of both Δr and ΔR. This kind of spatial measurements can be used to separate the C_1, C_2 and C_3 terms (in addition to spectral measurements, similar to those used in [26-28] for isolating the C_1 and C_2 contributions). The multiplicative character of C_1 and the additive character

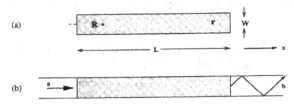

Figure 5. (a) Measurement of the local intensity: points $\mathbf{R} = (x_0, y_0, z_0)$ and $\mathbf{r} = (x, y, z)$ are the positions of the source and the observation point, respectively. (b) Measurement of t_{ba} in the waveguide geometry: a source and a detector of radiation are located outside the sample. The source produces a plane wave injected into the incoming channel a, and the intensity in outgoing channel b is measured.

of C_2 have been recently demonstrated in microwave experiments [34]. The measurement of the C_3 term in these experiments turns out to be rather difficult because of absorption.

The intensity distribution.   One of the characteristics of the speckle pattern is the intensity distribution $\mathcal{P}(I)$ at some point $\mathbf{r}$. More than a century ago Lord Rayleigh, using simple statistical arguments, proposed a distribution which bares his name:

$$\mathcal{P}_o(I) = \exp[-I/\langle I \rangle] . \tag{44}$$

The Rayleigh distribution has moments $\langle I^n \rangle = \langle I \rangle^n n!$ and it provides, in many cases, a rather accurate fit to experimental data, as long as I is not too large with respect to its average value [1]. For large I, however, the data show large deviations from Eq. (44) [1, 35-37]. Various extensions of Eq. (44) have been proposed in the literature. Jakeman and Pusey [38] proposed to fit the data with the K-distribution. It contains a phenomenological parameter κ and its moments are given by $\langle I^n \rangle = \langle I \rangle^n n! \kappa^{-n} \Gamma(n + \kappa)/\Gamma(\kappa)$. The experimentally relevant situation corresponds to $\kappa \gg 1$ and in this case moments up to $n \lesssim \kappa$ can be approximated as

$$\langle I^n \rangle \simeq \langle I \rangle^n n! \exp[n(n - 1)/\kappa] . \tag{45}$$

Therefore, only low moments, $n \ll \sqrt{\kappa}$, are close to the Rayleigh value.

Theoretical support to Eq. (45) has been given by Dashen [39], who considered smooth disorder with the typical size of inhomogeneities larger than the wavelength. More recently, the opposite case of short-range disorder in quasi-one dimensional systems have also been studied [40-43, 33] and significant deviations from Rayleigh statistics - for the distribution of the transmission coefficients $,t_{ba}$, or the local intensity inside the sample, $I(\mathbf{r}, \mathbf{R})$ - have been found. In particular, it was shown in [41, 42] that the distribution of the normalized transmission coefficients $s_{ba} \equiv t_{ba}/\langle t_{ba} \rangle$ crosses over from the Rayleigh distribution $\mathcal{P}(s_{ba}) = e^{-s_{ba}}$ to a stretched-exponential one $\mathcal{P}(s_{ab}) = e^{-2\sqrt{gs_{ba}}}$. The distribution of the local intensity, on the other hand, can exhibit different asymptotics [43]:

- if the source and detector are placed sufficiently far from each other, $\mathcal{P}(I)$ crosses over from Rayleigh distribution to a stretched-exponential.

- if both the source and the detector of radiation are embedded deep in the bulk of the medium, the stretched-exponential disappears and the Rayleigh distribution crosses over directly to a log-normal tail.

In what follows we shall restrict ourselves to the first case in which the local intensity and the transmission coefficients share the same statistics, i.e., $x - x_0 \approx L$ as shown in Fig. 5. The purpose is then to derive Eq. (45) for short-range disorder, identify the parameter κ phenomenologically introduced in [38] and obtain the corresponding intermediate asymptotics of the intensity distribution.

In the diagrammatic approach the intensity distribution, $\mathcal{P}(I)$, is obtained by calculating the moments $\langle I^n \rangle$. In the leading approximation [10], one should draw n-retarded and n-advanced Green's functions and insert diffusion ladders between pairs $\{\langle G \rangle, \langle G^* \rangle\}$ in all possible ways. This leads to $\langle I^n \rangle = n! \langle I \rangle^n$, and thus, to Eq. (44). Corrections to the Rayleigh result come from diagrams with intersecting ladders, which describe interaction between diffusons. The leading correction is due to a pairwise interaction as in Eq. (43). Setting $\Delta R = \Delta r = 0$ in this equation, and assuming $g \gg 1$, yields

$$\frac{\langle I^2 \rangle}{\langle I \rangle^2} = 2! \left(1 + \frac{2}{3g} \right) \tag{46a}$$

$$\frac{\langle \tilde{I}^2 \rangle}{\langle \tilde{I} \rangle^2} = \left(1 + \frac{2}{3g} \right), \tag{46b}$$

where $\tilde{I}$ is the total intensity, integrated over the cross section of the tube, and $\langle \tilde{I} \rangle = A \langle I \rangle$. It is easy to see now that any skeleton diagram contributing to $\langle \tilde{I}^n \rangle$ generates $n!$ diagrams for $\langle I^n \rangle$, which correspond to all possible permutations of the external vertices $\{r_1, r_2, \dots, r_n\}$. Therefore, for any n,

$$\langle I^n \rangle \simeq \frac{n!}{A^n} \langle \tilde{I}^n \rangle. \tag{47}$$

This relation holds in all orders of perturbation theory - even in the presence of high order interactions of diffusons - and it becomes an exact relation in the wide waveguide limit ($Ak_0^2 \to \infty$, $\ell/L \to 0$, fixed g). Normalizing both I and $\tilde{I}$ to their average values and introducing, respectively, the moments generating functions $M(z) = \langle \exp(-Iz) \rangle$ and $\tilde{M}(z) = \langle \exp(-\tilde{I}z) \rangle$, one immediately realizes that these functions are related by a Borel transformation:

$$M(z) = \int_0^\infty dI \mathcal{P}(I) e^{-Iz} = \sum_{n=0}^\infty \frac{(-z)^n}{n!} \langle I^n \rangle \tag{48a}$$

$$\tilde{M}(z) = B_1(z) = \sum_{n=0}^\infty \frac{(-z)^n}{n!} \frac{\langle I^n \rangle}{\Gamma(n+1)}. \tag{48b}$$

Therefore, representing $\mathcal{P}(I)$ and $\Gamma(n+1)$ as contour integrals which avoid the singularities of $M(z)$, one obtains

$$\mathcal{P}(I) = \oint \frac{dz}{2\pi i} \int_0^\infty \frac{dv}{z} \exp\left[I/z - zv - \phi(v) \right] \tag{49a}$$

$$\langle \tilde{I}^n \rangle = \frac{\langle I^n \rangle}{n!} = -(n!) \oint \frac{dw}{2\pi i} \frac{e^{-\phi(w)}}{(-w)^{n+1}}, \; n = 0, 1, 2, \dots, \tag{49b}$$

where all contours are taken around the origin and $\phi(z) \equiv -\ln \tilde{M}(z)$ is the cummulants generating function of the total intensity. In quasi-one-dimensional systems, $\phi(z)$ can be evaluated unperturbatively, using several methods [41-43], and the resulting expression for $g \gtrsim 1$ is

$$\phi(z) = g \ln^2 \left(\sqrt{1 + z/g} + \sqrt{z/g} \right) . \tag{50}$$

The representations of $\mathcal{P}(I)$ and its moments in terms of contour integrals are then suitable for making saddle point approximations. Note that Eq. (49a) is different than the corresponding expressions in Refs. [40, 41]. Here, for example, the integral over dz is recognized as Katalan's representation of the Bessel function which, instead of $\mathcal{P}(I) = \int_{-i\infty}^{i\infty} \frac{dv}{\pi i} K_0(2\sqrt{-Iv})e^{-\phi(v)}$, leads to

$$\mathcal{P}(I) = \int_0^\infty dv J_0(2\sqrt{Iv})e^{-\phi(v)} . \tag{51}$$

This expression reproduces the Rayleigh distribution by setting $g \to \infty$, i.e. $\phi(z) \to \phi_0(z) = z$. For large but finite g, the integral can be evaluated numerically as shown in Fig. 6a.

In the leading approximation $\phi(z)$ equals to $\phi_1(z) = z - z^2/3g$. Thus, keeping terms up to first order in $1/g$, one finds small corrections to Rayleigh statistics [41, 44]:

$$\mathcal{P}(I) = e^{-I} \left[1 + \frac{1}{3g} \left(I^2 - 4I + 2 \right) \right] , \qquad I \ll \sqrt{g} \tag{52a}$$

$$\langle I^n \rangle = n![1 + (n^2 - n)/3g] , \qquad n \ll \sqrt{g} . \tag{52b}$$

It is clear from Eqs. (49a and b), however, that these corrections are dominated by saddle point contributions and, thus, get to be exponentiated for sufficiently large values of I or n. Indeed, the intermediate asymptotics, still accessible by $\phi_1(z)$, takes the form

$$\mathcal{P}(I) \simeq \left(1 - \frac{4I}{3g} \right) \exp \left[-I + I^2/3g \right] , \quad \sqrt{g} \ll I \ll g \tag{53a}$$

$$\langle I^n \rangle \simeq n! \exp(n^2/3g) , \qquad \sqrt{g} \ll n \ll g , \tag{53b}$$

with a saddle point at $(u_c, z_c) \simeq -(I, 1)$. The far asymptotics of $\mathcal{P}(I)$ is determined by a different saddle point, $(u_c, z_c) \simeq -(g, \sqrt{I/g})$, which leads to the stretched-exponential tail [41, 42]:

$$\mathcal{P}(I) \sim \exp[-2\sqrt{gI}] , \qquad I \gg g \tag{54a}$$

$$\langle I^n \rangle \sim n! \exp[n \log(n/g)] , \qquad n \gg g . \tag{54b}$$

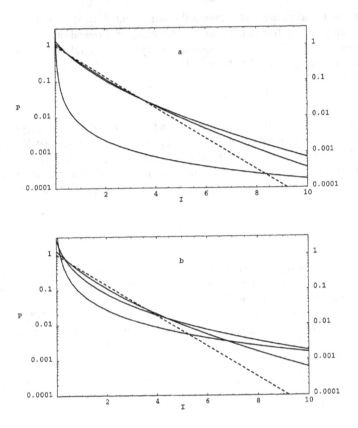

Figure 6.   The intensity distribution, $\mathcal{P}(I)$, obtained from Eqs. (51) and (55a). (a) Non-absorbing samples with $g = 0.1, 2, 4$ (lower to upper curves at $I = 2$). (b) Strongly absorbing samples with $g = 0.1, 0.5, 2$ as in [33]. The dashed line corresponds to Rayleigh distribution.

Eq. (45) is thereby justified for non-absorbing samples in the diffusive regime with $\kappa \equiv 3g \gg 1$ and $n \lesssim g$. It breaks down (i) in the localized regime, $g \lesssim 1$, and (ii) for strongly absorbing samples, $\alpha L \gg 1$:

- For non-absorbing samples $\widetilde{\mathcal{P}}(\tilde{I})$ crosses over from a Gaussian distribution, with non-Gaussian tails at $g \gg 1$, to a log-normal distribution as $g \to 0$. In the strongly localized regime [42]

$$\mathrm{var}\,\log \tilde{I} = -2\langle \log \tilde{I} \rangle = 2/g$$

and, thus, $\kappa \equiv g \ll 1$ for all values of n.

- For strongly absorbing samples $\tilde{I}$ is distributed log-normally for all the values of g with var $\log \tilde{I} = -2\langle \log \tilde{I} \rangle = 1/2g$. As a result [33],

$$\mathcal{P}(I) = \sqrt{\frac{g}{\pi}} e^{-1/16g} \int_{-\infty}^{\infty} dx \exp[-(gx^2 + 3x/2 + Ie^{-x})] \quad (55a)$$

$$\langle I^n \rangle = n! \exp[n(n-1)/4g] \, , \, n = 0, \pm 1, \pm 2 \cdots . \quad (55b)$$

The distribution $\mathcal{P}(I)$ for strongly absorbing samples is plotted in Fig.6b.

Conclusions

We have reviewed the diagrammatic technique for waves propagating in random media and showed how this technique is applied for calculating fluctuations in various physical quantities: fluctuations in transmission through disordered slabs, fluctuations in optical conductance, and the intensity distribution in quasi-one-dimensional samples. In our calculations we have used the Langevin scheme, as proposed by Zyuzin and Spivak [20], and concentrated on two experimentally relevant effects: the effect of absorption and the effect of geometry.

Fluctuations in transmission and optical conductance are sensitive to the width of the incident beam. For a Gaussian beam of radius R_0 impinging on a slab of width L, it was found that $\langle \delta T^2 \rangle = (\ell/k_0^2 L^3) F(L/R_0)$, where $F(x)$ is a monotonically increasing function of its argument: for small x it increases as x^2, for large x it is proportional to x. Fluctuations in optical conductance, $\langle \delta g^2 \rangle$, are expected to show qualitatively similar enhancement in restricted geometries [29].

In strongly absorbing quasi-1D samples $\langle \delta T^2 \rangle / \langle T \rangle^2 = 1/2g$, where $g = A k_0^2 \ell / 3\pi L$. Thus, the relative fluctuations in transmission assume an absorption-independent value, whereas $\langle \delta g^2 \rangle$ decreases exponentially.

The intensity distribution, $\mathcal{P}(I)$, shows significant deviations from Rayleigh statistics both in the diffusive and in the localized regimes. In strongly absorbing samples $\mathcal{P}(T)$ is distributed log-normally for all values of g.

Acknowledgments

Useful discussions with Eric Akkermans, Azi Genack, Ad Lagendijk, Frank Scheffold, Boris Shapiro and Patrick Snabre are gratefully acknowledged. Special thanks to Patrick Sebbah and Jean Michel Tualle who organized the school. The research was partially supported by the United States-Israel Binational Science Foundation (BSF).

References

[1] A. Ishimaru, *Wave Propagation and Scattering in Random Media* (Academic, New York, 1978).

[2] J. Goodman, in *Laser Speckle and Related Phenomena*, edited by J. C. Dainty (Springer-Verlag, Berlin, 1984), p. 9-75.

[3] For recent reviews see e.g. M. C. W. van Rossum and Th. Nieuwenhuizen, Rev. Mod. Phys **71**, 313 (1999); R. Berkovits and S. Feng, Phys. Rep. **238**, 135 (1994).

[4] B. L. Altshuler, P. A. Lee and R. A. Webb (Eds.), *Mesoscopic Phenomena in Solids*, (North-Holland, Amsterdam, 1991).

[5] For a comprehensive discussion of enhanced backscattering, with references to earlier work, see E. Akkermans, P. E. Wolf, R. Maynard, G. Maret, J. Phys. France **49**, 77 (1988); M. B. van der Mark, M. P. van Albada, A. D. Lagendijk, Phys. Rev. B **37**, 3575 (1988).

[6] S. Feng in *Scattering and Localization in Classical Waves*, edited by P. Sheng (Word Scientific, 1990); M. J. Stephen, in Ref. [4] p. 81-107; S. Feng and P. A. Lee, Science **251**, 633 (1991).

[7] A vector electromagnetic field was considered by M. J. Stephen and G. Cwilich, Phys. Rev. B **34**, 7564 (1986); K. Arya, Z. B. Su and J. L. Birman, Phys. Rev. Lett. **57**, 2725 (1986).

[8] A more realistic assumption would be to allow for a finite correlation length in $\langle \mu(\mathbf{r}) \mu(\mathbf{r}') \rangle$ or use discrete short-range scatterers. However, if one is not interested in the corrections to the speed of propagation, one can as well use the white-noise model.

[9] A. A. Abrikosov, L. P. Gorkov and I. E. Dzyaloshinskii, *Methods of Quantum Field Theory in Statistical Physics* (Pergamon, 1965), Sec 39; U. Frisch, in *Probabilistic Methods in Applied Mathematics*, edited by A. T. Barucha-Reid (Academic, 1968), Vol. 1, p. 76-198; B. Shapiro, in *Recent Progress in Many-Body Theories*, edited by Y. Avishai (Plenum Press, 1990), Vol. 2, p. 95-104.

[10] B. Shapiro, Phys. Rev. Lett. **57**, 2168 (1986).

[11] M. J. Stephen and G. Cwilich, Phys. Rev. Lett. **59**, 285 (1987).

[12] S. Hikami, Phys. Rev. B **24**, 2671 (1981); K. B. Efetov, A. I. Larkin and D. E. Khmel'nitskii, Sov. Phys. - JETP **52**, 568 (1980).

[13] S. Feng, C. Kane, P. A. Lee and A. D. Stone, Phys. Rev. Lett. **61**, 834 (1988).

[14] P. A. Mello, E. Akkermans and B. Shapiro, Phys. Rev. Lett. **61**, 459 (1988).

[15] P. A. Lee and A. D. Stone, Phys. Rev. Lett. **55**, 1622 (1985); B. L. Altshuler, JETP Lett. **41**, 648 (1985); P. A. Lee, A. D. Stone and H. Fukuyama, Phys. Rev. B **35**, 1039 (1986).

[16] B. Shapiro, Phys. Rev. Lett. **83**, 4733 (1999).

[17] S. Skipetrov and R. Maynard, unpublished.

[18] M. C. W. van Rossum, Th. M. Nieuwenhuizen and R. Vlaming, Phys. Rev. E **51**, 6158 (1995).

[19] C. L. Kane, R. A. Serota and P.A. Lee, Phys. Rev. B **37**, 6701 (1988).

[20] A. Yu. Zyuzin and B. Z. Spivak, Sov. Phys. - JETP **66**, 560 (1987); B. Z. Spivak and A. Yu. Zyuzin, Solid-State Comm. **65**, 311 (1988).

[21] This is, admittedly, a somewhat arbitrary choice and other types of boundary conditions for the diffusion equation occur in the literature [1]. For a discussion of the improved diffusion approximation and the effects of skin layers see A. Lagendijk, R. Vreeker and P. de Vries, Phys. Lett. A **136**, 81 (1989); Th. M. Nieuwenhuizen and J. M. Luck, Phys. Rev. E **48**, 569 (1993); and Ref. [30].

[22] R. Pnini and B. Shapiro, Phys. Rev. B **39**, 6986 (1989).

[23] R. Pnini and B. Shapiro, Phys. Lett. A **157**, 265 (1991).

[24] B. Z. Spivak and A. Yu. Zyuzin, in Ref. [4] p. 37-80.

[25] A. A. Burkov and A. Yu. Zyuzin, Phys. Rev. B **55**, 5736 (1997).

[26] N. Garcia and A. Z. Genack, Phys. Rev. Lett. **63**, 1678 (1989); A. Z. Genack, N. Garcia, W. Polkosnik, Phys. Rev. Lett. **65**, 2129 (1990).

[27] M. P. van Albada, J. F. de Boer and A. Lagendijk, Phys. Rev. Lett. **64**, 2787 (1990); J. F de Boer, M. P. van Albada and A. Lagendijk, Phys. Rev. B **45**, 658 (1992).

[28] N. Garcia, A. Z. Genack and A.A. Lisyansky, Phys. Rev. B **46**, 14475 (1992); N. Garcia, A. Z. Genack, R. Pnini and B. Shapiro, Phys. Lett. A **176**, 458 (1993).

[29] F. Scheffold, W. Härtl, G. Maret and E. Matijević, Phys. Rev. B **56**, 10942 (1997); F. Scheffold and G. Maret, Phys. Rev. Lett. **81**, 5800 (1998).

[30] A modified diffuson for the treatment of internal reflections was considered by M. C. W. van Rossum and Th. M. Nieuwenhuizen, Phys. Lett. A **177**, 452 (1993). This diffuson contains the injection depth, z_0, as additional parameter. In Eq. (31) it is assumed that $z_0 \ll L$.

[31] E. Kogan and M. Kaveh, Phys. Rev. B **45**, 1049 (1992).

[32] R. L. Weaver, Phys. Rev. B **47**, 1077 (1993).

[33] P. W. Brouwer, Phys. Rev. B **57**, 10526 (1998).

[34] P. Sebbah, R. Pnini and A. Z. Genack, to appear in Phys. Rev. E (2000).

[35] N. Garcia and A. Z. Genack, Opt. Lett. **16**, 1132 (1991); A. Z. Genack and N. Garcia, Europhys. Lett. **21**, 753 (1993).

[36] J. F. de Boer, M. C. W van Rossum, M. P. van Albada, Th. M. Nieuwen-huizen and A. Lagendijk, Phys. Rev. lett. **73**, 2567, (1994).

[37] M. Stoytchev and A.Z. Genack, Phys. Rev. Lett. **79**, 309 (1997); Opt. Lett. **24**, 262 (1999).

[38] E. Jackman and P. Pusey, Phys. Rev. Lett. **40**, 546 (1978).

[39] R. Dashen Opt. Lett. **10**, 110 (1984).

[40] E. Kogan, M. Kaveh, R. Baumgartner and R. Berkovits, Phys. Rev. B **48**, 9404 (1993); Physica A **200**, 469 (1993).

[41] Th. M. Nieuwenhuizen, M. C. W. van Rossum, Phys. Rev. Lett. **74**, 2674 (1995); E. Kogan, M. Kaveh, Phys. Rev. B **52**, R3813 (1995).

[42] S. A. van Langen, P. W. Brouwer and C. W. J. Beenakker, Phys. Rev. E **53**, R1344 (1996).

[43] A. D. Mirlin, R. Pnini, B. Shapiro, Phys. Rev.E **57**, R6285 (1998).

[44] N. Shnerb and M. Kaveh, Phys. Rev. B **43**, 1279 (1991).

Chapter 2

DYNAMIC SPECKLE CORRELATIONS

F. Scheffold

Fakultät für Physik, Universität Konstanz
Postfach 5560, D-78547 Konstanz, Germany
Physikalisches Institut, Universtät Freiburg
CH-1700 Freiburg, Switzerland
Frank.Scheffold@unifr.ch

G. Maret

Fakultät für Physik, Universität Konstanz
Postfach 5560, D-78547 Konstanz, Germany
Georg.Maret@uni-konstanz.de

Abstract   Long- and infinite range correlations C_2 and C_3 in the optical speckle pattern represent one of the most interesting phenomena in multiple scattering of light. Despite the strong scattering these correlations survive the averaging process of light diffusion and are even enhanced with increased randomness. In this article we are going to discuss the microscopic origin of these particular correlations which are explained in the simple picture of one and twofold crossing of multiple scattering paths. We present a comprehensive experimental study of *dynamic* speckle correlations, $C_2(t)$ and $C_3(t)$, where the phase shift between the multiple scattering paths is caused by the Brownian motion of the scattering particles. The shape and amplitude of the correlation functions $C_2(t)$ and $C_3(t)$ are in good overall agreement with theory. Deviations are found in the case of $C_2(t)$ when correlations are generated close to the incoming surface which can be explained by single scattering contributions.

Keywords: speckle correlations, diffusing wave spectroscopy, conductance fluctuations

Introduction

Light propagation in random media has attracted considerable attention over the last decade. In analogy to electronic transport in disordered metals, funda-

413

P. Sebbah (ed.), Waves and Imaging through Complex Media, 413–434.
© 2001 *Kluwer Academic Publishers. Printed in the Netherlands.*

mental issues such as localization of light have been addressed [1, 2]. It has been found that despite the randomness of the medium, various interference effects are essential for the light propagation in the multiple scattering regime. Since the discovery of weak localization of light [3, 4], a precursor of light localization, a much deeper understanding on wave propagation in random media has been achieved [5-7]. Most recently much attention has been paid to the reported observation of strong localization of light, which has been discussed controversially [8-10].

Another most interesting phenomenon in multiple scattering of classical waves is the appearance of correlations and fluctuations in the transmission speckle pattern. In this article we discuss recent experimental results about these correlations in the dynamic speckle pattern of laser light transmitted through a turbid colloidal suspension. Due to particular interference effects caused by crossing of scattering paths inside the random medium, two types of correlations between different speckle spots build up [11-19]. (1) Long range correlations in the scattered fields give rise to fluctuations in the angular integrated transmission. (2) Infinite range correlations cause fluctuations in the total transmission, independent both of the incoming and transmitted wave mode. The latter fluctuations are considered the optical analogue of "universal conductance fluctuations (UCF)" in electronic systems [13, 20].

After a brief review of the physical origin and theory, we discuss the temporal shape of the correlation function $C_2(t)$ and its amplitude dependence on sample thickness, beam spot size, and transport mean free path l^*. Universal conductance fluctuations of light will be the subject of the final part of this article.

The experiments show that due to of the inherently small noise level in dynamic light scattering experiments, photon correlation spectroscopy provides access to an unprecedented accuracy in the study of optical speckle correlation phenomena.

1.   THEORY

1.1.   THE PHYSICAL PICTURE

Both classical and electronic conductance fluctuations can be described in an appealing simple physical picture (Fig. 1) as further outlined below [13, 18, 19]. (C_1) : Interferences between waves scattered along independent paths give rise to short range angular fluctuations, in optics known as speckles. These are due to non-intersecting scattering paths which give rise to short range temporal and angular speckle fluctuations because of scatterers motion. There are no correlations between fields scattered along different paths.(C_2) : One crossing of scattering paths builds up correlations between different paths. Temporal decorrelation like in (C_1) occurs along the active section of the paths located

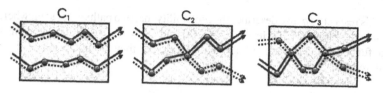

Figure 1.   The physical origin of speckle correlations can be explained in an appealing simple physical picture of independent (C_1) and crossing light paths (C_2 and C_3). One crossing of paths generates correlations of all output speckle spots (C_2) while two crossings cause UCF (C_3). (——) wave fields at correlation time $t = 0$, (- - -) phase shifted wave fields at $t > 0$ scattered along the same sequence of scatterers.

before the crossing event, while after the crossing the fields remain totally correlated (no mutual phase shifts) at all t and all output directions (b, b'). (C_3) : Twofold crossings generate universal conductance fluctuations (UCF). t-dependent phase shifts occur only between the crossing events, the intensity fluctuations are therefore insensitive to input (a, a') and output (b, b') wave modes.

1.2.   LONG AND INFINITE RANGE CORRELATIONS

The cylindrical waveguide.   The most simple case from a theoretical point of view is the diffuse transmission of classical waves through a cylindrical waveguide with perfectly reflecting walls (length L, width D). The average intensity, which is transmitted from an incoming plane light wave mode a to an outgoing plane wave mode b, is called $\langle T_{ab} \rangle$. The (dimensionless) conductance g of the sample is then defined as the sum over all incoming and outgoing modes:

$$g \equiv \sum_{a,b} \langle T_{ab} \rangle = \frac{Nl^*}{L} \sim D^2 . \tag{1}$$

N is the number of modes inside the waveguide of length L. N is proportional to D^2 and thus g is proportional to the surface area of the sample.

Feng *et al.* found that the intensity autocorrelation function $C(x) = \langle I(0)I(x) \rangle^2 / \langle I(0) \rangle^2 - 1$ can be written in terms of three leading contributions $C_1(x)$, $C_2(x)$ and $C_3(x)$, x being some quantity, such as frequency shift $\Delta\omega$ or correlation time t, which introduces phase shifts between optical fields [11, 13]:

$$C(x) = C_1(x) + C_2(x) + C_3(x) . \tag{2}$$

The amplitude of the different contributions was found to scale with

$$C_1(0) \simeq 1, C_2(0) \simeq g^{-1}, C_3 \simeq g^{-2} . \tag{3}$$

In this respect $1/g$ also describes the probability that two paths cross somewhere inside the sample [13]. Here $C_2 = C_2(0)$ and $C_3 = C_3(0)$ are independent of the functional behavior of $C_2(x)$ and $C_3(x)$, i.e. they are independent of the nature of x.

Slab geometry.   In practice it is difficult to realize a small optical waveguide for diffuse light propagation with perfectly reflecting walls. Therefore already Feng *et al.* suggested to investigate the transmission through a slab instead [11]. Pnini and Shapiro extended the theory of Feng to the general case of a finite beam spot incident on a slab [12]. They calculated the amplitude C_2 for a homogeneous beam spot of size W, with $I(r) = 1$ for $0 \leq r < W/2$ and $I(r) = 0$ otherwise. The result for $W \gg L$ is the same as for a cylindrical waveguide while for the case $W \ll L$ they find that the amplitude scales linearly with the inverse beam spot size $1/W$ [21]:

$$C_2 = \frac{4}{k_0^2 W^2} \frac{L}{l^*}; \ W \gg L \qquad (4)$$

$$C_2 = \frac{3}{2k_0^2 W l^*}; \ W \ll L. \qquad (5)$$

Later, de Boer *et al.* generalized this result for an incident gaussian beam, beamspot size w, usually encountered in optical experiments [22]. Although they considered frequency correlations $C(\Delta\omega)$ their results for the amplitude C_2 apply to the case of temporal correlations as well because of the insensitivity of C_2 on the phase shift introducing parameter:

$$C_2 = \frac{1}{\alpha} \times \frac{L}{w^2} \times F\left(\frac{w}{L}\right), \qquad (6)$$

$$\alpha = \frac{k_0^2 l^*}{3} = \frac{l^*}{3}\left(\frac{2\pi n}{\lambda}\right)^2, \qquad (7)$$

$$F\left(\frac{w}{L}\right) = \int_0^\infty dx \left(\frac{w}{L}\right)^2 \exp\left[-\left(\frac{w}{L}\right)^2 \frac{x^2}{32} \frac{x\left(\frac{\sinh(x)}{x} - 1\right)}{8\left(\cosh(x) - 1\right)}\right], \qquad (8)$$

where n is the refractive index, and λ is the wavelength of the incident light. In the limit $w \gg L$, $F(\infty) \longrightarrow 2/3$, hence $C_2 \sim w^{-2}$.

1.3.   SHAPE OF THE CORRELATION FUNCTION

Diffusing wave spectroscopy.   Light transmission through a slab containing Brownian particles shows strong fluctuations in the transmission speckle pattern. The fluctuations of the individual speckle spots are determined by the autocorrelation of electromagnetic fields scattered along individual paths, where the phase shift along a single path is, on average, directly proportional to the length of the path. For this case the leading contribution to the intensity-intensity correlation function $\langle I(0)I(t)\rangle / \langle I(0)\rangle^2 - 1 \simeq C_1(t)$ can be derived from diffusion theory [23]. Assuming light propagation on independent scattering paths (Fig. 1), it is possible to calculate the actual distribution of light paths and therefore calculate $C_1(t)$:

$$C_1(t, L) \cong \exp\left(-2\left(\frac{L}{l^*}\right)^2 \frac{t}{\tau_0}\right). \tag{9}$$

It was shown experimentally that this relation holds very well for samples of thickness L larger than 10 transport mean free paths l^* [24]. Here $\tau_0 = 1/Dk_0^2$ denotes the single scattering decay time and D_0 the translational diffusion constant of the scatterers. The correlation function is dominated by a typical path length L^2/l^* of the diffusing light where each scattering event contributes on average by $\exp(-t/\tau_0)$ to the decay of the correlation function. Measurements of $C_1(t)$ are widely exploited as the so called diffusing wave spectroscopy (DWS), which has become a powerful tool to study dynamics of colloids, emulsions, and other turbid soft matter [23, 25, 26]

Long range correlations.   Unlike the amplitude C_2, the dynamical part of the long range correlations $C_2(\Delta\omega)$ cannot be directly transferred to the time domain $C_2(t)$. To our knowledge the only theoretical treatment of long range $C_2(t)$ correlations for Brownian scatterers is presented in a paper by Berkovits and Feng [14]. Using a diagrammatic technique they derive the intensity-intensity correlation function for the case $w \gg L$ and find:

$$C_2(t) = \frac{3C_2}{2\sqrt{6\frac{t}{t_0}}}\left[\coth\left(\sqrt{6\frac{t}{t_0}}\right) - \frac{\sqrt{6\frac{t}{t_0}}}{\sinh^2\left(\sqrt{6\frac{t}{t_0}}\right)}\right], \tag{10}$$

$$t_0 \approx \left(\frac{l^*}{L}\right)^2 \tau_0. \tag{11}$$

The $C_2(t)$ correlation function decays over a much broader time scale than in the case of short range $C_1(t)$ correlations. In the long time limit an algebraic $t^{-1/2}$ behavior is predicted.

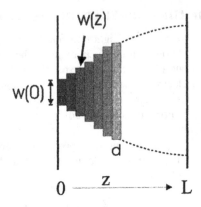

Figure 2.   Broadening of the photon cloud for the slab geometry. The increasing width $w(z)$ leads to a sharp decline of the "crossing probability" $C_2(z)$ with increasing depth z. If the beam spot size $w \ll L$, crossing events close to the incoming surface have a much higher weight and dominate C_2 and the temporal shape of the correlation function since the effective path lengths before a crossing event are much shorter as compared to a cylindrical waveguide of identical thickness L. With increasing beam spot size w the decay of the crossing probability becomes less sharp until, in the case $w \gg L$, it is independent of the depth z.

1.4.   THE INTEGRAL APPROXIMATION

Most of the theoretical results described above are derived from diagrammatic calculations which are quite complicated and physically not always very instructive. Most of these calculations are restricted to ideal cases, like the cylindrical waveguide. Often however the theoretical assumptions do not match the experimental conditions, e.g. the sample geometry or the influence of the boundary. How these deviations influence the amplitude and the decay of $C(t)$ cannot be easily derived from standard theory without doing the whole calculation from scratch. On the other side we have seen that the physics of long and infinite range correlations can be understood within the simple picture of crossing light paths [section 1.1], where the crossing probability is of the order $C_2 \approx 1/g$. While in the cylindrical waveguide the crossing probability is the same throughout the sample this is not true for other geometries. Figure 2 shows the conically shaped photon cloud in the case of a slab geometry. Here the crossing probability decreases with increasing depth z. Based on the simple picture of crossing light paths and the exact result for a cylindrical waveguide we have derived an approximate theory which can easily be adapted to the experimental conditions [27, 28]. As a starting point we consider the sample as a succession of Q thin slabs i of thickness d $[Q \cdot d = L]$. The thickness of the slabs is chosen such that d is much smaller than the lateral extension $w(z)$

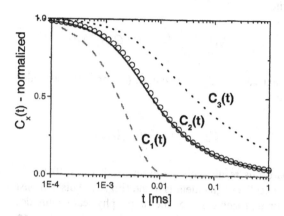

Figure 3.   Comparison of the shape of the correlation functions $C_1(t)$, $C_2(t)$ and $C_3(t)$ for $(L/l^*) = 20$, $\tau_0 = 2$ ms. The correlated contributions $C_2(t)$ and $C_3(t)$, Eq. (16), decay slower and over a much broader time range than $C_1(t)$. A comparison of the diagrammatic results for $C_2(t)$ [Eq. (10), solid line] and the result obtained from the integral approximation [Eq. (15), open symbols] shows excellent agreement.

of the photon cloud at a depth z inside the sample, hence $w(z) \gg d$. Each single slab can therefore be treated as a cylindrical waveguide with a crossing probability inside the slab i (4):

$$C_2^i \propto \frac{d}{w(z)^2} \,. \tag{12}$$

From this we find for the amplitude:

$$C_2 \simeq \sum_{i=1}^{Q} C_2^i \xrightarrow{d \to 0} \frac{1}{L} \int_0^L C_2(z) \, dz \,. \tag{13}$$

It is straightforward to extend this expression to describe the decay of the correlation function as well. Contributions to C_2 which are due to a crossing of paths in a depth z exhibit dephasing before the crossing event and are correlated afterwards. The path length distribution in this case is well approximated by the path length distribution of the uncorrelated function $C_1(t, z)$ (9). We can therefore write:

$$C_2(t) = \frac{1}{L} \int_0^L C_2(z) \cdot C_1(t,z)\, dz\,. \tag{14}$$

In the particular case of $C_2(z) = const.$ (cylindrical waveguide) we find the simple result:

$$C_2(t) = C_2 \frac{1}{L} \int_0^L C_1(t,z)\, dz\,. \tag{15}$$

If we compare this result with the exact result Eq. (10) from diagrammatic calculations, excellent agreement is found (Fig. 3). This demonstrates the consistency of our approach based on the simple physical picture described above. Equation (14) represents a complete description of long range speckle correlations $C_2(t)$ for a known distribution of $w(z)$.

The same approach can be used to determine the correlation function $C_3(t)$ [29]. The active path sections contributing to $C_3(t)$ are located between two crossing events (Fig. 1) resulting in a further broadening and slowing down compared to $C_2(t)$. In the integral approximation the correlation function for a waveguide geometry is given by a double integral over $C_1(t,z)$ or a single integral over $C_2(t,z)$:

$$C_3(t) = \frac{C_3}{L} \int_0^L C_2(t,z)/C_2\, dz\,. \tag{16}$$

2.   DYNAMIC LONG RANGE CORRELATIONS

2.1.   EXPERIMENT

Dynamic long range correlations have been studied by angular averaging of light transmitted through a slab containing a turbid colloidal suspension. The colloidal suspensions were prepared from monodisperse $BaTiO_3$ suspended in water [30]. Values of l^* where determined independently by static transmission measurement [24]. A minimal value of $l^* = 0.98$ μm was found at a volume fraction $\Phi = 27\%$. The fluctuations of the integrated transmission were measured with the setup illustrated in Fig. 4.

A gaussian laser beam (diameter roughly 1mm) from an Ar-laser operating in single frequency mode at 457.9 nm was focused onto a sample cell of variable thickness yielding a transverse intensity profile at the sample surface :

$$I(r) = \frac{2}{w^2\sqrt{\pi}} \exp\left(-\frac{4r^2}{w^2}\right)\,. \tag{17}$$

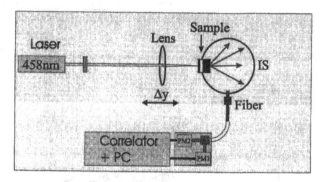

Figure 4. Experimental setup: The fluctuations of the integrated transmission through a slab were measured by angular averaging the transmitted light with an integrating sphere (IS). Detected by a photomultiplier unit (PM) the correlation function was subsequently analyzed using a digital correlator [18]. The incident laser beam is strongly focused by a lens.

The beam waist w is defined by the distance between the 1/e points of the transverse intensity distribution. To obtain small beam spot sizes, we used either an optical lens of a focal length of 5cm, which yields a minimum beam spot size of $w = 11.6$ µm, or in one case a microscope objective to obtain beam spots down to $w = 3.4$µm. This setup allowed us to change the actual beam spot size by variation of the sample-lens distance. The size of the beam spot w was determined by replacing the sample by a 10µm pinhole, or a 1 µm pinhole in the latter case, and scanning across the beam (accuracy ca. 5%). The glass cell was mounted in a sample holder and placed into an integrating sphere in order to average scattering intensities over all scattering angles of the transmitted light. A thick fiber bundle (diameter 5mm), positioned perpendicular to the incoming beam, was used to conduct the transmitted light from the integrating sphere to a photomultiplier. The fluctuations of the integrated transmission were analyzed using a commercial photon correlation setup . The detection limit for the intensity correlation function was determined to be lower than 10^{-6} over the whole range of correlation times 4×10^{-7} s$< \tau < 10^{-5}$ s considered. Details of the experimental setup can be found in [18]. Figure 5 shows the intensity-intensity correlation functions for three different beam spot sizes, with $l^* = 0.98$ µm. The maximum signal observed for this film thickness $L = 19.6$ µm is of the order of $\langle I(0)I(0) \rangle / \langle I(0) \rangle^2 - 1 \approx 2 \times 10^{-4}$ corresponding to a conductance of $g \approx 5000$.

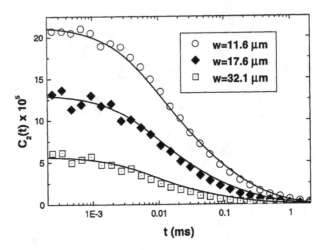

Figure 5.   Dynamic long range correlation function $C_2(t)$ for a slab of thickness L=19.6 μm [$l^* = 0.98$ μm]. The amplitude C_2 increases with decreasing beam spot size w. The solid lines are calculated within the integral approximation, Eq. (22), with no adjustable parameter ($z_0 = l^*$).

2.2.   INFLUENCE OF THE BOUNDARY LAYER

We first want to explore the limits of $w \approx l^*$, where an increased influence of the light propagating in a layer near the sample surface is expected. In the case of C_2 correlations, crossing of light paths near the incoming surface results in short "active" scattering paths, after which no further dephasing occurs. These short active paths are responsible for the long time tail of the correlation function $C_2(t)$ [see also section 2.4].

A significant influence of this boundary layer on the amplitude C_2 is expected when the beam size w is of the order of the transport mean free path l^*, which is the length scale over which the incident light is randomized. Figure 6 shows the dependence of C_2^{-1} on w for thick films ($L = 90 \pm 10$ μm) of colloidal suspensions of different l^*. The concentration Φ of the suspensions is in all cases 11% or lower, therefore the refractive index $n \simeq n_{water} \simeq const$. In this range, i.e. $w/L < 0.2$, C_2^{-1} scales in good approximation linearly with w :

$$C_2^{-1} \simeq \frac{8\alpha}{5}\, w \,. \tag{18}$$

Equation (18) can be derived by expanding Eq. (6) in the limit $w/L \to 0$. $C_2^{-1} \propto w$ follows also directly from the integral approximation (13) assuming diffuse linear spreading of the photon cloud inside the slab $w(z) \simeq w(0) + \beta z$ with β of order 1 [18].

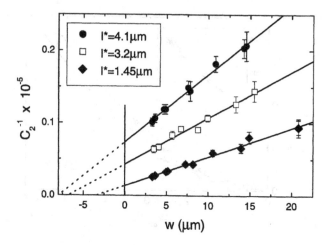

Figure 6. Measurements of C_2 vs. the beam spot size w show that C_2^{-1} values extrapolated for $w \to 0$ are non-zero and a distinct function of the transport mean free path l^*. A minimum beam spot size, $w_{min} = 2.4l^*$, can be defined by extrapolation of $C_2^{-1}(w) \to 0$ (dotted lines). The thickness of the films ($L = 90 \pm 10$ µm) is in all cases much larger than the beam spot size w.

In the experiments (Fig. 6) we clearly observe the linear dependence of C_2^{-1} on w, however C_2^{-1} does not tend to zero for small values of w, but reaches a well defined minimum value $(1/C_2)_{min}$. Apparently, the light incident on the slab does not contribute to the long range correlations before it is scattered at least once inside a surface layer, hence broadening the beam spot [18]. This result suggests that in fact the photon intensity distribution is broadened by scattering in a surface layer of thickness $(1 - 2)l^*$. We can account for this surface scattering by introducing an effective beam spot size:

$$w_{eff} = (w_{min} + w) = 2.4l^* + w. \tag{19}$$

We note that this value is somewhat larger than one would expect from single scattering contributions and also larger then the value determined from the shape of the correlation function (see section 2.4). Recent calculations suggest that finite size effects may account for this discrepancy since they lead to an increase of C_2^{-1}[31]. In the case $w/L \to 0$ the effective length L_{eff} of the sample (where the correlations are built up) is of the order $L_{eff} \approx w$ which means L_{eff} is comparable to l^* for small beamspot sizes w.

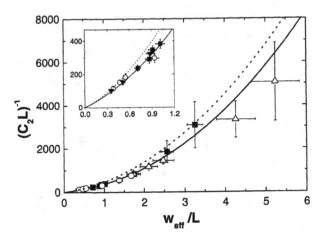

Figure 7.   Scaling dependence of $(C_2 L)^{-1}$ on the reduced beam spot size w_{eff}/L. The measured values for four different film thicknesses $L = 8, 19.6, 36.7, 40.2$ μm follow a master curve [Eq. (6), solid line] with $\alpha = 149 \pm 15$ the only adjustable parameter. The integral approximation yields fairly good agreement with the same set of parameters [Eq. (21), dotted line].

2.3.   AMPLITUDE SCALING OF C_2

According to Eq. (6) the product $C_2 \cdot L$ should depend solely on the ratio of the beam spot size and sample thickness w/L, independently of the actual values of w and L. For this reason, the measured values of $(C_2 L)^{-1}$ are predicted to follow a master curve. We expect the rescaled amplitude $(C_2 L)^{-1}$ to increases linearly with w/L for $w \ll L$, whereas for large ratios $w/L \gg 1$ the quadratic dependence should be recovered: $(C_2 L)^{-1} \sim (w/L)^2$. The amplitude C_2 for films of different thicknesses ($L = 8, 19.6, 36.7, 40.2$ μm) and effective beam spot sizes from 14 to about 60 μm was determined for a single minimum transport mean free path of $l^* = 0.98$ μm. Over the explored range of w_{eff}/L, the values $(C_2 L)^{-1}$ are found to be in good agreement with the scaling prediction Eq. (6) (Fig. 7) with only one adjustable parameter $\alpha = 149 \pm 15$ [1/μm]. This value is in quantitative agreement with theory, $\alpha = 138 \pm 27$ [1/μm], and static measurements in the frequency domain [22].

Fairly good agreement with the same set of parameters is also obtained from Eq. (13) with the approximate intensity distribution $w(z) \simeq w_{eff} + \beta z, \beta = 16/15$ [27, 32]:

$$C_2(z) = \frac{2}{3\alpha} \frac{1}{w(z)^2}, \tag{20}$$

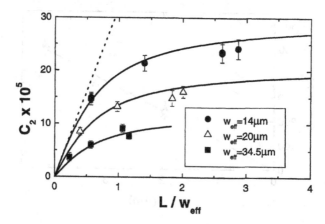

Figure 8. Dependence of C_2 on the slab thickness L. The values for three different beam spot sizes $w_{eff} = 14, 20, 34.5$ µm are plotted with the corresponding theoretical lines, Eq. (6), using $\alpha = 149$ [1/µm]. The dotted line shows the result for a corresponding cylindrical waveguide with $D = 14$ µm Eq. (4).

$$C_2 L \simeq \int_0^L C_2(z)\, dz = \frac{2}{3\alpha} \frac{1}{[(w_{eff}/L)^2 + \beta(w_{eff}/L)]}. \qquad (21)$$

Another feature is illustrated in Fig. 8. For $L/w_{eff} \gg 1$ the magnitude of the C_2 correlations becomes independent of L. This is due to the broadening of the beam inside the sample (Fig. 2). If L is much larger than w_{eff}, the width of the photon cloud deep inside the sample becomes so large, that the crossing probability is very small in most of the sample except the region of thickness of order $L_{eff} \approx w_{eff}$ near to the entrance surface. Increasing the thickness yields only asymptotically small increases in C_2 (Fig. 8).

Localization of light.   Since $1/g \simeq C_2$ we find that for a slab geometry (by increasing L and decreasing w) the value of the dimensionless conductance g cannot be reduced below a certain value g_{min}. In fact, the maximum amplitude C_2 or the minimal g is determined by the transport mean free path (18, 19): $g_{min} \simeq (8/5)\, \alpha\, (2.4 l^*) = 1.3\, (k_0 l^*)^2$ [18].

The dimensionless conductance g is also an important quantity with respect to the transition from diffusion to localization of light. For a waveguide geometry $g < 1$ implies a localization transition while the role of g for the localization transition in a slab geometry is still under discussion.

From our experiments we extrapolate that $g < 1$ can be achieved at $kl^* \approx 1$, a value that is of the same order as the Ioffe-Regel criterion for the localization

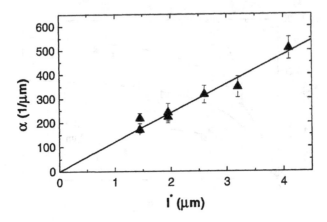

Figure 9.   The values of α show the expected linear dependence on l^*. The α values are determined from the slope of measured $C_2^{-1}(w)$ curves (Fig. 6).

transition: $kl^* \approx 1$ [33]. However for optical wavelengths $k \approx 15/\mu m$ ($\lambda/n \approx 400nm$) this is only realized for (unphysically) small beamspot sizes $w \ll l^* \leq 1/k_0 \approx 70$ nm [27].

Scaling with l^*.   We were able to confirm the predicted linear dependence of $C_2^{-1} \propto \alpha$ on the transport mean free path l^* (6). Figure 9 shows the values of α determined from the slope of the $(C_2(w))^{-1}$-curves (Fig. 6). A linear fit yields $\alpha/n^2 = (71 \pm 9) * l^*$, compared to $\alpha/n^2 = 63 * l^*$ from theory (7).

2.4.   SHAPE OF THE CORRELATION FUNCTION

Finally we want to discuss the time dependence of the correlation function $C_2(t)$. Crossing of light paths can occur at any point inside the sample, its probability being determined only by the effective lateral extension of the photon cloud. In the case of a cylindrical wave the crossing probability is independent of the depth z which leads to Eq. (15), or equivalently Eq. (10). $C_2(t)$ therefore decays much slower and broader than $C_1(t)$. The semi-logarithmic plots in Fig. 3 and Fig. 10 clearly reveal this behavior.

For a complete description for any combination of w and L we use again the integral approximation for the correlation function $C_2(t)$ (14). Using Eq. (21) with $w(z) \simeq w_{eff} + \beta z, \beta = 16/15$ we find [27, 32]:

$$C_2(t) = \frac{C_2}{L} \int_{z_0}^{L} \frac{2}{3\alpha} \frac{1}{w(z)^2} \cdot \exp\left(-2\left(\frac{z}{l^*}\right)^2 \frac{t}{\tau_0}\right) dz. \qquad (22)$$

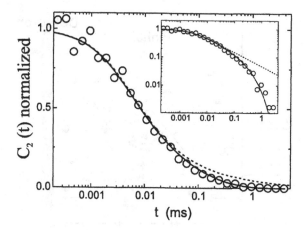

Figure 10.   Normalized correlation function $C_2(t)$ for $w > L$ [$L = 19.6$ μm, $w = 32$ μm]. A numerical fit with Eq. (10), dashed line, yields good overall agreement with $L = 19.4$ μm in agreement with $L = 19.6$ μm determined from measurements of $C_1(t)$. For long correlation times deviations from the predicted $t^{-1/2}$ behavior show up. We find perfect agreement (solid line) when introducing a lower bound $z_0 = 1.3l^*$, in the integral approximation, Eq. (22). Inset: log-log plot of the same data set.

For the cylindrical waveguide limit, $w_{eff} \gg L$ and $z_0 = 0$, Eq. (22) reduces to Eq. (15) (see also Fig. 3). We furthermore introduced a lower bound z_0 for the integral which can be non-zero. This allows us to take into account single scattering contributions close to the boundary. In fact for long correlation times $C_2(t)$ does not show the expected algebraic decay $t^{-1/2}$ but decays much faster [18]. The suggested explanation is that before a crossing of two light paths can take place there has to be at least one single scattering event close to the surface. This sets a lower bound to the minimum "active" path length. We take account for this by setting $z_0 \simeq l^*$. In Fig. 5 and Fig. 10 it is shown that Eq. (22) perfectly describes the experiments over the whole range of correlation times both in the limit $w \ll L$ and $w \approx L$. The good agreement for longer correlation times gives further evidence for the suggested single scattering contributions.

3.   UNIVERSAL CONDUCTANCE FLUCTUATIONS OF LIGHT

3.1.   THE EXPERIMENTAL REALIZATION

The setup to measure $C_3(t)$, the optical analog of universal conductance fluctuations in disordered metals [13], is schematically displayed in Fig. 11. It

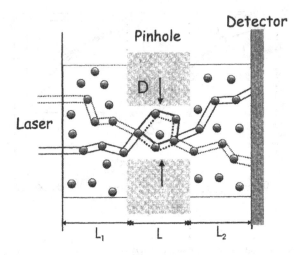

Figure 11.   Sketch of the experimental setup. A small cylindrical pinhole (diameter D, length L) sandwiched between two layers L_1 and L_2 is filled with a turbid colloidal suspension. Photon paths which cross twice inside the pinhole give rise to UCF.

was designed in analogy to a mesoscopic wire in two lead configuration. The prelayer of variable thickness L_1 enables the separation of $C_2(t)$ and $C_3(t)$ in the time domain. The active path sections of $C_2(t)$ are located before the (single) crossing events which occur almost exclusively within the pinhole. A sufficiently thick prelayer L_1 therefore leads to a rapid decay of $C_2(t)$ very similar to $C_1(t)$ for $L_1 \gg L$, as can be seen when replacing the integration limits in Eq. (15) by $[L_1, L_1 + L]$:

$$C_2'(t) \simeq C_2' \frac{1}{L} \int\limits_{L_1}^{L_1+L} C_1(t,z) dz \qquad (23)$$

$$\simeq C_2' \exp[-2 \left(\frac{L_1}{l^*}\right)^2 \frac{t}{\tau_0}]; \, L_1 \gg L \, .$$

To distinguish it from the broad and slow decay of $C_2(t)$ we call this rapidly decaying function $C_2'(t) \simeq C_1(t, L_1)$. On the other hand, according to Eq. (16), $C_3(t)$ is expected to show an algebraic decay independent of the thickness of both layers L_1 and L_2 since the two crossing events occur essentially only within the pinhole. Physically the colloidal prelayer scrambles the incoming modes very rapidly thereby creating an effective multi-mode illumination of the pinhole on the time scales of interest for $C_3(t)$.

3.2.   SETUP

Samples were prepared from commercial colloidal TiO_2 suspended in water. After stabilization with polyacrylic acid and filtering the suspension was found fairly monodisperse with an average particle diameter of $d \simeq 290$ nm, determined by single scattering photon correlation spectroscopy. Concentration was about 7 % by volume. Using a cell of known thickness (L=100μm) we find from DWS in transmission (9) $l^* = 1.35 \pm 0.1$ μm, ($\tau_0 \simeq 3$ ms).

A cylindrical pinhole (laser drilled in a disc shaped stainless steel foil of thickness $L = 13$ μm) was embedded in the suspension providing a liquid reservoir on both sides of the sample. The thickness of both layers L_1, L_2 sealed by glass windows was varied using different spacers. The sample was illuminated with a laser beam ($\lambda = 514.5$ nm) focused down to 150-200 μm beam diameter at intensities < 100 mW. We performed measurements of the autocorrelation function of the angular integrated transmitted intensity collected with a thick multi-mode fiber (*detector*). Multiple runs of typically 500 × 3 *sec* were carried out at photon count rates of 500 − 2000 kHz. In this geometry the contribution $C_1(t)$ is time independent due to the angular averaging of the outgoing light over many speckle spots.

3.3.   LONG RANGE CORRELATIONS

We characterized the setup by measuring $C_2'(t)$ for different pinhole sizes using moderately thick surrounding layers (L_1, $L_2 \approx 50$ μm). The inset in Fig. 12 shows the C_2' values determined from the amplitude of $C_2'(t)$. For a quantitative comparison with theory it is necessary to take also into account contributions outside the pinhole where the effective lateral confinement of photon cloud spreads out linearly. We can write:

$$C_2' \simeq \sum_{i=1}^{2} C_2^i.$$ (24)

The contributions inside the pinhole are given by the expression for a cylindrical waveguide $C_2^1 = (4L)/(k_0^2 D^2 l^*)$, Eq. (4), whereas the contributions outside the pinhole are due to an intensity step profile spreading out in a semi-infinite sample $C_2^2 = 3/(2k_0^2 D l^*)$, Eq. (5), hence

$$C_2' \simeq \frac{4}{k_0^2 D^2} \frac{L}{l^*} + \frac{3}{2k_0^2 D l*} = \frac{4}{k_0^2 D^2} \frac{L + (3/8)D}{l^*}.$$ (25)

Hence we can take account for this additional contribution by introducing an effective length of the pinhole $L + (3/8)D$.

The experimental results are in excellent agreement with this prediction. It can be readily seen that due to the quadratic dependence of C_2' on D the expected

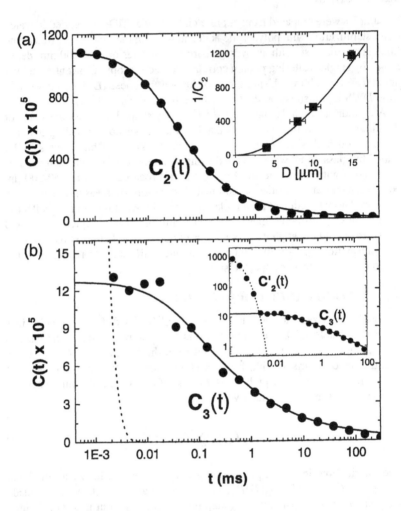

Figure 12.   (a) Inset: Inverse amplitude $1/C_2'$ as a function of the pinhole diameter D. Solid line: theoretical prediction , Eq. (25), with no adjustable parameter. Main figure: $C_2(t)$ correlation function for the smallest pinhole $D = 4$ μm (full circles). Solid line: best fit by Eq. (15) with $C_2 = 1.1 \cdot 10^{-2}$, $L = 13.1 \pm 1.3$ μm. (b) Universal conductance fluctuations $C_3(t)$ in comparison with $C_2'(t)$. For $t > 2 \times 10^{-3}$ ms, $C_2'(t)$ (dotted line) has decayed and $C_3(t)$ clearly shows up. Solid line: theory (16) with $C_3 = 1.3 \cdot 10^{-4}$, $L = 13.1$ μm. Inset: log-log plot of the same data set.

amplitude of $C_3(t)$ is very small, $C_3 \approx (C_2')^2 < 10^{-5}$, for all pinhole sizes except for the smallest one $D = (4 \pm 0.5)$ μm.

3.4.   OBSERVATION OF OPTICAL UCF

To achieve an effective separation of time scales between $C_2'(t)$ and $C_3(t)$ we used a prelayer thickness $L_1 \approx 100$ μm. To increase transmission, we replaced the colloidal suspension in (L_2) by pure water. Due to the absence of scattering in L_2 we were now able to measure the full $C_2(t)$ correlation function (15) of a cylindrical waveguide by illuminating the sample from the L_2-side. This provides additional information about the dynamics of the particles inside the pinhole which is difficult to obtain otherwise. The measured correlation function is shown in Fig. 12. Since we expect $C_2 = (1 \pm 0.2) \cdot 10^{-2} \approx 1/g$ from theory for the pinhole foil thickness $L = 13$ μm, it is in excellent agreement with the theoretical prediction in amplitude, shape and characteristic decay time. These results demonstrate that the dynamics of the particles inside the pinhole are largely unaffected by the lateral confinement and that the distribution of path lengths is not significantly altered by residual absorption at the pinhole walls.

In order to measure UCF the *identical* sample ($D = 4$ μm) was used which was now simply illuminated from the opposite side (L_1). As seen in Fig. 12, the contribution of $C_2'(t)$ now decays very fast. For longer correlation times we observe $C_3(t)$ which decays over more than four decades in time. The amplitude $C_3 \simeq 1.3 \cdot 10^{-4}$ is found in good agreement with the value $C_2 = 1.1 \cdot 10^{-2}$ obtained from the $C_2(t)$ measurement. Equally good agreement is found by comparing the shape of the $C_3(t)$ correlation function with the theoretical curve (16) without any adjustable parameter.

Summary and Conclusions

It has been shown that the use of coherent laser sources combined with accurate time correlation techniques allows to study very precisely the higher order correlation functions $C_2(t)$ and $C_3(t)$. Based on a series of measurements with different sample thicknesses L, beam spot sizes w, and transport mean free paths l^*, it has been possible to quantitatively confirm the scaling predictions for C_2. The time dependent correlation function $C_2(t)$ shows a good overall agreement with diagrammatic calculations. However the predicted long time $t^{-1/2}$ algebraic tail has not been observed. Quantitative agreement can be achieved by introducing a cut-off for the contribution of short scattering paths to $C_2(t)$. The study of the amplitude C_2 for extremely small values of $w \approx l^*$ delivers further evidence that light has to be scattered at least once before correlations can be built up by crossing of light paths.

Universal conductance fluctuations (UCF) in the transmission of classical waves have been observed using very small samples of concentrated colloidal suspensions. The experimental results provide a complete picture of the microscopic origin of UCF in disordered conductors in general. This demonstrates that the (quantum) wave interference can be quantitatively described by the simple model of diffusing waves crossing at locations inside the sample where the lateral confinement is high. Like weak and strong Anderson localization, UCF are a direct consequence of wave interference effects on a macroscopic scale. These interference corrections increase with randomness resulting in the breakdown of classical transport theory.

Acknowledgments

F. S. would like to thank Juanjo Saenz for useful comments and many stimulating discussions.

References

[1] P. W. Anderson, Phil. Mag. **52** 505 (1985); S. John, Phys. Rev. B **31** 304 (1985).

[2] S. John, Physics Today, May 1991, p.32-40.

[3] M. P. van Albada and A. Lagendjik, Phys. Rev. Lett. **55**, 2692 (1985).

[4] P. E. Wolf and G. Maret, Phys. Rev. Lett. **55**, 2696, (1985).

[5] P. Sheng (Ed.): *Scattering and localization of classical waves in random media*, World Scientific, Singapore 1990, Ping Sheng,*Introduction to Wave Scattering, Localization, and Mesoscopic Phenomena*, Academic Press, Boston, 1995.

[6] C. M. Soukoulis (Ed.): *Photonic band gaps and localization,* Nato ASI Series B, Physics **308** (Plenum, N.Y., 1993).

[7] G. Maret in: *Mesoscopic Quantum Physics*, E. Akkermans, G. Montambaux, J-L. Pichard and J. Zinn-Justin, eds (Elsevier Science B. V., North Holland, 1995), p.147.

[8] D. S. Wiersma, P. Bartolini, A. Lagendijk and R. Righini, Nature **390**, 671 (1997).

[9] F. Scheffold, R. Lenke, R. Tweer and G. Maret, Nature **398**, 206 (1999).

[10] F. J. P. Schuurmans, M. Megens, D. Vanmaekelbergh and A. Lagendijk, Phys. Rev. Lett **83**, 2183 (1999).

[11] S. Feng, C. Kane, P. A. Lee and A. D. Stone, Phys. Rev. Lett. **61**, 834 (1988).

[12] R. Pnini and B. Shapiro, Phys. Rev. B **39** , 6986 (1989).

[13] S. Feng and P. A. Lee, Science **251**, 633 (1991).

[14] R. Berkovits and S. Feng, Phys. Rep. **238**, 135(1994).

[15] M. C.W. van Rossum and T. M. Nieuwenhuizen, Rev. Mod. Phys. **71**,313 (1999).

[16] A. Z. Genack, N. Garcia, W. Polkosnik, Phys. Rev. Lett. **65**, 2129 (1990).

[17] M. P. van Albada, J. F.de Boer and A. Lagendijk, Phys. Rev. Lett. **64**, 2787 (1990).

[18] F. Scheffold, W. Härtl, G. Maret and E. Matijević, Phys. Rev. B **56**, 10942 (1997).

[19] F. Scheffold and G. Maret, Phys. Rev. Lett. **81**, 5800 (1999).

[20] C. P. Umbach *et al.*, Phys. Rev. B. **30** 4048 (1984); R. A. Webb *et al.*, Phys. Rev. Lett. **54** 2696 (1985).

[21] This differs by a factor of 2 from the original results for scalar waves [12] due to the two polarization states of electromagnetic waves [22].

[22] J. F. de Boer, M. P. van Albada and A. Lagendijk, Phys. Rev. B **45**, 658 (1992).

[23] G. Maret and P. E. Wolf, Z. Phys. B **65** 409 (1987); D. J. Pine, D. A. Weitz, P. M. Chaikin and E. Herbolzheimer, Phys. Rev. Lett. **60** 1134 (1988); D. A. Weitz and D. J. Pine in *Dynamic Light Scattering*, W. Brown Ed. (Oxford Univ. Press, New York, 1993), pp 652-720.

[24] P. D. Kaplan, M. H. Kao, A. G. Yodh and D. J. Pine, Appl. Opt. **32**, 21, 3828(1993).

[25] G. Maret, Curr. Opin. Coll. Int. Sci. **2**, 251-257 (1997).

[26] S. Romer, F. Scheffold and P. Schurtenberger, submitted to Phys. Rev. Lett.

[27] F. Scheffold and G. Maret, in preparation.

[28] F. Scheffold, PhD thesis, University of Konstanz (1998).

[29] We note that a first diagrammatic calculation of $C_3(t)$ has been published by Berkovits and Feng [14] but we were not able to reproduce their plots which are based on a complex combination of different diverging functions. Recently van Rossum and Nieuwenhuizen have presented an new calculation of $C_3(t)$ which has not been compared yet to the results of our simple integral approximation, Eq. (16) [15]. Attempts are now underway to do so [27].

[30] Y-S. Her, E. Matijević and M. C. Chon, J. Mater. Res. **10**, 12 , 3106 (1995).

[31] A. Garcia-Martin, F. Scheffold, M. Nieto-Vesperinas and J. J. Saenz, in preparation.

[32] The prefactor to Eq. (12) and the value of $\beta = 16/15$ where determined by comparing the limiting cases [$w \ll L, w \gg L$] to the results of Eq. (6).

[33] Mott, N. F. *Metal Insulator Transitions* (Taylor and Franzis, London, 1974).

Chapter 3

SPATIO-TEMPORAL SPECKLE CORRELATIONS FOR IMAGING IN TURBID MEDIA

S. E. Skipetrov

Laboratoire de Physique et Modélisation des Milieux Condensés
Université Joseph Fourier, Maison des Magistères — CNRS
B.P. 166, 38042 Grenoble Cedex 9, France
sergey@belledonne.polycnrs-gre.fr

Abstract   We discuss the spatio-temporal correlations of waves in a turbid medium. A hidden heterogeneous region (inclusion), characterized by a distinct scatterer dynamics, is assumed to be embedded in the medium. We show that the spatio-temporal correlation is affected by the inclusion which suggests a new method of imaging in turbid media. Our results allow qualitative interpretation in terms of diffraction theory: the cross-correlation of scattered waves behaves similarly to the intensity of a wave diffracted by an aperture.

Keywords: speckle correlations, correlation imaging, turbid media

A considerable progress has been made during the recent years in the understanding of wave transport in disordered media [1]. Very similar phenomena are shown to exist in multiple scattering of electrons and classical waves (e.g., light) under particular circumstances [2]. Some of the concepts developed first theoretically, and then studied in model experiments, are now very close to practical applications. One of the important fields where the physics of multiple-scattered waves is currently finding its applications is the (medical) imaging of disordered, turbid media [3]. The light waves scattered inside a turbid medium (e.g., human tissue) carry information on the properties of the medium. The information can be considered as being "encoded" in the statistics of the waves. Analysis of the latter statistics allows one to reconstruct (or "image") the scattering medium.

P. Sebbah (ed.), Waves and Imaging through Complex Media, 435–443.
© *2001 Kluwer Academic Publishers. Printed in the Netherlands.*

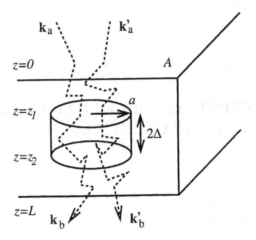

Figure 1.   A cylindrical inclusion of height $2\Delta = z_2 - z_1 \gg \ell$ and radius $a \gg \ell$ (ℓ is a photon transport mean free path) is embedded at $z_0 = (z_1 + z_2)/2$ inside a turbid slab of width L and surface area $A = W^2$, $W \gg L$, $W \gg a$. $\mathbf{k}_a$ and $\mathbf{k}'_a$ denote the wave vectors of incident waves, while $\mathbf{k}_b$ and $\mathbf{k}'_b$ —the wave vectors of transmitted waves. The scatterers in the medium undergo Brownian motion with diffusion coefficients D_{in} (inside the inclusion) and D_{out} (outside the inclusion).

A simplified version of a typical geometry considered in connection with imaging problems is shown in Fig. 1. A slab of turbid medium occupies the space between the planes $z = 0$ and $z = L$, and some region (a cylinder-shaped inclusion) inside the slab is assumed to have somewhat different properties as compared to the surrounding medium. If "different properties" means different scattering μ'_s and/or absorption μ_a coefficients, one can image the inclusion by measuring the spatial distribution of the average intensity $I(\mathbf{r}) = \langle E(\mathbf{r}, t) E^*(\mathbf{r}, t) \rangle$ of transmitted (or reflected) wave [4, 5]. Here $E(\mathbf{r}, t)$ is the amplitude of scattered wave at spatial position $\mathbf{r}$ at time t. If μ'_s and μ_a are constant throughout the medium, and the contrast between the inclusion and surrounding medium is provided by the scatterer dynamics (different types and/or intensities of scatterer motion inside and outside the inclusion), the methods of diffusing-wave spectroscopy [6, 7] can be applied to visualize the inclusion [8]. In the latter case, one measures the time autocorrelation function $C_1(\mathbf{r}, \tau) = \langle E(\mathbf{r}, t) E^*(\mathbf{r}, t + \tau) \rangle$ of scattered wave field at multiple positions $\mathbf{r}$, which allows visualization of the inclusion.

In the present contribution, we propose to use the *spatio-temporal cross-correlation* function $C_1(\mathbf{r}, \Delta\mathbf{r}, \tau) = \langle E(\mathbf{r}, t) E^*(\mathbf{r} + \Delta\mathbf{r}, t + \tau) \rangle$ for the purpose of imaging in turbid media. Since the time autocorrelation function $C_1(\mathbf{r}, \tau)$ carries more information about the turbid medium than the average intensity $I(\mathbf{r})$, we suggest that the information contents of the spatio-temporal correlation function $C_1(\mathbf{r}, \Delta\mathbf{r}, \tau)$ should be even richer. If the points $\mathbf{r}$ and $\mathbf{r} + \Delta\mathbf{r}$ are taken far enough from the medium (in the far-field of scattered wave), $C_1(\mathbf{r}, \Delta\mathbf{r}, \tau)$ is equivalent to the angular-temporal correlation function $\langle E(\mathbf{k}_b, t)E^*(\mathbf{k}_b', t + \tau)\rangle$ [where $E(\mathbf{k}, t)$ is the spatial Fourier transform of $E(\mathbf{r}', t)$ with $\mathbf{r}'$ taken at the plane where the scattered waves leave the medium].

We start with a macroscopically homogeneous turbid medium (no inclusion), and assume that scatterers in the medium undergo Brownian motion with a diffusion coefficient D. In addition, we assume a weak-scattering limit $k\ell \gg 1$ (where $\ell = 1/\mu_s'$ is the photon transport mean free path), and neglect the absorption of light in the medium ($\mu_a = 0$). As depicted in Fig. 1, a plane wave is incident upon a turbid slab at time t with a wave vector $\mathbf{k}_a$. The transmitted wave leaves the slab with a wave vector $\mathbf{k}_b$. Similarly, for an incident wave with a wave vector $\mathbf{k}_a'$ at time $t + \tau$, the transmitted wave has a wave vector $\mathbf{k}_b'$. Assuming unit amplitudes of incident waves, we calculate the correlation function of transmitted fields $C_1(\mathbf{k}_a, \mathbf{k}_b; \mathbf{k}_a', \mathbf{k}_b'; \tau) = \langle E(\mathbf{k}_a, \mathbf{k}_b, t)E^*(\mathbf{k}_a', \mathbf{k}_b', t + \tau)\rangle$ using the standard diagrammatic techniques [9], following the general calculation scheme developed in Ref. [10]. In the leading order in a small parameter $1/k\ell$, we obtain:

$$C_1 = \frac{\ell^2}{4k^2A^2} \int\int d^2\mathbf{R}_1\, d^2\mathbf{R}_2 \, \exp\left(-i\Delta\mathbf{q}_a\mathbf{R}_1 + i\Delta\mathbf{q}_b\mathbf{R}_2\right) \times$$
$$\times \quad P\left(\{\mathbf{R}_1, \ell\}, \{\mathbf{R}_2, L - \ell\}, \tau\right), \tag{1}$$

where $\mathbf{q}$'s denote projections of $\mathbf{k}$'s onto the plane $z = const$, $\Delta\mathbf{q}_a = \mathbf{q}_a - \mathbf{q}_a'$, $\Delta\mathbf{q}_b = \mathbf{q}_b - \mathbf{q}_b'$, $\mathbf{r} = \{\mathbf{R}, z\}$ with $\mathbf{R}$ being a two-dimensional vector perpendicular to the z-axis, and we assume the first and the last scattering events to occur at $z = \ell$ and $z = L - \ell$, respectively. If $L \gg \ell$ and $|\mathbf{r}_1 - \mathbf{r}_2| \gg \ell$, the reduced ladder propagator P entering into Eq. (1), obeys the diffusion equation:

$$\left[\nabla^2 - \alpha^2(\tau)\right] P(\mathbf{r}_1, \mathbf{r}_2, \tau) = -\frac{3}{\ell^3}\delta(\mathbf{r}_1 - \mathbf{r}_2), \tag{2}$$

where $\alpha^2(\tau) = 3\tau/(2\tau_0\ell^2)$ with $\tau_0 = (4k^2D)^{-1}$. The solution of Eq. (2) with Dirichlet boundary conditions at $z = 0$ and $z = L$ ($P = 0$ if $z_1 = 0, L$ or $z_2 = 0, L$) is readily found [11]:

$$P_0(\mathbf{r}_1, \mathbf{r}_2, \tau) = \frac{12\pi}{\ell^3} \int d^2\mathbf{p} \, \frac{\sinh[\beta_a(L - z_>)]\sinh(\beta_a z_<)}{\beta_a \sinh(\beta_a L)} \times$$
$$\times \quad \exp\left[i(\mathbf{R}_1 - \mathbf{R}_2)\mathbf{p}\right]. \tag{3}$$

Here $\beta_a^2 = \mathbf{p}^2 + \alpha^2(\tau)$, $z_> = \max\{z_1, z_2\}$, $z_< = \min\{z_1, z_2\}$, and the subscript "0" of P_0 denotes macroscopically homogeneous case. Inserting Eq.

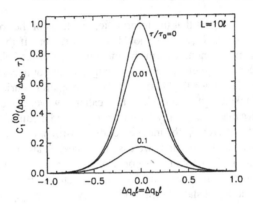

Figure 2.   Normalized angular-temporal correlation of a wave transmitted through a macroscopically homogeneous (no inclusion) slab of width $L = 10\ell$ for three different time delays $\tau/\tau_0 = 0, 0.01, 0.1$. This correlation function vanishes identically for $\Delta\mathbf{q}_a \neq \Delta\mathbf{q}_b$, which corresponds to the memory effect [12].

(3) into Eq. (1), we get

$$C_1^{(0)}(\Delta\mathbf{q}_a, \Delta\mathbf{q}_b, \tau) = \delta_{\Delta\mathbf{q}_a, \Delta\mathbf{q}_b} \frac{3\pi}{k^2 A} \frac{\sinh^2(\beta_a\ell)}{\beta_a\ell\,\sinh(\beta_a L)} \qquad (4)$$

with $\beta_a^2 = \Delta\mathbf{q}_a^2 + \alpha^2(\tau)$. For $\alpha^2(\tau) = 0$, Eq. (4) reduces to the angular correlation function [10], while for $\Delta\mathbf{q}_a = \Delta\mathbf{q}_b = 0$ the time autocorrelation function of transmitted light [7] is recovered. The Kronecker delta symbol in Eq. (4) describes the memory effect [12].

Now we turn to the case of macroscopically heterogeneous medium, assuming that the scatterer diffusion coefficient D_{in} inside a cylindrical region depicted in Fig. 1 is not the same as D_{out} in the surrounding medium (while ℓ is assumed to be constant throughout the whole sample). The correlation function of transmitted waves can be again described by Eqs. (1), (2) but with $\alpha^2(\tau) = \alpha_0^2(\tau) + \alpha_1^2(\tau) = 3\tau/(2\tau_{in}\ell^2)$ inside the inclusion and $\alpha^2(\tau) = \alpha_0^2(\tau) = 3\tau/(2\tau_{out}\ell^2)$ outside it. Here $\tau_{in,out} = (4k^2 D_{in,out})^{-1}$ and $\alpha_1^2(\tau) = 3\tau/(2\ell^2)[1/\tau_{in} - 1/\tau_{out}]$.

Assuming $\left|\alpha_1^2(\tau)\right| \ll \alpha_0^2(\tau)$, we can write an approximate solution of Eq. (2) as a sum of P_0 corresponding to the macroscopically homogeneous medium with $\tau_0 = \tau_{out}$ [see Eq. (3)], and P_1 which describes the influence of inclusion

[11]:

$$P_1(\{\mathbf{R}_1, \ell\}, \{\mathbf{R}_2, L - \ell\}, \tau) \simeq$$

$$\simeq -\alpha_1^2(\tau) \frac{\ell^3}{3} \int d^3\mathbf{r}\, P_0(\{\mathbf{R}_1, \ell\}, \mathbf{r}, \tau) P_0(\mathbf{r}, \{\mathbf{R}_2, L - \ell\}, \tau) =$$

$$= -\frac{144\pi^3}{3} \alpha_1^2(\tau) a^2 \frac{\Delta}{\ell} \int \int d^2\mathbf{p}\, d^2\mathbf{s}\, F_T(\mathbf{p}, \mathbf{s}, \tau) \times$$

$$\times \exp(i\mathbf{p}\mathbf{R}_1 - i\mathbf{s}\mathbf{R}_2), \tag{5}$$

where the integration of the second line is taken over the volume of inclusion, and F_T is a form factor:

$$F_T(\mathbf{p}, \mathbf{s}, \tau) = \frac{2J_1(a|\mathbf{p} - \mathbf{s}|)}{a|\mathbf{p} - \mathbf{s}|} \frac{\sinh(\beta_a \ell)\sinh(\beta_b \ell)}{(\beta_a \ell)(\beta_b \ell)\sinh(\beta_a L)\sinh(\beta_b L)} \times$$

$$\times \left\{ \frac{\sinh[(\beta_a - \beta_b)\Delta]}{(\beta_a - \beta_b)\Delta} \cosh[\beta_a(L - z_0) + \beta_b z_0] - \right.$$

$$\left. - \frac{\sinh[(\beta_a + \beta_b)\Delta]}{(\beta_a + \beta_b)\Delta} \cosh[\beta_a(L - z_0) - \beta_b z_0] \right\}. \tag{6}$$

Here J_1 is the Bessel function of the first order, $\beta_a^2 = \mathbf{p}^2 + \alpha_0^2(\tau)$, $\beta_b^2 = \mathbf{s}^2 + \alpha_0^2(\tau)$, $\Delta = (z_2 - z_1)/2$, and $z_0 = (z_1 + z_2)/2$ (see Fig. 1).

Inserting $P = P_0 + P_1$ into Eq. (1), we obtain the angular-temporal correlation function corresponding to the macroscopically heterogeneous medium as a sum of two contributions: $C_1 = C_1^{(0)} + C_1^{(1)}$, where $C_1^{(0)}$ is given by Eq. (4) with $\alpha(\tau) = \alpha_0(\tau)$, and

$$C_1^{(1)}(\Delta\mathbf{q}_a, \Delta\mathbf{q}_b, \tau) = -4\pi\alpha_1^2(\tau)\ell^2 \frac{3\pi}{k^2 A} \frac{\pi a^2}{A} \frac{\Delta}{\ell} F_T(\Delta\mathbf{q}_a, \Delta\mathbf{q}_b, \tau). \tag{7}$$

Equation (7) is the main result of the paper. We now compare the $C_1^{(0)}$ correlation function, corresponding to a macroscopically homogeneous slab [Eq. (4)], and the $C_1^{(1)}$ correlation function [Eq. (7)], originating from the presence of a dynamically heterogeneous region (inclusion). As follows from Eq. (4), $C_1^{(0)}$ vanishes identically if $\Delta\mathbf{q}_a \neq \Delta\mathbf{q}_b$, which is a manifestation of the memory effect [12]. If $\Delta\mathbf{q}_a = \Delta\mathbf{q}_b$, $C_1^{(0)}$ decays to zero for $\Delta q_a > 1/L$ (see Fig. 2). In contrast, $C_1^{(1)}$ correlation is not necessarily zero for $\Delta\mathbf{q}_a \neq \Delta\mathbf{q}_b$ (see Fig. 3). The memory effect is still present for the $C_1^{(1)}$ correlation function, as it is peaked near $\Delta\mathbf{q}_a = \Delta\mathbf{q}_b$ due to $J_1(a|\Delta\mathbf{q}_a - \Delta\mathbf{q}_b|)/(a|\Delta\mathbf{q}_a - \Delta\mathbf{q}_b|)$ term in $F_T(\Delta\mathbf{q}_a, \Delta\mathbf{q}_b, \tau)$. The memory effect for $C_1^{(1)}$ is considerably less sharp than for $C_1^{(0)}$, as one can see from Fig. 3.

Let us consider the simplest and practically important case of a single incident plane wave ($\Delta\mathbf{q}_a = 0$). For a macroscopically homogeneous slab, the

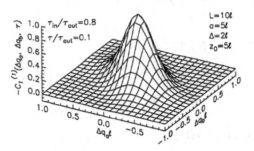

Figure 3.   Normalized angular-temporal cross-correlation function $C_1^{(1)}$ corresponding to the setup of Fig. 1. We assume $\Delta\mathbf{q}_a \parallel \Delta\mathbf{q}_b$ for this plot. In contrast to $C_1^{(0)}$ (see Fig. 2), $C_1^{(1)}$ is not necessarily zero for $\Delta\mathbf{q}_a \neq \Delta\mathbf{q}_b$. $C_1^{(1)} \neq 0$ only for $\tau \neq 0$.

angular-temporal cross-correlation vanishes if $\Delta\mathbf{q}_b \neq 0$, i.e. the waves scattered in different directions are uncorrelated.[1] If a heterogeneous region is embedded inside the slab, the $C_1^{(1)}$ term appears and correlation between the waves scattered along different directions is not necessarily zero. The $C_1^{(1)}$ term is plotted in Fig. 4 for three different radii a of the inclusion (solid lines). As is seen from the figure, the correlation range can be estimated as $\Delta\mathbf{q}_b \sim 1/a$. It is worthwhile to note that the correlation between the waves scattered along different directions ($\Delta\mathbf{q}_b \neq 0$), introduced by the inclusion, exists only for $\tau \neq 0$. If $\tau = 0$, $C_1^{(1)} = 0$ and the correlation function is given by $C_1^{(0)}$ which is identically zero for $\Delta\mathbf{q}_b \neq \Delta\mathbf{q}_a = 0$.

Some qualitative insight into the behavior of the $C_1^{(1)}$ correlation can be gained by comparing Eq. (7) with the angular distribution of the wave field $\mathcal{E}$ (wave number $K = 2\pi/\Lambda$) diffracted by a circular aperture of radius b [13] (see Fig. 4, dashed lines):

$$\mathcal{E}(K_\perp) \propto b^2 \frac{J_1(K_\perp b)}{K_\perp b}, \qquad (8)$$

where $K_\perp$ is the projection of $\mathbf{K}$ onto the plane of the aperture. In the case of $a \gg \Delta$, which corresponds to a "pill-shaped" inclusion, the correlation function given by our Eq. (7) and the diffraction pattern of Eq. (8) are remarkably close (see, e.g., the curves corresponding to $a = 20\ell$ and $b = 20\Lambda$ in Fig. 4). In this case, one can explain the appearance of correlation between the waves scattered in different directions in a macroscopically heterogeneous medium using the classical diffraction theory [13], and assuming $\Lambda = \ell$. As a consequence, some

---

[1]In reality, correlation persists as long as $\Delta q_b < 1/W$, and our result (4) corresponds to the limit $W \to \infty$.

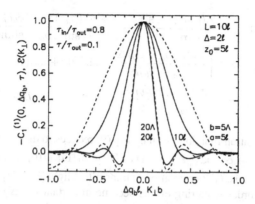

Figure 4. Normalized angular-temporal correlation functions of transmitted waves for a single plane wave ($\Delta\mathbf{q}_a = 0$) incident upon a slab with a cylindrical inclusion inside (solid lines). Dashed lines show the (normalized) angular distribution of the wave field $\mathcal{E}$ diffracted by a circular aperture of radius b ($b = 5\Lambda$ and $b = 20\Lambda$ with Λ being the wavelength). For $a = 20\ell \gg \Delta$ and $b = 20\Lambda$, correlation and diffraction curves are remarkably close, suggesting that correlation is "diffracted" by the inclusion.

theorems known for diffraction of waves (e.g., the Babinet's principle), apply directly to the angular-temporal correlation function of light transmitted through a macroscopically heterogeneous turbid medium. It is worthwhile to note that this holds for any shape of inclusion, provided that the transverse extent of inclusion is significantly greater than its extent along the z-axis ($a \gg \Delta$ in our notation). For $a \sim \Delta$, the quantitative agreement between Eqs. (7) and (8) is absent, although their overall behavior is similar (see, e.g., the curves corresponding to $a = 5\ell$ and $b = 5\Lambda$ in Fig. 4).

Up to now, our analysis has been devoted to the correlation functions of transmitted light. In experiments, however, it could be more convenient to work with diffusely reflected waves. Calculation of the angular-temporal correlation function of reflected waves is performed similarly to that of transmitted ones. If $|\mathbf{q}_a + \mathbf{q}_b|$, $|\mathbf{q}'_a + \mathbf{q}'_b|$, $|\mathbf{q}_a + \mathbf{q}'_b|$, $|\mathbf{q}'_a + \mathbf{q}_b| \gg 1/\ell$, we can ignore the time-reversal symmetry and obtain:

$$C_1^{(0)}(\Delta\mathbf{q}_a, \Delta\mathbf{q}_b, \tau) = \delta_{\Delta\mathbf{q}_a, \Delta\mathbf{q}_b} \frac{3\pi}{k^2 A} \frac{\sinh[\beta_a(L - \ell)]\sinh(\beta_a\ell)}{\beta_a\ell\sinh(\beta_a L)}, \quad (9)$$

$$C_1^{(1)}(\Delta\mathbf{q}_a, \Delta\mathbf{q}_b, \tau) = -4\pi\alpha_1^2(\tau)\ell^2 \frac{3\pi}{k^2 A} \frac{\pi a^2}{A} \frac{\Delta}{\ell} F_R(\Delta\mathbf{q}_a, \Delta\mathbf{q}_b, \tau), \quad (10)$$

$$F_R(\mathbf{p}, \mathbf{s}, \tau) = \frac{2 J_1\left(a\left|\mathbf{p}-\mathbf{s}\right|\right)}{a\left|\mathbf{p}-\mathbf{s}\right|} \frac{\sinh(\beta_a \ell) \sinh(\beta_b \ell)}{(\beta_a \ell)(\beta_b \ell) \sinh(\beta_a L) \sinh(\beta_b L)} \times$$

$$\times \left\{ \frac{\sinh\left[(\beta_a + \beta_b)\Delta\right]}{(\beta_a + \beta_b)\Delta} \cosh\left[(\beta_a + \beta_b)(L - z_0)\right] - \right.$$

$$\left. - \frac{\sinh\left[(\beta_a - \beta_b)\Delta\right]}{(\beta_a - \beta_b)\Delta} \cosh\left[(\beta_a - \beta_b)(L - z_0)\right] \right\}. \qquad (11)$$

The $C_1^{(0)}$ correlation function given by Eq. (9) reduces to the result of Ref. [14] for $\tau = 0, \beta_a L \to \infty$ and $\beta_a \ell \to 0$. The $C_1^{(1)}$ term [Eq. (10)] has the same qualitative features as the $C_1^{(1)}$ correlation of transmitted light [Eq. (7)]. In reflection, however, low-order scattering events become important for $\Delta q_a, \Delta q_b \sim 1/\ell$, and thus our results (9)–(11) make sense only for $\Delta q_a \ell, \Delta q_b \ell \ll 1$. If $\left|\mathbf{q}_a + \mathbf{q}_b\right|$, $\left|\mathbf{q}_a' + \mathbf{q}_b'\right|, \left|\mathbf{q}_a + \mathbf{q}_b'\right|$, or $\left|\mathbf{q}_a' + \mathbf{q}_b\right|$ are of order of or smaller than $1/\ell$, the time-reversal symmetry of the problem cannot be ignored any more. This significantly complicates calculations even in the case of macroscopically homogeneous medium [15].

In conclusion, we have calculated and discussed the angular-temporal cross-correlation functions of waves scattered in a turbid, dynamically heterogeneous medium. Our analysis demonstrates that the considered correlation functions can be used to image a hidden dynamic inclusion embedded in an otherwise homogeneous medium. Comparison of our results with diffraction patterns obtained for a wave diffracted by an aperture suggests that the angular-temporal correlation function can be considered as being "diffracted" by inclusion. Such an interpretation is particularly successful for "pill-shaped" inclusions which are much more extended in the transverse directions (i.e., in the directions parallel to the surfaces of the slab where they are embedded) than in the longitudinal one.

References

[1] M. C. W. van Rossum and Th. M. Nieuwenhuizen, Rev. Mod. Phys. **71**, 313 (1999); see also E. Akkermans and G. Montambaux in this volume.

[2] A. Lagendijk and B. A. van Tiggelen, Phys. Rep. **270**, 143 (1996).

[3] A. Yodh and B. Chance, Phys. Today **10**, No. 3, 34 (1995).

[4] M. A. O'Leary, D. A. Boas, B. Chance, and A. G. Yodh, Phys. Rev. Lett. **69**, 2658 (1992).

[5] P. N. den Outer, T. M. Nieuwenhuizen, and A. Lagendijk, J. Opt. Soc. Am. A **10**, 1209 (1993).

[6] G. Maret and P. E. Wolf, Z. Phys. B **65**, 409 (1987).

[7] D. J. Pine, D. A. Weitz, P. M. Chaikin, and E. Herbolzheimer, Phys. Rev. Lett. **60**, 1134 (1988).

[8] See G. Maret and M. Heckmeier (this volume) and references therein.

[9] U. Frisch, in: *Probabilistic Methods in Applied Mathematics,* ed. A. T. Bharucha-Reid (Academic, New York, 1968).

[10] R. Berkovits and S. Feng, Phys. Rep. **238**, 135 (1994).

[11] S. E. Skipetrov, Europhys. Lett. **40**, 382 (1997).

[12] I. Freund, M. Rosenbluh, S. Feng, Phys. Rev. Lett. **61**, 2328 (1988).

[13] M. Born and E. Wolf, *Principles of Optics* (Pergamon Press, Oxford, 1965).

[14] L. Wang and S. Feng, Phys. Rev. B **40**, 8284 (1989).

[15] R. Berkovits and M. Kaveh, Phys. Rev. B **41**, 2635 (1990); R. Berkovits, Phys. Rev. B **42**, 10750 (1990).

Chapter 4

SPECKLE CORRELATIONS AND COHERENT BACKSCATTERING IN NONLINEAR RANDOM MEDIA

R. Bressoux and R. Maynard

LPM2C - Université Joseph Fourier / CNRS

Maison des magistères - BP 166, 38000 Grenoble France.

bressoux@belledonne.polycnrs-gre.fr

maynard@polycnrs-gre.fr

Abstract   We calculate field-field and intensity-intensity correlation functions for light transmitted and reflected by a weakly nonlinear random medium containing randomly distributed scatterers. In a nonlinear Kerr medium, these correlation functions involve a characteristic length for phase decorrelation due to a change of the refractive index along the diffusive path of light. A new approximate expression for the nonlinear backscattering cone is obtained.

Keywords: multiple scattering, nonlinearities, correlations

Introduction

When high laser intensity pulses of very short duration illuminate a sample, the nonlinear response of the system must be considered. Multiple scattering of waves in such media offers a new field of research. Very few studies in the regime of nonlinear multiple scattering have been carried out. The speckle correlation function of the second harmonic (generated by 90 ps pulses on LiNbO$_3$ particles in quartz window) has been measured and analysed [1]. An anomalous linewidth shape of the backscattering cone is predicted for a random and Kerr medium in a frame of perturbated multiple scattering by nonlinear interactions [2, 3]. Several predictions have been made also in the general context of weak nonlinearities and modest beam intensity [4, 5]. Here, we calculate the intensity speckle correlation function in the weakly nonlinear

P. Sebbah (ed.), Waves and Imaging through Complex Media, 445–453.

regime for a random Kerr medium with a nonlinear susceptibility of the third order (χ_3). We will start from a perturbation scheme where the nonlinearity produces a correction to the ladder diagrams for the average intensity. We propose a new expression of the short range correlation function of the speckle pattern, as a function of angle and intensity, by extending its linear functional dependence to the nonlinear regime. The main output of this approach is a new characteristic length that collects both nonlinearity and randomness. An extension to long range correlation function is also proposed.

1. THE NONLINEAR CORRELATION FUNCTION

We consider a slab of scatterers imbedded in a homogeneous nonlinear medium (see Fig. 1) and we are interested in the field-field and the intensity-intensity correlation function of the light emerging from this slab.

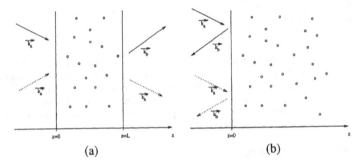

(a)   (b)

Figure 1.   Illumination of a slab by two beams of wave vectors k_a or k_a': the scattered light is detected in either the direction k_b or k_b'. The speckle correlation function is parameterized by all four wave vectors as well as the input amplitudes A and A'. (a) Transmission of a slab of length L. (b) Reflection on an semi-infinite slab.

First we will calculate the field-field correlation function, which depends on the incident directions k_a, k_a', the emerging directions k_b, k_b' and the amplitudes A and A' of the two incident beams :

$$C^{\mathrm{E}}(A, A', k_a, k_a', k_b, k_b') = < E_{k_a}(k_b) E_{k_a'}^*(k_b') > . \qquad (1)$$

It describes the correlation between two outgoing fields for two incident beams with different amplitudes. In the following, we will assume that A' is small. In order to simplify the calculation, we use the scalar approximation for the electromagnetic field. This simplification is justified in the regime of strong multiple scattering where the light is totally depolarized. The wave equation is,

$$\nabla^2 E(\mathbf{r}) + \frac{\omega^2}{c^2} \left\{ \epsilon(\mathbf{r}) + \alpha \, |E(\mathbf{r})|^2 \right\} E(\mathbf{r}) = J(\mathbf{r}), \qquad (2)$$

where $J(\mathbf{r})$ is the external source of radiation, $\epsilon(\mathbf{r}) = \epsilon_0 + \delta\epsilon(\mathbf{r})$ is the homogeneous linear dielectric constant ϵ_0 plus its random part $\delta\epsilon(\mathbf{r})$. Finally $\alpha = 4\pi\chi^{(3)}$ is the nonlinear susceptibility contributing to the dielectric constant. We suppose that there is no absorption in the linear regime and that $\epsilon(\mathbf{r})$ is real. In order to solve Eq. (2), we use a perturbational approach and expand the electric field as $E(\mathbf{r}) = E^{(0)}(\mathbf{r}) + E^{(1)}(\mathbf{r}) + \ldots$, where $E^{(0)}(\mathbf{r})$ and $E^{(1)}(\mathbf{r})$ are solutions of, respectively,

$$\begin{cases} \nabla^2 E^{(0)}(\mathbf{r}) + \dfrac{\omega^2}{c^2}\epsilon(\mathbf{r})E^{(0)}(\mathbf{r}) = J(\mathbf{r}). \\[2mm] \nabla^2 E^{(1)}(\mathbf{r}) + \dfrac{\omega^2}{c^2}\epsilon(\mathbf{r})E^{(1)}(\mathbf{r}) = -\dfrac{\omega^2}{c^2}\alpha \mid E^{(0)}(\mathbf{r}) \mid^2 E^{(0)}(\mathbf{r}). \end{cases} \quad (3)$$

In the second equation, the nonlinear term $-\alpha\,\omega^2/c^2 \mid E^{(0)}(\mathbf{r}) \mid^2 E^{(0)}(\mathbf{r})$ acts as a nonlinear source inside the volume of the slab. For the field-field correlator, we have to lowest order in α:

$$C^{\mathrm{E}} = < E^{(0)}_{\mathbf{k}_a}(\mathbf{k}_b)E^{(0)*}_{\mathbf{k}'_a}(\mathbf{k}'_b) > + < E^{(1)}_{\mathbf{k}_a}(\mathbf{k}_b)E^{(0)*}_{\mathbf{k}'_a}(\mathbf{k}'_b) > . \quad (4)$$

2. NONLINEAR CORRELATIONS IN TRANSMISSION

In transmission through a slab of length L, the expression of the linear part of the field correlator is given by (see formula (20) of ref [6]),

$$< E^{(0)}_{\mathbf{k}_a}(\mathbf{k}_b)E^{(0)*}_{\mathbf{k}'_a}(\mathbf{k}'_b) > = \delta_{\Delta\mathbf{q}_a,\Delta\mathbf{q}_b}\frac{3AA'}{4k^2S}\frac{l}{L}F_T(\Delta\mathbf{q}_a L), \quad (5)$$

where $F_T(x) = x/\sinh(x)$.

The expression of the nonlinear field correlator $< E^{(1)}_{\mathbf{k}_a}(\mathbf{k}_b)E^{(0)*}_{\mathbf{k}'_a}(\mathbf{k}'_b) >$ can be obtained by diagramatic expansion. In Fig. 2, we have represented the corresponding diagram in real space, which contributes to lowest order in $1/kl$. Its mathematical expression is [2, 3]:

$$< E^{(1)}_{\mathbf{k}_a}(\mathbf{k}_b)E^{(0)*}_{\mathbf{k}'_a}(\mathbf{k}'_b) > = \delta_{\Delta\mathbf{q}_a,\Delta\mathbf{q}_b}\frac{3AA'}{4k^2S}\frac{27i\alpha A^2}{2n_0^2}\frac{kl}{l^4}\times$$

$$\int_0^L D(L-l,z,\Delta\mathbf{q}_b)\, D(z,l,\Delta\mathbf{q}_a)\, D(z,l,0)\, dz, \quad (6)$$

where $D(z,z',\mathbf{q})$ is the intensity propagator from plane z' to plane z solution of $(q^2 - \frac{\partial}{\partial z^2})D(z,z',\mathbf{q}) = \frac{1}{2\pi}\delta(z-z')$. We use the solution with the Dirichlet boundary conditions $D(\mathbf{r} - \mathbf{r}') = 0$ on the plane $z = 0$ and $z = L$. (In this transmission geometry, the results obtained using the Dirichlet boundary

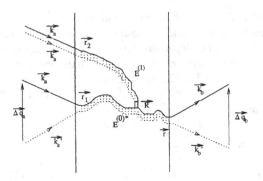

Figure 2.   The diagram, in real space, used to calculate the first nonlinear correction to the correlation function. The nonlinear vertex is represented by a square that connects three different propagators of intensity.

conditions are equivalent, in the leading order of l/L to those obtained using the radiative boundary conditions). Upon doing the integral over z in Eq. (6), we obtain:

$$< E_{\mathbf{k}_a}^{(1)}(\mathbf{k}_b) E_{\mathbf{k}_a'}^{(0)*}(\mathbf{k}_b') > = \delta_{\Delta q_a, \Delta q_b} \frac{3AA'}{4k^2 S} \frac{l}{L} (-i\gamma_{NL}^2 L^2 \frac{F_T'(\Delta q_a L)}{4\Delta q_a L}), \quad (7)$$

where $F_T'(x)$ is the derivative of $F_T(x)$ with respect to x and γ_{NL}^{-1} is the characteristic length defined by:

$$\gamma_{NL}^{-1} = (\frac{27}{4\pi} \frac{\alpha A^2}{n_0} \frac{\omega}{c_0 l})^{-1/2}. \quad (8)$$

The presence of $\delta_{\Delta q_a, \Delta q_b}$ in Eq. (7) means that the "memory effect" [6] still exists since the nonlinear vertex doesn't break the average transverse translational invariance of the slab. By collecting the expressions (5) and (7), we obtain:

$$C^E = \delta_{\Delta q_a, \Delta q_b} \frac{3AA'}{4k^2 S} \frac{l}{L} \left\{ F_T(\Delta q_a L) - i\frac{\gamma_{NL}^2 L^2}{\Delta q_a L} F_T'(\Delta q_a L) + \dots \right\}. \quad (9)$$

Expression (9) results from a perturbational expansion valid when $\gamma_{NL}^2 L \ll \Delta q_a$. We recognise the first term of the Taylor serie of the function F_T if we would replace the variable $\Delta q_a L$ by $\sqrt{\Delta q_a^2 L^2 - 2i\gamma_{NL}^2 L^2}$. This suggests to extrapolate Eq. (9) to:

$$C^E = \delta_{\Delta q_a, \Delta q_b} \frac{3AA'}{4k^2 S} \frac{l}{L} \ F_T(\sqrt{\Delta q_a^2 L^2 - 2i\gamma_{NL}^2 L^2}). \quad (10)$$

This "ansatz" can be justified, in the case of a nonlinear Kerrmedium with a real coefficient α [7]; γ_{NL}^{-1} is the characteristic length for phase decorrelation along the diffusive paths, due to the dependence of the index of refraction on the amplitudes A and A' of the incident beams.

We turn now to the calculation of the intensity-intensity correlation function. In the so-called $C^{(1)}$ approximation [6], this function is given by the simple formula: $C_T^{(1)}(A, A', \mathbf{k}_a, \mathbf{k}_a', \mathbf{k}_b, \mathbf{k}_b') = \left| C^E(A, A', \mathbf{k}_a, \mathbf{k}_a', \mathbf{k}_b, \mathbf{k}_b') \right|^2$. This approximation, valid when $kl \gg 1$, ignores possible crossings of multiple scattering paths. Here, we will also consider the weakly nonlinear regime where nonlinearities do not affect the scattering properties of the particles (e.g. the mean free path). Within this approximation we have:

$$C_T^{(1)} = \delta_{\Delta q_a, \Delta q_b} < I_{\mathbf{k}_a}(\mathbf{k}_b) > < I_{\mathbf{k}_a'}(\mathbf{k}_b') > \left| F_T(\sqrt{\Delta q_a^2 L^2 - 2i\gamma_{NL}^2 L^2}) \right|^2.$$

(11)

The function $|F_T(\sqrt{\Delta q_a^2 L^2 - 2i\gamma_{NL}^2 L^2})|^2$ is plotted as a function of $\Delta q_a L$ and $\gamma_{NL} L$ in Fig. 3. We note that a successful experimental observation of the

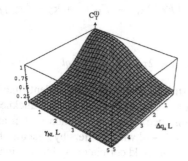

Figure 3.   The speckle correlation function in a nonlinear random Kerr medium. The decrease of the correlation function is presented as a function of: angle (Δq_a) and intensity (γ_{NL}).

present nonlinear effect should reach values of $\gamma_{NL} L \sim 2$, which corresponds, for a slab of length $L = 10^{-2}m$ and a mean free path of $l = 10^{-3}m$, to a change of the refractive index equal to $\Delta n = n_2 A^2 \sim 2 \times 10^{-5}$. This value is not beyond the short pulse regime in usual nonlinear materials.

Beyond this short range $C^{(1)}$ contribution, long range correlations [6] may be expected, particularly in confined geometries. These contributions arise from the crossing of diffusive paths in the random medium. The expression for the long range $C^{(2)}$ correlations in the frequency domain, found using a

Langevin method [8], can be translated to our nonlinear problem by substituting $\gamma_{\Delta\omega} = \sqrt{3n_0\Delta\omega/2c_0l}$ (called α in [8]) by γ_{NL}.

3.   NONLINEAR CORRELATIONS IN REFLECTION

Let us calculate the correlation function for the light reflected from a slab of infinite length. Taking into account the role of the time reversal symmetry, the linear field correlator is given by:

$$< E^{(0)}_{\mathbf{k}_a}(\mathbf{k}_b)E^{(0)*}_{\mathbf{k}'_a}(\mathbf{k}'_b) >= \delta_{\Delta\mathbf{q}_a,\Delta\mathbf{q}_b} \frac{3AA'}{4k^2S} \frac{1}{l^2} \int_0^\infty\int_0^\infty e^{-z'/l} \times$$

$$\left\{ D(z,z',\Delta\mathbf{q}_a) + D(z,z',\mathbf{k}_a+\mathbf{k}_b+\Delta\mathbf{q}_a) \right\} e^{-z/l} dz dz' \quad . \quad (12)$$

Using the intensity propagator $\tilde{D}(z,z',\Delta\mathbf{q}_a)$ obtained with the radiative boundary condition: $\tilde{D}(0,z',\Delta\mathbf{q}_a) - z_0 \frac{d}{dz}\tilde{D}(0,z',\Delta\mathbf{q}_a) = 0$, where $z_0 = \tau_0 l$ is the extrapolation length and $\tau_0 = 2/3$ in the diffusion approximation (see [9]), we have:

$$< E^{(0)}_{\mathbf{k}_a}(\mathbf{k}_b)E^{(0)*}_{\mathbf{k}'_a}(\mathbf{k}'_b) >= \delta_{\Delta\mathbf{q}_a,\Delta\mathbf{q}_b} \frac{3AA'}{4k^2S} \frac{l}{2} \times$$

$$\{ F_R(\Delta\mathbf{q}_a) + F_R(\mathbf{k}_a+\mathbf{k}_b+\Delta\mathbf{q}_a) \} , \quad\quad (13)$$

where

$$F_R(\mathbf{q}) \sim (1+2\tau) - 2(1+\tau)^3 ql + \ldots \text{ for } ql \ll 1. \quad (14)$$

Berkovits *et al.* [6, 10] have obtained an equivalent function, for a vanishing boundary conditions ($\tau = 0$) and for a slab of finite thickness. They considered sources located at a distance l from the boundary. For an infinite slab, their formula gives $F_R(\mathbf{q}) \sim 1 - ql$ which differs by a factor of 2 from Eq. (14).

The nonlinear part of the field correlator is obtained in the limit $\Delta q_a l \ll 1$:

$$< E^{(1)}_{\mathbf{k}_a}(\mathbf{k}_b)E^{(0)*}_{\mathbf{k}'_a}(\mathbf{k}'_b) >= \delta_{\Delta\mathbf{q}_a,\Delta\mathbf{q}_b} \frac{3AA'}{4k^2S} \frac{27i\alpha A^2}{n_0^2} l^4(1+\tau)^3 \times$$

$$\left\{ \frac{1}{\Delta q_a} + \frac{1}{|\mathbf{k}_a+\mathbf{k}_b+\Delta\mathbf{q}_a|} \right\} + \ldots \quad\quad (15)$$

With the same "ansatz" as was proposed in the previous section, one can collect the two expressions (13) and (15) to get the $C^{(1)}$ correlation function:

$$C^{(1)}_R = \delta_{\Delta\mathbf{q}_a,\Delta\mathbf{q}_b} < I_{\mathbf{k}_a}(\mathbf{k}_b) >< I_{\mathbf{k}'_a}(\mathbf{k}'_b) > \times$$

$$\left| F_R(\sqrt{\Delta q_a^2 - 2i(1+\tau)\gamma_{NL}^2}) + F_R(\sqrt{|\mathbf{k}_a+\mathbf{k}_b+\Delta\mathbf{q}_a|^2 - 2i(1+\tau)\gamma_{NL}^2}) \right|^2 ,$$

with γ_{NL} given by the formula (8). We have represented the $C_R^{(1)}$ function in Figure 4. For $\gamma_{NL} = 0$, it exhibits two peaks which correspond to the two conditions $\Delta\mathbf{q}_a = 0$ ($\mathbf{k}_a' = \mathbf{k}_a$ and $\mathbf{k}_b' = \mathbf{k}_b$) like in the correlation in transmission

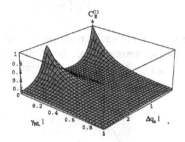

Figure 4.   The reflection correlation function in a nonlinear medium as a function of two variables: the backscattering angle through $\Delta\mathbf{q}_a$ and the intensity through γ_{NL}. The two peaks occur respectively for $\Delta\mathbf{q}_a = 0$ and $\mathbf{k}_a + \mathbf{k}_b + \Delta\mathbf{q}_a = 0$.

and for $\mathbf{k}_a + \mathbf{k}_b + \Delta\mathbf{q}_a = 0$ ($\mathbf{k}_a' = -\mathbf{k}_b$ and $\mathbf{k}_b' = -\mathbf{k}_a$) which corresponds to the time reversed directions. As for the correlation in transmission, a nonlinear decorrelation occurs through the coefficient γ_{NL}.

4.   NONLINEAR COHERENT BACKSCATTERING

The nonlinear backscattering from a random and nonlinear absorbing medium has been studied by Agranovitch *et al.* [2] and Heiderich *et al.* [3]. They used the same perturbational approach and obtained an expression for the coherent backscattering cone to first order in α. Beyond this first order of perturbation, one can use the "ansatz" described previously. It gives, for $\mathbf{q} = \mathbf{k}_a + \mathbf{k}_b$,:

$$< I_{\mathbf{k}_a}(\mathbf{k}_b) > \ = \ \delta_{\Delta\mathbf{q}_a, \Delta\mathbf{q}_b} \frac{3A^2}{4k^2 S} \frac{l}{2} \left\{ F_R\left(\sqrt{\frac{1}{L_a^2} - 2(1+\tau)Re(i\gamma_{NL}^2)} \right) + \right.$$

$$\left. F_R\left(\sqrt{q^2 + \frac{1}{L_a^2} - 2(1+\tau)Re(i\gamma_{NL}^2)} \right) \right\} , \qquad (16)$$

where $L_a = \sqrt{ll_a/3} = l/\sqrt{12\pi k_0 l Im\chi^{(1)}}$ is the linear absorption length of the medium. The backscattered intensity depends only on the nonlinear coefficient of absorption through $Re(i\gamma_{NL}^2)$ and it is instructive to define a generalized absorption length L_a^{NL} by:

$$\left(\frac{1}{L_a^{NL}}\right)^2 = \frac{12\pi \, k_0 l \, Im(\chi^{(1)}) + 27(1+\tau) \, k_0 l \, Im(\chi^{(3)})A^2}{l^2} . \qquad (17)$$

A more complete expression of the backscattering cone is now obtained as:

$$\frac{<I(q)>}{<I>_{inc}} = \frac{F_R\left(\frac{1}{L_a^{NL}}\right) + F_R\left(\sqrt{q^2 + (\frac{1}{L_a^{NL}})^2}\right)}{F_R\left(\frac{1}{L_a^{NL}}\right)}. \tag{18}$$

All nonlinear characteristics of the medium are put into absorption length L_a^{NL}. The lineshape of the backscattering cone, shown in Fig. 5, presents a usual angular dependence, without the dip predicted by Agranovitch [2] and Heiderich [3].

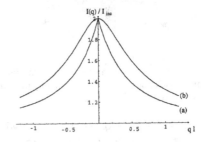

Figure 5.   The enhanced backscattering cone for two values of the nonlinear absorbing coefficient. (a) No absorption: $L_a^{NL} = \infty$. (b) $L_a^{NL} = 10l$. L_a^{NL} has been defined in Eq. (17).

Conclusion

In conclusion, we have developed a perturbational approach for calculating the correlation function of the intensity in a weakly nonlinear random media. The short range as well as long range correlation functions of the speckle pattern have been obtained as function of the input intensity and the angles of scattering. The result was analyzed in a simple way by introducing a new characteristic length γ_{NL} which incorporates the ingredients of the nonlinear susceptibilities, the multiple scattering via the light mean free path and the incident beam intensity. By using this approach, a new expression for the nonlinear backscattering cone has been proposed.

References

[1] J. de Boer, A. Lagendijk, and R. Sprik, Phys. Rev. Lett. **71**, 3947 (1993).

[2] V. Agranovitch and V. Kravtsov, Phys. Rev. **B43**, 13691 (1991).

[3] A. Heiderich, R. Maynard, and B.A. van Tiggelen, Opt. Comm. **115**, 392 (1995).

[4] V. Kravtsov, V. Yudson, and V. Agranovitch, Phys. Rev. **B41**, 2794 (1990).

[5] V. Kravtsov, V. Agranovitch, and K. Grigorishin, Phys. Rev. **B44**, 4931 (1991).

[6] R. Berkovits and S. Feng, Phys. Rep. **238**, 135 (1994).

[7] R. Bressoux and R. Maynard, Europhys. Lett. to be published.

[8] R. Pnini and B. Shapiro, Phys. Rev. **B39**, 6986 (1989).

[9] A. Ishimaru, *Wave Propagation and Scattering in Random Media* (Academic press Inc., San Diego, 1978).

[10] R. Berkovits and M. Kaveh, Phys. Rev. **B41**, 2635 (1990).

Index

Author Index